ENCYCLOPEDIA OF ENVIRONMENTAL ISSUES

ENCYCLOPEDIA OF ENVIRONMENTAL ISSUES

Volume II

Environmental justice and environmental racism—
Population growth

Editor

Craig W. Allin

Cornell College

Project Editor

Robert McClenaghan

Salem Press, Inc.

Pasadena, California   Hackensack, New Jersey

Managing Editor: Christina J. Moose
Research Supervisor: Jeffry Jensen
Acquisitions Editor: Mark Rehn
Photograph Editor: Karrie Hyatt

Project Editor: Robert McClenaghan
Copy Editor: Doug Long
Production Editor: Janet Alice Long
Layout: William Zimmerman

Library of Congress Cataloging-in-Publication Data

Encyclopedia of environmental issues / editor, Craig W. Allin; project editor, Robert McClenaghan.
    p. cm.
    Includes bibliographical references and index.
    ISBN 0-89356-994-1 (set : alk. paper). — ISBN 0-89356-995-X (v. 1 : alk. paper) — ISBN 0-89356-996-8 (v. 2 : alk. paper) — ISBN 0-89356-997-6 (v. 3 : alk. paper)
    1. Environmental sciences—Encyclopedias. 2. Pollution—Encyclopedias. I. Allin, Craig W. (Craig Willard) II. McClenaghan, Robert, 1961-
  GE10 .E52 2000
  363.7′003—dc21

    99-046373
    CIP

Second Printing

PRINTED IN THE UNITED STATES OF AMERICA

Contents

List of Articles by Category

Indoor air quality
London smog disaster
Montreal Protocol
Particulate matter
Radon
Smog
Sudbury, Ontario, emissions

BIOTECHNOLOGY AND GENETIC ENGINEERING
Bacillus thuringiensis
Biofertilizers
Biopesticides
Biotechnology and genetic engineering
Cloning
Diamond v. Chakrabarty
Dolly the sheep
Genetically altered bacteria
Genetically engineered foods
Genetically engineered organisms
Genetically engineered pharmaceuticals
Wilmut, Ian

ECOLOGY AND ECOSYSTEMS
Antarctica Project
Aral Sea destruction
Balance of nature
Bioassays
Biodiversity
Biomagnification
Bioremediation
Biosphere concept
Biosphere II
Brundtland, Gro Harlem
Carson, Rachel Louise
Club of Rome
Darwin, Charles
Dubos, René
Earth Day
Earth Summit
Ecology
Ecosystems
Environmental policy and lobbying
Environmental Protection Agency
Food chains
Global Biodiversity Assessment
Global Environment Facility
Global 2000 Report, The

Gore, Albert, Jr.
Group of Ten
Hardin, Garrett
International Biological Program
Leopold, Aldo
Limits to Growth, The
Lovelock, James
Marsh, George Perkins
National Environmental Policy Act
Restoration ecology
Sale, Kirkpatrick
Snyder, Gary
Soviet Plan for the Transformation of Nature
Spaceship Earth metaphor
Sustainable development
Tansley, Arthur G.
United Nations Environment Programme
United Nations Environmental Conference
Vernadsky, Vladimir Ivanovitch

ENERGY
Alternative energy sources
Alternative fuels
Alternatively fueled vehicles
Biomass conversion
Corporate average fuel economy standards
Energy-efficiency labeling
Energy policy
Fossil fuels
General Motors solar-powered car race
Geothermal energy
Hydroelectricity
Lovins, Amory
Oil crises and oil embargoes
Oil drilling
Power plants
Refuse-derived fuels
Siberian pipeline
Solar energy
Solar One
Sun Day
Synthetic fuels
Tennessee Valley Authority
Tidal energy
Trans-Alaskan Pipeline
Wind energy

Debt-for-nature swaps
Earth First!
Echo Park Dam proposal
Ecotourism
Foreman, Dave
Gila Wilderness Area
Glen Canyon Dam
Grand Canyon
Grand Coulee Dam
Green Plan
Hetch Hetchy Dam
International Union for the Conservation
    of Nature/World Wildlife Fund
Kings Canyon and Sequoia
    National Parks
Lake Baikal
Los Angeles Aqueduct
Mather, Steven T.
Muir, John
National parks
Nature preservation policy
Nature reserves
Pinchot, Gifford
Powell, John Wesley
Preservation
Roosevelt, Theodore
Scenic Hudson Preservation Conference v.
    Federal Power Commission
Serengeti National Park
Sierra Club
Sierra Club v. Morton
Tellico Dam
Three Gorges Dam
Watt, James
Wetlands
Wilderness Act
Wilderness areas
Wilderness Society
World Heritage Convention
Yosemite
zapovednik system

RESOURCES AND RESOURCE MANAGEMENT
Accounting for nature
Conservation
Conservation policy
Energy conservation
Environmental economics

Environmental engineering
Renewable resources

THE URBAN ENVIRONMENT
Bookchin, Murray
Mumford, Lewis
Olmstead, Frederick Law, Sr.
Planned communities
Shanty towns
Urban parks
Urban planning
Urbanization and urban sprawl

WASTE AND WASTE MANAGEMENT
Citizen's Clearinghouse for Hazardous Waste
Composting
Disposable diapers
Garbology
Hazardous waste
Landfills
Mobro barge incident
Nuclear and radioactive waste
Ocean dumping
Recycling
Resource recovery
Seabed disposal
Septic systems
Sewage treatment and disposal
Sludge treatment and disposal
Solid waste management policy
Waste management
Waste treatment
Welsh mining disaster

WATER AND WATER POLLUTION
Acid mine drainage
Amoco Cadiz oil spill
Aquifers and aquifer restoration
Argo Merchant oil spills
Braer oil spill
Brent Spar occupation
Chlorination
Clean Water Act and amendments
Cultural eutrophication
Cuyahoga River fires
Desalination
Dredging
Drinking water

ENCYCLOPEDIA OF ENVIRONMENTAL ISSUES

Environmental justice and environmental racism

CATEGORY: Philosophy and ethics

Advocates of environmental justice express concern for social justice in terms of such issues as the location of toxic waste dumps and the enforcement of environmental regulations. They maintain that everyone has the right to enjoy a clean environment. Environmental racism typically refers to the charge that minority groups are most often the victims of pollution and government failure to enforce environmental laws.

During the 1960's most Americans involved in environmental activities were white and upper or middle class. Issues such as conservation and preservation were of little interest to members of low-income or minority groups, who were often more concerned with civil rights and improving economic conditions. However, as concern over lead poisoning, hazardous waste dumps, and soil and water pollution mounted, minority leaders took notice. Research indicated that dumps and contaminated sites were more likely to be located near communities with a higher-than-average percentage of minorities. A 1983 study of landfill and incinerator sites in Houston, Texas, found that these facilities were usually found near African American neighborhoods. This and other studies led to a grassroots movement during the 1980's that attacked the problem of environmental racism.

Activists contend that racism has prompted governments to issue permits for waste facilities in low-income and minority areas. In some cases, communities have welcomed the facilities as a source of employment. In addition, members of these communities are less likely to possess the knowledge and the resources to oppose regulatory decisions. Through public protests and political pressure, activists seek to bring national attention to the problem and influence the decisions of local and national policymakers.

Grassroots organizations can point to some successes. Residents of one California community successfully pressured their town council to implement a program to screen for lead poisoning. When citizens in Halifax County, Virginia, learned that the federal government was considering their community as the location for a nuclear waste depository, they formed a group to fight the proposal. More than 1,400 African American and white residents voiced their opposition at a public meeting. Shortly thereafter, government officials dropped the county as a potential depository site. In 1986 the residents of Revilletown, Louisiana, received cash settlements and relocation assistance from a nearby chemical manufacturer. These and other victories indicate the potential power of grassroots movements that seek environmental justice.

Such successes do not mean that the movement is without its critics, who argue that environmental justice, as a concept, is so broad and vague that it cannot serve as a guide for policymakers. Moreover, they maintain that the available evidence regarding environmental racism is flawed. They contend that the studies fail to determine if the facilities were located in already existing minority communities or if the communities coalesced around the facilities.

Proponents of environmental justice reject these arguments as further evidence of injustice and racism. They claim that major corporations that hope to maximize profits have a vested interest in attacking the movement, which, if successful, would significantly raise the cost of production and waste disposal. Activists also complain that the national media consistently ignores environmental racism, favoring instead sensational stories that do not examine the deeper institutional causes of the environmental disasters featured in the headlines. Finally, proponents also fault the mainstream environmental movement for its fixation on preservation issues.

Despite the continuing debate over the meaning and reality of environmental justice and racism, some politicians have perceived the issues as deserving of legislation. In 1992 two members of the U.S. Congress sponsored the Environmental Justice Act, which required the Environmental Protection Agency (EPA) to identify and monitor areas with high levels of toxic chemicals. The measure failed but brought heightened attention to the issue. Two years later President Bill Clinton issued Executive Order 12898,

In 1994 residents of Highland Park, a mostly African American community in Michigan, protested plans to build a medical waste incineration facility in their neighborhood. (Jim West)

which required federal agencies to pursue environmental justice and acknowledged the existence of environmental racism. The order has had limited impact but has drawn more attention to the issue.

One major problem regarding accusations of environmental racism centers on the issue of proof. Courts require evidence that the alleged racism is intentional, which in most cases is impossible to prove. However, in 1997 a federal judge ruled that suits can be filed on the basis of "disparate impact," which means that the effect of racial discrimination, regardless of the intent, can be used to assess responsibility. The judgement was a victory for activists, who saw increased opportunities to pursue remedies in the courts. However, some observers noted that the ruling would prompt industrial interests, many of which

would not be guilty of polluting, to avoid minority communities out of fear of expensive lawsuits.

The issues of environmental justice and environmental racism have garnered support because their stated goals, a clean environment for all and an end to racist industrial and government practices, are attractive to many people. However, achieving those goals has proven difficult. People tend to desire both justice and the manufactured goods that cause environmental degradation. While most people oppose racism, they understandably have no desire to relocate polluting industries into their own neighborhoods. As the debate over disparate impact indicates, conflicts over environmental racism may have the unintended consequence of denying low-income people jobs that they desire.

Thomas Clarkin

SUGGESTED READINGS: Richard Hofrichter, editor, *Toxic Struggles: The Theory and Practice of Environmental Justice* (1993), includes articles on gender and international issues. Christopher H. Foreman, Jr., *The Promise and Peril of Environmental Justice* (1998), takes a critical view of the movement. Robert D. Bullard, editor, *Confronting Environmental Racism: Voices from the Grassroots* (1993), offers nine case studies of community efforts aimed at ending environmental racism. Stephen B. Huebner, *Are Storm Clouds Brewing on the Environmental Justice Horizon?* (1998), is a brief but thoughtful assessment of the movement's future. Andrew Szasz, *EcoPopulism, Toxic Waste, and the Movement for Environmental Justice* (1995), analyzes environmental justice as a mass movement.

SEE ALSO: Environmental policy and lobbying; Hazardous and toxic substance regulation; Landfills.

Environmental policy and lobbying

CATEGORY: Philosophy and ethics

Environmental policy is legislation passed on the national and state level intended to preserve and protect the environment and its natural resources. Environmental organizations or concerned individuals can influence policy by lobbying for new laws or changes in existing legislation.

Adequate enforcement of environmental laws, combined with the compliance of businesses, industry, and individuals, can lead to dramatic changes in the state of the environment. However, if these laws are not properly enforced by federal and state governments, they are ineffective. Many environmental laws are inspired by the depletion of natural resources, the pollution of air and waterways, and the extinction of animal or plant species. In many cases, the future of a proposed law is determined by the influence that lobbying groups have on legislators and their decisions.

HISTORY OF ENVIRONMENTAL LAW

Natural resource management first became a serious issue in the United States after the Civil War. In the late nineteenth century rapid settlement of the frontier land in the West, along with the increased use of railroads, caused the destruction of thousands of acres of forest and wildlife habitat. This resulted in an escalating need for legislation that would control the use of forestland. Several laws, such as the Forest Reserve Act and the Forest Management Act, were passed in the 1890's to create forest reserves that would protect remaining resources. However, it was not until the early twentieth century, during the presidency of Theodore Roosevelt, that environmental issues came to the fore. Roosevelt had always been known as a man of action, and one of his main interests was conservation. Even without the support of Congress, Roosevelt successfully created or expanded thirty-two national forests, passed laws that dealt with historic preservation, and created the first wildlife refuge. Subsequent administrations were not as concerned with environmental problems until Franklin D. Roosevelt's New Deal emerged. Among other things, Franklin Roosevelt fought for land and resource restoration and reforestation to replenish the forests and prevent further soil erosion.

The 1950's and 1960's saw the passage of many environmental laws, including the 1955 Air Pollution Control Act, the 1956 Federal Pollution Control Act, the 1963 Clean Air Act, and the 1965 Water Quality Act. Although legislators recognized that there were problems with air and water pollution, lack of proper regulation made the laws nearly ineffective. Much of this changed in the mid- to late 1960's during President Lyndon B. Johnson's administration. Many laws aimed at the protection of the nation's wilderness and rivers were passed, and they were followed with much stricter enforcement. The attention of American citizens and legislators was becoming focused on the environment. All of this led to the explosion of environmental laws and protection groups across the country during the 1970's.

The first annual Earth Day was celebrated nationwide on April 22, 1970. This decade also saw the passing of much key legislation, including the 1970 Clean Air Act amendments, the 1970 Occupational Safety and Health Act (aimed at regulating workplace conditions), the 1972 Clean Water Act, the 1973 Endangered Species Act, and the 1974 Safe Drinking Water Act. The vast amount of new legislation led President Richard Nixon to combine five different federal departments and agencies to create the Environmental Protection Agency (EPA). The EPA is a federal agency that establishes and enforces standards intended to protect the nation's air, water, and soil. Among the many thousands of environmental laws that have been passed, three of the most influential and well-known are the Clean Air Act, the Clean Water Act, and the Endangered Species Act.

CLEAN AIR ACT

The Air Pollution Act of 1955 was the first national legislation that dealt with air pollution. It gave control of air pollution regulations to state governments and concentrated on air pollution research. The Clean Air Act of 1963 expanded the research from the previous act and gave more regulatory power to the federal government. The act also increased research and development of alternatives that would decrease the amount of industrial and vehicle pollutants. Congress was dissatisfied with the progress of the state governments and consequently passed the 1967 Air Quality Act. Although most regulatory responsibility continued to reside with the states, federal air-quality standards were established for the first time. In 1970 Congress passed the most effective legislation dealing with air pollution thus far, the Clean Air Act amendments of 1970, thereafter referred to as simply the Clean Air Act (CAA).

The CAA regulated air emissions from area, stationary, and mobile sources and gave the EPA the authority to establish national ambient air-quality standards (NAAQS) in all states. The NAAQS were designed to protect public health and the environment. The original goal of the CAA was for each state to submit, by 1975, a state implementation plan (SIP) to the EPA outlining the state's regulations for air-emissions control. However, the CAA had to be amended in 1977 to

extend the deadline for state compliance with the NAAQS. The CAA was again amended in 1990 to address problems such as acid rain, ground-level ozone, depletion of the ozone layer, and toxic air pollution. The 1990 CAA amendments also required the EPA to set limits on the emissions of 189 specific toxic air pollutants.

The effects of the Clean Air Act are, at times, difficult to determine. Much of the emissions data are reported by individual industries and corporations. According to annual reports by the EPA, the CAA has had a positive impact on the environment. During the twenty-five years following the passing of the CAA, the total reported emissions of the six criteria pollutants—lead, nitrogen dioxide, sulfur dioxide, carbon monoxide, particulate matter, and ground-level ozone—were reduced by 24 percent. Individually, lead emissions were reduced by 98 percent, ozone by 23 percent, sulfur dioxide by 32 percent, carbon monoxide by 23 percent, and particulate matter by 78 percent. Only emissions of nitrogen dioxide, which increased by 14 percent, failed to improve. This rise was caused by an increase in coal burning at power plants. There was also a 70 to 80 percent decrease in the emission of pollutants from typical cars, and the annual release of the 189 toxic air pollutants regulated by the EPA was reduced by 900,000 tons. In urban cities where smog causes constant health problems, the pollution standard index (PSI) also showed dramatic improvement.

Although these reports are promising, the EPA recognizes the need for more improvement in many areas. In 1994 62 million Americans lived in areas that did not meet air-quality standards, and 92 million lived in areas that did not meet ozone standards. There continues to be a need to develop more efficient ways to travel in order to decrease the amount of nitrous oxide emissions.

CLEAN WATER ACT

According to the EPA, the Clean Water Act (CWA) of 1977 is actually an amendment to the 1972 Federal Water Pollution Control Act. The CWA has two main foci. First, the CWA controls the types and amounts of pollutants an industry may release into its waste streams. This area of control is known as technology-based effluent limitations. Effluent limitation refers to the restrictions on the amount, rate, and concentration of discharged substances. Second, the EPA requires states to categorize their navigable waters according to primary use, such as recreational, agricultural, or industrial. Standards of water quality are then established and enforced by the EPA. This type of control is known as water-quality-based control.

Under the CWA, it is illegal for any person to discharge pollutants from a point source without obtaining a permit from the EPA. A point source is any location or entity that releases pollutants directly into a waterway. The CWA also regulates the nonpoint sources of pollution, such as agricultural runoff and drainage from streets. However, the amount of pollutants released from nonpoint sources is more difficult to determine.

Like the Clean Air Act, the nationwide effects of the Clean Water Act are also difficult to measure. The Toxic Release Inventory (TRI) is an annual publication that compiles industrial reports of toxic chemical releases. The TRI is used by the EPA as a method of determining the level of water pollution contributed by industries. However, the accuracy of the TRI depends on proper industrial reports. There are also many loopholes that must be considered. The TRI only covers approximately three hundred chemicals and those industries that produce in excess of 25,000 pounds of the chemicals each year. Assuming that the TRI is highly accurate, the 1990 report showed that the amount of toxic chemicals released into the nation's waterways was reduced from 4.1 billion pounds in 1987 to just under 2 billion pounds in 1990.

According to the EPA's 1994 annual report, 20 percent more of the nation's surveyed rivers, lakes, and estuaries were safe for fishing and swimming compared to 1974. As a result of the CWA, ocean disposal of sewage sludge, industrial wastes, plastic debris, and medical wastes has been banned. The EPA estimates that established standards for more than fifty industries prevented more than one billion pounds of toxic pollution from entering the nation's waterways.

ENDANGERED SPECIES ACT

In 1966 the first legislation was passed that dealt primarily with the threat of species extinction. This law was replaced in 1973 with a more efficient version known as the Endangered Species Act (ESA). The ESA required all federal

---

Measurable Effects of Environmental Legislation

LAW	ENVIRONMENTAL EFFECTS
Clean Air Act amendments (1970)	Between 1970 and 1994, emissions of criteria pollutants were reduced by 24 percent; lead emissions were reduced by 98 percent.
Clean Water Act amendments (1977)	Between 1977 and 1994, 20 percent more of the surveyed water became safe for swimming and fishing.
Endangered Species Act (1973)	The extinction rate for listed species is less than 1 percent; since 1973, gray wolves, peregrine falcons, bald eagles, whooping cranes, California condors, and black-footed ferrets have recovered from endangered status.

Sources: Environmental Protection Agency National Air Quality and Emissions Report, 1994; Environmental Protection Agency National Quality Inventory, 1994; National Wildlife Federation, *EnviroAction* (December-January, 1999).

agencies to take part in species conservation but placed primary responsibility with the U.S. Fish and Wildlife Service and the National Marine Fisheries Service, known collectively as the Services. The Services must maintain a list of all species classified as endangered or threatened. The ESA defines an endangered species as being in danger of extinction throughout all or a large part of its range. A threatened species is one for which the threat of endangered status is imminent. The ESA prohibits any activity that harasses, collects, harms, or destroys the habitat of a listed species. Any federal project or any project taking place on federal land must be reviewed first by the Services to ensure that no listed species will be put in danger as a result.

According to the National Wildlife Federation (NWF), in 1973 109 species were listed as threatened or endangered. In 1998 listed species numbered 1,500. There are many success stories related to the ESA. One of the most significant is that of the peregrine falcon. Dichloro-diphenyl-trichloroethane (DDT), an organochlorine pesticide, was used throughout the 1940's, 1950's, and 1960's to control insects. Its use had damaging effects on many of the raptors, including the peregrine falcon. The pesticide resulted in significant thinning of the falcon's eggshells and caused unsuccessful reproduction. It was estimated that by the middle of the 1970's only 10 to 20 percent of the falcon's population had survived. DDT was banned in the United States in 1972. The falcon was placed on the endangered species list, its habitat was placed under protection, and six thousand captive-bred young were released into the wild. By 1998 the peregrine falcon had made a dramatic recovery and was being considered for removal from the endangered list.

There are many more cases of species recovering from near extinction, including gray wolves, whooping cranes, California condors, black-footed ferrets, and red wolves. According to the NWF, of all the species ever placed on the endangered list, only two have become extinct.

HISTORY OF ENVIRONMENTAL LOBBYING

For as long as legislators have been passing laws, individuals or groups have been trying to influence their decisions. The idea of lobbying has existed for centuries, although it was not always referred to as such. The first documented legislative use of the word "lobby" was in the 1808 annals of the tenth Congress. The 1946 Federal Regulation of Lobbying Act set many of the standards lobbying groups must follow. The law defined a lobbyist as any person who collects or receives money intended for the influencing of legislation by way of direct contact with Congress. One of the basic stipulations of the law required all lobbyists to register with the federal government each year in order to provide a concise documentation of the active lobbying groups and the organizations they represented. However, because of the unclear nature of the act, many individuals or groups that participated in lobbying either did not regularly register or did not report the name of the represented organization. Therefore, in 1995 Congress passed the Lobbying Disclosure Act. These amendments to the original law clarified the issue of registration.

Lobbying groups exist for a vast range of purposes and in support of many different causes. There are many groups and organizations that lobby to influence legislation concerning the environment. While there are many groups who represent industries and press for the lowering of air- and water-quality standards, there are also environmental groups that demand higher standards. Environmental and conservation groups have been actively participating in the legislative process since the early twentieth century. The Sierra Club was one of the first environmental groups to become involved in political matters.

In order for a lobbying group or a lobbyist to be effective, several features are necessary. A group needs to have financial support in order to make campaign contributions and gain access to adequate research. It is also beneficial if the represented organization has many members and a good relationship with their lobbyist. Most important, the lobbyists must have a thorough understanding of the issue for which they are lobbying. Many of the legislators know little or nothing about the particular issue being considered. Therefore, they depend on lobbyists to provide a clear, concise description of the issue. Lobbying groups can influence legislation in many different ways, such as providing testimony at hearings, meeting directly with members of Congress, or organizing grassroots campaigns. Grassroots campaigns are designed to involve not only the members of an organization but the general public as well.

ENVIRONMENTAL LOBBYING GROUPS

There are thousands of environmental lobbying groups in the United States. Some are established organizations that date back to the nineteenth century, while others have only been in existence for a few decades. In general, the primary focus of such groups is not lobbying, but rather conservation. However, either through grassroots campaigns or direct contact with Congress, many of these groups or their representatives influence legislation. Among the most influential groups have been the Sierra Club, the National Resources Defense Council (NRDC), and the Environmental Defense Fund (EDF).

The Sierra Club, led by John Muir, was organized in 1892 in the San Francisco Bay area. The original goal of the club was to protect and promote the Yosemite and Sierra Nevada regions. In 1900 the club had 384 members. In 1998, nearly one century later, the membership had grown to over 500,000. The Sierra Club is one of the oldest and largest environmental organizations and has had a dramatic impact on the environment. A few of its success stories include the establishment of Glacier National Park in 1910, the enlargement of Sequoia National Park in 1926, and the addition of 47,000 acres of land to the Olympic National Park in 1953. After a lengthy campaign by the Sierra Club, Congress passed the Wilderness Act in 1964 to establish the National Wilderness Preservation System. In 1975 the club convinced Congress to enlarge Grand Canyon National Park and in 1994 lobbied Congress to pass the California Protection Act, which provided protection for more than 7 million acres of national forests.

The National Resources Defense Council (NRDC) also has an impressive success record. Founded in 1971, the NRDC has over 400,000

members and is composed of eight different programs. In 1998 the NRDC's expenses totaled $27.6 million, $376,000 of which was devoted to legislative action. With such large membership and financial support, the council's impact on the environment has been extensive. In 1978 the NRDC won the fight to remove chlorofluorocarbons (CFCs), which damage the ozone layer, from aerosol cans. The council helped to obtain federal protection for one million acres of land in Alaska in 1980. In 1991, with the help of the NRDC, a bill was defeated that would have allowed oil drilling in the Arctic National Wildlife Refuge.

A final example of an influential environmental lobbying group is the Environmental Defense Fund (EDF). Organized in 1967 by several Long Island scientists, the original goal of the EDF was to ban the use of DDT as a pesticide. The charter members of the EDF won their fight, and the organization has continued defending the environment. With more than 300,000 members, the EDF plays an active role in writing environmental laws on the federal and state level, working directly with businesses to reduce the amount of wastes they produce, and protecting endangered wildlife. Among the EDF's accomplishments were adding all hunted whales to the endangered species list in 1970, fighting for the ban of asbestos in hair dryers in 1979, and convincing federal legislators to cut the amount of lead in gasoline by 90 percent in 1985.

Carolyn Simmons and Massimo D. Bezoari

SUGGESTED READINGS: *The Clean Water Act: 20 Years Later* (1993), by Robert W. Adler, Jessica C. Landman, and Diane M. Cameron, depicts the effects of the Clean Water Act using many graphs, charts, and tables for easy comprehension. Derek Elsom's *Atmospheric Pollution* (1987) describes the air pollution problems in the United States and other countries. For an overall understanding of how Congress operates, see Edward V. Schneier and Bertram Gross, *Legislative Strategy: Shaping Public Policy* (1993). A detailed and well-documented history of lobbying can be found in James Deakin's *The Lobbyists* (1966). An easy-to-read synopsis of each environmental law can be found in *Environmental Compliance in Alabama* (1998), edited by Sandra Romano. *Environmental Law: From Resources to Recovery* (1993), edited by Celia Campbell-Mohn, is a thorough discussion of the history of environmental law and its role in the future.

SEE ALSO: Air pollution policy; Clean Air Act and amendments; Clean Water Act and amendments; Endangered Species Act; Environmental Protection Agency; Hazardous and toxic substance regulation.

Environmental Protection Agency

DATE: established December 2, 1970
CATEGORY: Ecology and ecosystems

The Environmental Protection Agency (EPA) is an agency of the United States government charged with administering various environmental regulatory and distributive programs as well as conducting various environmental research activities.

From its inception in 1970, the EPA has been beset by differing conceptions of its mission as the primary federal environmental regulatory agency. The EPA expanded during the 1970's but came under severe attack in the 1980's, particularly during Anne Gorsuch Burford's tenure as director. The agency was revitalized, and its mission expanded in the late 1980's and early 1990's. The EPA has directed its attention to protecting the health of the human population by reducing the pollution of the environment. It has attempted to achieve this goal primarily through enforcing congressional legislation and issuing regulations.

The EPA consists of a director (appointed by the president), a headquarters in Washington, D.C., and ten regional offices, each headed by a regional administrator. In some cases the implementation of the EPA's regulatory burden is entrusted to state environmental agencies. The major regulatory tasks assigned to the agency center on air quality, water quality, disposal of hazardous and radioactive wastes, regulation of chemicals (including pesticides), and setting noise levels for construction equipment, transportation equipment, motors, and electronic equipment. Legislation that the EPA is charged with implementing includes the Clean Air Act (1963), the Clean Water Act (1965); the Resource Conservation and Recovery Act (1976); the Toxic Substances Act (1976); the Federal Insecticide, Fungicide, and Rodenticide Act (1947); and Superfund (1980) legislation.

The EPA often suffers from overload and finds it difficult to carry out its various mandates. This inability stems, in part, from congressional action. Congress has often specified unrealistic deadlines for action with various statutory penalties should the EPA not comply, coupled to detailed management instructions. In addition, some members of Congress have used environmental legislation as political pork barrels so that the EPA has not always been able to allocate its funds in the most efficient fashion. Also, as Marc Landy, Marc Roberts, and Stephen Thomas point out in *The Environmental Protection Agency: Asking the Wrong Questions from Nixon to Clinton* (1994), the EPA itself has often "asked the wrong questions" concerning its mission. By concentrating on enforcement, the EPA has not always been able to establish its scientific credentials. The EPA has also missed opportunities to educate the public regarding environmental issues; therefore, citizens often have unrealistic expectations regarding the agency's ability to deal with environmental problems. The leadership of the EPA has not always been attentive to the need to develop a strategic perspective in dealing with environmental issues, a failing that is fostered by leaving conceptual ambiguities unresolved. At times factionalism within the agency has detracted from the accomplishment of its mission.

Protecting the public's health is at the core of the EPA's mission. Regulating human contact with dangerous pesticides and improving water quality are two good examples of this mission. Cancer prevention, while not always a stated goal, has been one of the key concerns of the agency. In the late 1970's the EPA, the Food and Drug Administration (FDA), the Occupational Safety and Health Administration (OSHA), and the Consumer Product Safety Commission worked together through the Interagency Regulatory Liaison Group to formulate standards for dealing with chemicals that might possibly cause cancer. However, the resulting document did not enunciate a clear standard of risk assessment or establish a sound level of cancer risk. It also failed as an attempt to educate the public concerning cancer risks and scientific uncertainty.

The risk of cancer also underlies the public controversy concerning the implementation of Superfund legislation. Part of the problem with Superfund was Congress's inability to set cleanup priorities or cleanup levels for Superfund sites. The EPA was deficient in providing Congress with the detailed information necessary to make these decisions. Because Superfund was conceptually deficient, the EPA often had to react to what the public perceived as crisis situations such as the Times Beach, Missouri, dioxin cleanup in the early 1980's. Such actions were not always based on reliable research and often displayed an inability to educate the public regarding environmental risk.

Although public health concerns have been the stated rationale for much of the EPA's actions, a further concern has been protecting the environment and resource conservation. Enhancing water quality, for example, has obvious benefits for aquatic life. Controlling the negative impact of pesticides has an impact all the way along the food chain. The threat of acid deposition to forest products and water quality in some regions of the country is substantial, and the EPA's efforts at establishing air-quality standards under the 1990 revisions of the Clean Air Act are an effort to deal with this issue. Indirectly, the EPA's regulations dealing with improving automobile mileage have decreased the consumption of steel and oil. Waste reduction has been the goal of much of the EPA's regulatory efforts. Waste reduction reduces the amount of natural resources consumed, providing a decreased environmental impact.

The 1984 revisions of the Resource Conservation and Recovery Act (RCRA) directed the EPA to advocate conservation as a means of dealing with hazardous materials. The RCRA directed that the placement of hazardous wastes in landfills was the least favored option in dealing with these materials. The most favored approach was for an industry to generate less of the material, thus practicing resource conservation. In the mid-1990's this ideal was still unrealized, although some industries were finding substitutes for hazardous materials.

John M. Theilmann

SUGGESTED READINGS: A fundamental starting point for an examination of the EPA is Marc K. Landy, Marc J. Roberts, Stephen R. Thomas, *The Environmental Protection Agency: Asking the Wrong Questions from Nixon to Clinton* (1994). Other useful studies include Peter C. Yeager, *The Limits of Law* (1991), Walter A. Rosenbaum, *Environmental Politics and Policy* (1995), and Paul R. Portney, editor, *Public Policies for Environmental Protection* (1990).

SEE ALSO: Clean Air Act and amendments; Clean Water Act and amendments; Environmental policy and lobbying; Hazardous waste and toxic substance regulation; Superfund.

Erosion and erosion control

CATEGORY: Land and land use

Erosion is the loss of topsoil through the action of wind and water. Erosion control is vital because soil loss from agricultural land is a major contributor to nonpoint-source pollution and desertification and represents one of the most serious threats to world food security.

In the United States alone some two billion tons of soil erode from cropland on an annual basis. About 60 percent, or 1.2 billion tons, is lost through water erosion, while the remainder is lost through wind erosion. This is equivalent to losing 0.3 meters (1 foot) of topsoil from two million acres of cropland each year. Although soil is a renewable resource, soil formation occurs at rates of just a few inches per hundred years, which is much too slow to keep up with erosive forces. The loss of soil fertility is incalculable, as are the secondary effects of polluting

Antierosion fences have been placed along several stretches of the Massachusetts shoreline in an effort to stop erosion caused by rising sea levels and strong currents. In Plymouth, such currents threaten to destroy houses built near eroding cliffs. (AP/Wide World Photos)

surrounding waters and increasing sedimentation in rivers and streams.

Erosion removes the topsoil, the most productive soil zone for crop production and the plant nutrients it contains. Erosion thins the soil profile, which decreases a plant's rooting zone in shallow soils, and can disturb the topography of cropland sufficiently to impede farm equipment operation. It carries nitrates, phosphates, herbicides, pesticides, and other agricultural chemicals into surrounding waters, where they contribute to cultural eutrophication. Erosion causes sedimentation in lakes, reservoirs, and streams, which eventually require dredging.

There are several types of wind and water erosion. The common steps in water erosion are detachment, transport, and deposition. Detachment releases soil particles from soil aggregates, transport carries the soil particles away and, in the process, scours new soil particles from aggregates. Finally, the soil particles are deposited when water flow slows. In splash erosion, raindrops impacting the soil can detach soil particles and hurl them considerable distances. In sheet erosion, a thin layer of soil is removed by tiny streams of water moving down gentle slopes. This is one of the most insidious forms of erosion because the effects of soil loss are imperceptible in the short term. Rill erosion is much more obvious because small channels form on a slope. These small channels can be filled in by tillage. In contrast, ephemeral gullies are larger rills that cannot be filled by tillage. Gully erosion is the most dramatic type of water erosion. It leaves channels so deep that even equipment operation is prevented. Gully erosion typically begins at the bottom of slopes where the water flow is fastest and works its way with time to the top of a slope as more erosion occurs.

Wind erosion generally accounts for less soil loss than water erosion, but in states such as Arizona, Colorado, Nevada, New Mexico, and Wyoming, it is actually the dominant type of erosion. Wind speeds 0.3 meters (1 foot) above the soil that exceed 16 to 21 kilometers per hour (10 to 13 miles per hour) can detach soil particles. These particles, typically fine-to medium-size sand fewer than 0.5 millimeters (0.02 inches) in diameter, begin rolling and then bouncing along the soil, progressively detaching more and more soil particles by impact. The process, called saltation, is responsible for 50 to 70 percent of all wind erosion.

Larger soil particles are too big to become suspended and continue to roll along the soil. Their movement is called surface creep.

The most obvious display of wind erosion is called suspension, when very fine silt and clay particles detached by saltation are knocked into the air and carried for enormous distances. The Dust Bowl of the 1930's was caused by suspended silt and clay in the Great Plains of the United States. It is also possible to see the effects of wind erosion on the downward side of fences and similar obstacles. Wind passing over these obstacles deposits the soil particles it carries. Other effects of wind erosion are tattering of leaves, filling of road and drainage ditches, wearing of paint, and increasing incidence of respiratory ailments.

The four most important factors affecting erosion are soil texture and structure, roughness of the soil surface, slope steepness and length, and soil cover. There are several passive and active methods of erosion control that involve these four factors. Wind erosion, for example, is controlled by creating windbreaks, rows of trees or shrubs that shorten a field and reduce the wind velocity by about 50 percent. Tillage perpendicular to the wind direction is also a beneficial practice, as is keeping the soil covered by plant residue as much as possible.

Water erosion is controlled by similar cultural practices. Highly erosive, steeply sloped land can be protected by placing it in the U.S.-government-sponsored Conservation Reserve Program. Tillage can be done along the contour of slopes. Long slopes can be shortened by terracing, which also reduces the slope steepness. Permanent grass waterways can be planted in areas of cropland that are prone to water flow. Likewise, grass filter strips can be planted between cropland and adjacent waterways to impede the velocity of surface runoff and cause suspended soil particles to sediment and infiltrate before they can become contaminants.

Conservation tillage practices such as minimal tillage and no-tillage are being widely adapted by farmers as a simple means of erosion control. As the names imply, these are tillage practices in which as little disruption of the soil as possible occurs and in which any crop residue remaining after harvest is left on the soil surface to protect the soil from the impact of rain and wind. The surface residue also effectively impedes water flow, which causes less suspension of soil particles. Because the soil is not disturbed, practices such as no-tillage also promote rapid water infiltration, which also reduces surface runoff. No-tillage is rapidly becoming the predominant tillage practice in southeastern states such as Kentucky and Tennessee, where high rainfall and erodible soils occur.

Mark Coyne

Leading No-Tillage States, 1994-1997

STATE	PERCENT OF CROPLAND PLANTED USING NO-TILL PRACTICES		
	1994	1996	1997
Kentucky	44	51	48
Maryland	41	46	45
Tennessee	42	44	43
West Virginia	37	39	39
Delaware	37	38	38
Ohio	35	37	36

Source: Data adapted from G. R. Haszler, "No Tillage Use for Crop Production in Kentucky Counties in 1997," *Soil Science News and Views* 19 (1998).

SUGGESTED READINGS: Several texts concerning soil science and management thoroughly cover the mechanisms of erosion and methods of erosion control. *Soil Science and Management* (1997), by Edward Plaster, has an excellent chapter on erosion, while *Soil and Water Conservation Engineering* (1993), by Glen Schwab et al., addresses erosion control measures. *Plowman's Folly* (1943), by Edward Faulkner, is a classic diatribe against the effects of conventional agriculture and agricultural practices on soil erosion and environmental quality.

SEE ALSO: Deforestation; Desertification; Dust Bowl; Logging and clear-cutting; Runoff: agricultural; Soil conservation; Strip farming.

Experimental Lakes Area

DATE: established 1968
CATEGORY: Water and water pollution

Founded in 1968, the Experimental Lakes Area (ELA) is a research facility that studies environmental problems, especially acidification, in lakes in eastern Canada. ELA scientists conduct research aimed at solving water pollution problems and maintaining the health of freshwater lakes in Canada.

Increasing acidification in lakes in Ontario and Nova Scotia led to significant declines of fish populations during the 1960's through the 1980's. As a result, Canadian environmentalist David Schindler founded the Experimental Lakes Area research facility in 1968 in order to investigate acidification and other environmental problems that were impacting the lakes of eastern Canada. The goals of the ELA are fourfold: to develop better understanding of global threats to the environment through knowledge gained from ecosystem, experimental, and scientific research; to monitor and demonstrate the impacts of human activities on watersheds and lakes; to develop appropriate environmental responsibility for the preservation, restoration, and enhancement of ecosystems; and to educate and promote environmental protection and conserva-

tion for ecosystems. The ELA includes fifty-eight small lakes and their drainage basins, plus three additional stream segments.

The ELA is operated by the Central and Arctic Region of the Canadian Department of Fisheries and Oceans from its Freshwater Institute in Winnipeg. Since the ELA research facility is located in a sparsely inhabited region of southern Ontario, it is relatively unaffected by external human influences and industrial activities. As such, it serves as a natural laboratory for the study of physical, chemical, and biological processes and interactions operating on an ecosystem over a large area and a multiyear time scale.

With renewed operating support in the late 1990's, the ELA took on several new experimental studies in addition to whole-lake acidification experiments and eutrophication recovery studies. For example, one ELA ecosystem study investigated additions of trace amounts of mercury to one of the ELA lakes. It is generally believed that high concentrations of methyl mercury, the most toxic form of mercury, in fish in remote lakes is caused by elevated inputs of atmospheric inorganic mercury deposited directly into lakes and indirectly through their watersheds. The ELA is investigating this hypothesis as well as an alternative that suggests that the most important source of mercury to remote lakes is geologic mercury that originates from the weathering of mineral deposits in lake basins.

ELA scientists are also focusing on the effects of climate change, dissolved organic carbon, and ultraviolet radiation on lakes. One study investigates the effects of experimentally deepening the mixed layer of a lake, thereby exposing more of the lake water and organisms to surface radiation and simulating the effects observed in natural ELA lakes during two decades of warming between 1970 and 1990. Another area of research involves reducing the natural color of an ELA lake to investigate the effects of increased ultraviolet radiation on lake life-forms. Other projects include studies of the effects of persistent toxic substances—such as cadmium and hydrocarbons—in lakes, contributions of forest materials to lake nutrient inputs, and the alteration of food chain processes caused by human intervention.

Alvin K. Benson

SEE ALSO: Acid deposition and acid rain; Ecosystems; Water pollution; Water quality.

Extinctions and species loss

CATEGORY: Animals and endangered species

Before humans appeared on the earth, animal and plant species were eliminated through extinction at an average rate of about one per year. However, the influence of human population growth and technology on the biological ecosystems of the world has caused extinction rates to skyrocket to between four thousand and fifty thousand species per year.

Extinctions that occurred during the past 4.6 million years were primarily caused by the earth's changing surface and climate. The rates of speciation (the creation of new species) and extinction have been fairly constant throughout history. However, there have been five mass extinctions in Earth's history. The most famous extinction occurred 65 million years ago at the end of the Mesozoic era. A large number of land and marine animals, including the dinosaurs, suddenly disappeared from the fossil record. Some scientists credit this extinction to a large meteor that hit the earth and subsequently caused climate changes that devastated some life-forms. Other scientists credit the demise of dinosaurs to the evolutionary superiority of mammals.

CAUSES OF RECENT EXTINCTIONS

As human populations grow, people clear forests and land at increasing rates to build homes, businesses, and farms. This process disrupts the habitats that are home to microbes, plants, and animals, which all have unique ecological relationships to one another. Habitat destruction is one of the most prominent causes of extinctions in recent history. In destroying one species, a secondary extinction can often occur. For example, the worldwide extinction of the dodo bird was followed by the extinction of the calveria tree because the dodo bird was instrumental in the calveria's repro-

duction. The dodo would eat the tree's seeds, digest the hard outer shells, and then excrete the exposed seed. The extinction of a food species may cause the extinction of the species that feeds on it. The panda bear of China exists largely on a diet of bamboo. As bamboo became scarce, the panda population began falling.

Highway development can cause fragmentation of a habitat, preventing migration between the two fragments. This can cause a species to die out if neither area is capable of supporting a viable population of the species. This is also true of habitats that are decreasing in size. In the United States, 99.8 percent of tall grass prairies have been destroyed, 50 percent of the wetlands are gone, and 98 percent of virgin and old-growth forests have been cut. At least five hundred native species have become extinct, with tremendous losses in the populations of animals such as wolves, black bears, bison, and cougars.

Tropical rain forests, which are home to more than one-half of the world's plants and animals, are estimated to lose approximately seventeen thousand species per year because of deforestation. Many of the lost species are unique to the tropical forests. This destruction has caused the near extinction of many migratory songbirds that winter in Central and South America. Meanwhile, the songbirds' northern summer homes are becoming fragmented, causing more losses to the population.

Even when a species is not completely obliterated by development, extinction can occur if the population becomes too small to recover. Such a population is said to fall into an extinction vortex. This problem may occur if there are too few females left in the population to breed, or, if the habitat is too fragmented, if individuals are not able to locate a partner with which to mate. If enough mates cannot be found, genetic inbreeding can destroy the viability of the species. The panther population in Florida, for example, has fallen to fewer than fifty animals. Because of genetic inbreeding, the male panthers suffer many testicular and other deformities.

Small populations that are vulnerable to environmental fluctuations may also fall into an extinction vortex. For example, extremely harsh

Estimated Loss of Species after Isolation of Five Tropical Land-bridge Islands

	NUMBER OF SPECIES LOST			PERCENT LOST IN FIRST CENTURY
ISLAND	In 10,000 Years	First 100 Years	First 1,000 Years	
Trinidad	144	2	22	0.6
Margarita	246	10	80	3.2
Coiba	172	5	45	2.2
Tobago	218	8	63	2.6
Rey	179	8	63	3.7

Source: Data adapted from John Terborgh, *Diversity and the Tropical Rain Forest.* New York: W. H. Freeman, 1992.

winter weather could wipe out an entire species that has already been reduced to a small population. On the other hand, larger groups are better able to survive adversity. The chance of being destroyed by environmental fluctuations increases exponentially with a decreasing population size.

The smallest population of a species that is able to stay above the extinction vortex is often called the minimum viable population (MVP). If a population declines below this size, it is usually only a matter of time before breeding problems and climatic fluctuations will destroy the whole population. Likewise, if a habitat is reduced to a point where it is unable to support the MVP in an adverse year, the population will vanish.

POLLUTION, OVERUSE, AND OVERHUNTING

Pollution can kill many plants and animals and at the same time alter and destroy habitats. Acid rain and air pollution are detrimental to forests and forest animals. Sediment and excessive nutrients that run off into lakes, rivers, and bays often have adverse effects on aquatic life. Pesticides that degrade at slow rates, such as dichloro-diphenyl-trichloroethane (DDT), have caused large losses and near extinction in some bird species. The effects of chemicals are amplified as they become more concentrated in the fatty tissue of organisms at higher trophic levels in the food chain.

Exotic species are animals or plants that have been introduced into an area to which they are not native. It is estimated that the introduction of exotic species has contributed to approximately 40 percent of all animal extinctions worldwide since 1600. Because the new exotic species may not have any natural predators or competitors in its new habitat, it can dominate the new ecosystem and reduce the population of many native species. Islands are particularly vulnerable to exotic species. The brown tree snake was introduced to the Pacific island of Guam during World War II. In the ensuing forty years, the snakes destroyed eleven of the eighteen species of birds native to the island. Because the brown snake has no natural predator on Guam, some areas have more than five thousand snakes per square mile.

During the nineteenth century whales were harvested at a rate of fewer than one hundred every three years. However, faster boats and more efficient weapons were developed, and by 1933 the whaling industry was killing thirty thousand whales per year; by 1967 that number had more than doubled. Interestingly, 2.5 million barrels of whale oil were harvested in 1933, as opposed to only 1.5 million barrels in 1967. This is because the larger whales, the blues and fins, had been hunted to the brink of extinction by 1967.

In the United States, overhunting has led to

the extinction or near extinction of many species, such as the American bison. Early in the nineteenth century the passenger pigeon was one of the most abundant birds on earth. Alexander Wilson, a renowned ornithologist, is said to have observed a flock of passenger pigeons that took several hours to fly by him. He estimated that the flock, which appeared to be approximately 1.6 kilometers (1 mile) wide and 386 kilometers (240 miles) long, was composed of two billion birds. The passenger pigeon became extinct in 1914, largely because of overhunting by market hunters. The hunters used nets, guns, and even dynamite to trap the birds, which were viewed as a culinary delicacy.

Many species are under government and international protection but are still hunted because the economic incentive of selling skins or horns outweighs the risk of a small fine or short prison sentence. A coat made from the fur of a Bengal tiger can sell for $100,000, and rhinoceros horns can sell for as much as $12,500 per pound. The horn is then ground into powder to make medicine and aphrodisiacs. It is estimated that the poaching industry brings in five billion to eight billion dollars per year.

PEST CONTROL, MONOCULTURES, AND WILD PLANTS

Species can become endangered or even extinct if they must compete with the human population for food. African elephants have been killed by farmers to prevent the elephants from eating or trampling food crops. In the United States the Carolina parakeet was exterminated by farmers in 1914 because it fed on fruit crops, while 98 percent of the prairie dog population has been exterminated with poisons so that horses and cattle would not break their legs stepping into burrows. The prairie dog's primary predator, the black-footed ferret, has come close to extinction with the massive reduction of the population of its food source.

Extinction and loss of species is a critical problem for the human population because human life is dependent on the biodiversity of species. One area in which this is apparent is agriculture. Farmers have developed high-yielding monocultures of approximately thirty species,

each with a minimum of genetic variation. These monocultures lack the vigor of wild plants, which are constantly developing new ways to adapt to adverse conditions and fend off the animals and microorganisms that attack them. When monoculture crops fail because of disease or other problems, plant breeders must go back to wild species to find the traits that their crops need to thrive and breed these characteristics into their crop. If wild species are not protected, the genetic gene pool will dwindle and eventually disappear.

Wild plants, especially in the tropical rain forests, have important uses as medicine. In 1960 the rosy periwinkle, which grows in Madagascar, was found to contain two chemicals that revolutionized the treatment of childhood leukemia and Hodgkin's disease. The use of these drugs brought about a 95 percent remission rate in children who previously had little chance of surviving leukemia. Taxol, a drug extracted from the bark of the Pacific yew tree, has been valuable in the treatment of breast, ovarian, and small-cell cancer. A plant related to the periwinkle, rauwolfia, provides an alkaloid that is widely used in the control of high blood pressure. Digitalis, which comes from the foxglove plant, is a highly effective drug used in the treatment of chronic heart failure. Of the world's 250,000 plant species, only about five thousand have been studied for medicinal use. It is believed that there may be thousands of plants in tropical forests that have cancer-fighting properties.

Toby Stewart and Dion Stewart

SUGGESTED READINGS: *Extinctions: The Causes and Consequences of the Disappearance of Species* (1981), by Paul Ehrlich and Anne Ehrlich, is one of the premier books on extinction throughout the world. *Environment 98/99* (1998), edited by John L. Allen, provides interesting articles on extinction and other aspects of the environment. Another book that contains thought-provoking articles on extinction is *The Last Extinction* (1987), edited by Les Kaufman and Kenneth Mallory.

SEE ALSO: Biodiversity; Dodo birds; Endangered species; International Whaling Ban; Passenger pigeon.

Exxon Valdez oil spill

DATE: March 24, 1989
CATEGORY: Water and water pollution

In March of 1989 the Exxon Valdez supertanker ran aground in a shallow stretch of Prince William Sound off the coast of Alaska, resulting in the largest oil spill in the history of the United States.

Oil spills are common occurrences in United States waters. According to a U.S. Coast Guard report, there are about ten thousand spills in and around U.S. waters each year that total 15 million to 25 million gallons of oil. These oil spills may result from drilling accidents such as the Santa Barbara, California, spill in 1969 or may be related to problems associated with supertankers such as the *Exxon Valdez*.

The *Exxon Valdez* was a single-hulled supertanker operated by the Exxon Company. The ship was 300 meters (987 feet) long and cost about $125 million to build. It was equipped with state-of-the-art instruments for depth sounding, guidance, and navigation. On March 23, 1989, the vessel was loaded with more than ten million barrels, or approximately 420 million gallons, of North Slope crude. The oil had been transported about 1,300 kilometers (800 miles) from Prudhoe Bay near the Arctic Circle to the Port of Valdez in southern Alaska. Shortly after midnight on Friday, March 24, the ship left port and traveled west and southwest down the Valdez fiord to the vicinity of Bligh Island in Prince William Sound, where it ran aground on Bligh Reef.

At the time the ship left port, conditions for sailing were ideal: light winds, calm seas, and good visibility. However, because of miscalculations by the officers in charge, the ship hit a chain of rocks about 4 kilometers (2.5 miles) west of Bligh Island. The rocks tore a gash in the tanker hull and allowed an estimated 10 million to 11 million gallons of crude oil to escape before noon. The oil spread across 4,600 square kilometers (1,776 square miles) of water in Prince William Sound and the Gulf of Alaska. Approximately 5,100 kilometers (3,169 miles) of shoreline received some oiling. The oiled areas included a number of fishing villages, a national forest, state and national parks, national wildlife refuges, critical habitat areas, and a state game sanctuary.

The damage to coastal areas in southeastern Alaska and marine life in Prince William Sound was enormous. The spill affected the livelihood of villagers along the west side of the sound, the Alaska and Kenai Peninsulas, the Kodiak Archipelago, and part of Cook Inlet. Both large and small marine mammals, birds, fish, mollusks, and plant life were devastated. Thousands of dolphins, sea lions, sea otters, and harbor seals were killed or had their environment fouled. An estimated 250,000 to 300,000 sea birds were also killed by the oil spill. These included harlequin ducks, pigeon quillemotes, common murres, and marbled murrelets. A significant number of bald eagles (more than one hundred), Canadian geese, and cormorants also perished. Economically important fish such as pink and sockeye salmon and Pacific herring, as well as a large number of small forage fish (capelin, pollock, sandlance, and smelt), died as a result of the spill. Shellfish, such as clams, crabs, oysters, and shrimp, were also killed or threatened.

In *Environmental Science* (1993), G. Tyler Miller, Jr., remarked that "most forms of marine life recover from exposure to large amounts of crude oil within three years." This statement was generally true for most marine life in the sound with the exception of harbor seals, which had not recovered by the mid-1990's.

Various techniques were used to consume or disperse the oil slick and beach coating. Attempts to ignite the oil met with limited success because they occurred after most of the lighter, more volatile components had evaporated, leaving behind emulsified, pancakelike layers of crude oil. Boat skimmers, which are designed to confine and collect the oil with floating booms, recovered less than 5 percent of the oil. These booms are most effective when the wave height is below 1 meter (3 feet). The application of chemical dispersants to the oil slick also was not effective. In cold water, the mousselike layers of oil residue are almost impossible to break up chemically.

Techniques used to wash beach rocks ranged from hand-applied cold water to steam cleaning. The latter was effective in some areas. Exxon workers substantially reduced the oil coating on beach rocks at some islands in the northern part of Prince William Sound. These include Lone, Naked, and Smith Islands. However, this method may have caused more harm than good because most of the small, beneficial organisms along the beach and in the tidal zone were killed by the hot waters. Meanwhile, tons of oiled gravel underlying mussel beds were removed from Prince William Sound and the Kenai Peninsula by recovery team workers and local residents. The oiled gravel was replaced with clean sediment.

Perhaps the most effective technique used to clean the oil was bioremediation. Workers sprayed a fertilizer solution along miles of the cobbled beaches in Prince William Sound in an effort to promote the growth of naturally occurring, oil-eating microbes. Such microorganisms primarily consist of spiral-type bacteria that can rapidly develop under ideal conditions. Most of these encapsulated cells reproduce asexually by splitting in half, a process known as fission. These oil-metabolizing bacteria can double their number in fewer than twenty-four hours.

As a result of the spill, both criminal charges and civil damage claims were filed against Exxon by the state of Alaska and the United States. In the civil settlement, Exxon was required to pay $900 million over a ten-year period. A state-federal trustee council consisting of six members was designated to administer the settlement and coordinate studies of the spill effects on wildlife and the environment. In an agreement concerning the criminal charges, Exxon was originally required to pay a fine of $250 million. However, because of Exxon's cooperation with

governmental agencies during the cleanup and the quick payment of most private claims, $125 million of the fine was forgiven.

Donald F. Reaser

SUGGESTED READINGS: Excellent case studies of the *Exxon Valdez* oil spill are provided in Carla Montgomery, *Environmental Geology* (1995), G. Tyler Miller, Jr., *Environmental Science* (1993), and Bernard W. Pipkin, *Geology and the Environment* (1994). Pipkin shows the track of the *Exxon Valdez* supertanker from the port of Valdez to the Bligh Reef. He also discusses the death toll of marine life from the disaster as well as from other supertanker spills. Keith Kvenvolden et al., "Ubiquitous Tar Balls with a California-source

On March 26, 1989, two days after the Exxon Valdez *ran aground in Prince William Sound, the* Exxon Baton Rouge *(smaller ship) pulled alongside in an attempt to offload crude oil. The accident was the worst oil spill in the history of the United States.* (AP/Wide World Photos)

Signature on the Shorelines of Prince William Sound, Alaska," *Environmental Science and Technology* (October, 1995), provides evidence for older (pre-1989) hydrocarbon residues along the shorelines of Prince William Sound. Jonathan Turk and Graham Thompson, *Environmental*

Geoscience (1995), contains an excellent discussion on the effects of hydrocarbons on living organisms.

SEE ALSO: *Amoco Cadiz* oil spill; *Argo Merchant* oil spill; Oil spills; *Sea Empress* oil spill; Tobago oil spill; *Torrey Canyon* oil spill.

F

Fish and Wildlife Service, U.S.

DATE: established 1940
CATEGORY: Animals and endangered species

The Fish and Wildlife Service was established in 1940 to protect fish and wildlife and to conserve their habitats in the United States.

During the late nineteenth century, resource use in the United States developed at such a rapid pace that it led to the abuse of the environment. One of the resources most depleted was the national fisheries. Major declines in fish populations and commercial fish catches led to the establishment of the Bureau of Fisheries in 1870. Its initial responsibilities were stream restoration, the improvement of fisheries, and the stocking of lakes and streams.

Concern for wildlife in relation to agriculture also led to the establishment of the U.S. Biological Survey in the late nineteenth century. Both of these agencies, through merger and reorganization, eventually became the Fish and Wildlife Service in 1940. During this evolution, a variety of federal acts were passed that strongly influenced the direction and authority of the Fish and Wildlife Service over environmental resources. The establishment of wildlife refuges was begun under President Benjamin Harrison. In 1900 the Lacey Act was passed, under which authority the federal government began to enforce laws regarding interstate and foreign commerce in wildlife. During Theodore Roosevelt's presidency, many more wildlife refuges were created from public domain lands. The Migratory Bird Treaty Act of 1918 significantly expanded the service's authority by involving two foreign governments (Canada and Mexico) in wildlife protection.

The National Wildlife Refuge System, which became a reality in 1929, eventually came under the stewardship of the Fish and Wildlife Service.

The Pittman-Robertson Act, which has been identified by some writers as the most significant wildlife act in U.S. history, was passed in 1937. It established an excise tax on firearms, archery equipment, and ammunition used in hunting, as well as a manufacturers' tax on handguns. These tax dollars, which are specified for wildlife and fisheries, are apportioned to state wildlife agencies through the Fish and Wildlife Service. This flow of money demonstrates and emphasizes the cooperative nature of Fish and Wildlife Service activities in environmental affairs.

The Fish and Wildlife Service's responsibilities include operating the national wildlife refuges, cooperating with state fish and game agencies, establishing university extension programs regarding fish and wildlife, and collecting, maintaining, and distributing statistical information on fisheries, wetlands, and estuaries. One of the most diverse areas of involvement is the Biological Services Program. Under the provision of this program, the Fish and Wildlife Service provides scientific information and methodologies on key environmental issues that impact fish and wildlife resources. To carry out all of these activities, the service has been organized into regional divisions. The output from this program reflects the wide range of regional environmental issues that have an impact on fish and wildlife resources.

Jerry E. Green

SEE ALSO: Wetlands; Wildlife management; Wildlife refuges.

Fish kills

CATEGORY: Animals and endangered species

A fish kill is a significant, sudden mass mortality of fishes. The cause of a fish kill is rarely obvious, and its determination often requires complex analyses of water quality, fish tissues, and environmental and human events that have occurred in the watershed.

Notification of a fish kill usually occurs after, rather than during, the event, in which case causative agents may have dissipated or become diluted, and fish tissue may have decomposed. Consequently, formal field investigation involves trained personnel, fishery biologists, fish pathologists, and conservation officers. Relevant field investigation requires collection of accurate, factual, and precise data and information. These are obtained through careful observation, interview of witnesses, and meticulous sampling of water and dead fish. Since fish kills caused by human activity may result in litigation, collected samples and other evidence must be processed according to prescribed methods and quickly transported to an analytical laboratory.

About one-half of reported fish kills are the result of a natural event. Natural causes include oxygen depletion; blooms of toxic algae; infec-

tious diseases caused by bacteria, viruses, protozoans, or parasites; stranding of fish schools; overabundant runoff of silt or ash from forest fires; toxic runoff water; and life-cycle events. In most cases a combination of causative factors occur. For example, during the summer when water temperatures exceed the optimal limits of some fish, the fish become physiologically stressed. If all other environmental factors remain optimal, fish tolerate the increase in temperature. However, oxygen solubility is lowered as temperature increases. Therefore, the concentration of dissolved oxygen is lower at a time when stress requires the fish to increase their oxygen uptake. Also, the increased metabolism of all other biological organisms puts additional demands on the limited supply of oxygen. The killer in this case is not temperature but oxygen depletion, the primary culprit of fish kills. Oxygen depletion can also be caused by input of high organic loads, algal blooms and their degradation, low water flow, and ice cover on a lake or pond during the winter. Snow cover reduces

Fish killed by siltation accumulate along the banks of the Potomac River near Alexandria, Virginia. (Archive Photos)

light penetration, which inhibits photosynthesis, the only source of dissolved oxygen in ice-covered lakes. This results in winter fish kills. A major cause of natural fish kills in coastal waters is blooms of microorganisms that generate red tides and other epidemiological effects.

A variety of human activities cause major fish kills. The dumping of treated and untreated sewage waste into natural waters artificially increases the nutrient load of the system. This in itself does not cause a fish kill, but it stimulates increased production and biodegradation processes, which reduce dissolved oxygen below critical levels. A more obvious cause is the introduction of toxic pollutants such as pesticides, heavy metals, and industrial wastes. Alterations of the landscape of a watershed increases water temperature and siltation; silt reduces the efficiency of oxygen uptake by the gills. Dams impede fish migration and, during low water levels, create pockets of deoxygenated water. Turbines of hydroelectric power plants and other water intake facilities are another detriment to fish. Disposal of undesirable fish during commercial fishing operations often generate fish kill reports.

Richard F. Modlin

SEE ALSO: Acid deposition and acid rain; Agricultural chemicals; Thermal pollution.

Flavr Savr tomato

CATEGORY: Agriculture and food

Flavr Savr tomatoes are grown on plants that have been genetically altered to delay fruit ripening and rotting. As the first genetically engineered food product on the market, the Flavr Savr tomato was closely monitored by food-safety advocates and environmentalists.

Tomatoes are an important agricultural crop for use as fresh produce and for the food-processing industry. To survive shipping and storage, tomatoes are typically harvested while still hard and green to avoid bruising and crushing, damage that makes them susceptible to rot. Treatment with ethylene gas—the natural ripening agent—

induces softening and a red color change, but such tomatoes do not taste like vine-ripened fruit.

In order to alter the ripening process and enhance flavor, scientists at Calgene in Davis, California, used genetic material known as antisense deoxyribonucleic acid (DNA). When inserted into tomato plants, this DNA interferes with a natural plant gene responsible for tomato softening. Since the normal genetic message is blocked, production of a critical protein that breaks down pectin is diminished, and the ripening process is inhibited. Flavr Savr tomatoes can be kept on the vine longer to turn red and develop better flavor, but the fruit remains firm. The fruit is later softened by ethylene treatment.

In 1991 Calgene asked the U.S. Food and Drug Administration (FDA) to examine data to determine the Flavr Savr tomato's safety. In May, 1994, the FDA approved the tomato; almost immediately, Calgene began marketing it under the brand name of MacGregor. Although not required, Calgene labeled their tomatoes as genetically engineered and provided information to consumers regarding the genetic alterations.

In 1992 the FDA ruled that genetically engineered foods did not require premarket approval or special labeling. This ignited fears that future genetically modified foods would not be as thoroughly tested as was the Flavr Savr tomato. Activists are especially concerned about the presence of bacterial antibiotic resistance genes in the foods. In approving the Flavr Savr tomato, the FDA concluded that these genes, which serve as markers to determine if an organism has been successfully modified, did not significantly differentiate the tomato from others. However, such markers might be allergenic and could potentially be transferred to intestinal bacteria in people who eat the tomato. The development of new antibiotic-resistant strains of bacteria is already a serious medical problem.

The environmental impact of genetically engineered crops is yet unknown, but there are two major concerns. Engineered plants could become "superweeds" that would damage ecosystems as well as agricultural lands; this, in turn, might necessitate additional herbicide use. The foreign genes present in engineered plants could also be transferred to wild relatives growing in the vicinity with unpredictable consequences.

Concerns raised by consumer groups and activists, a barrage of press using terms such as "Frankentomato," and general public distrust of technology all contributed to the poor market performance of the Flavr Savr tomato. Additionally, when grown commercially, the plants did not have acceptable yields or disease resistance, and the tomatoes did not withstand the shipping process as well as expected. All of these factors contributed to Calgene discontinuing production of the Flavr Savr tomato in 1997.

Diane White Husic

SEE ALSO: Biotechnology and genetic engineering; Genetically engineered foods.

Flood control

CATEGORY: Water and water pollution

Flood control involves the containment or diversion of water flows to reduce flood damage. It is only intended to provide a specified level of flood protection and not to eliminate flooding altogether. The Army Corps of Engineers and the Natural Resources Conservation Service are the two governmental agencies responsible for flood control in the United States.

Structural flood control relies on the construction of dams, levees, and diversion channels. Reservoirs formed by the dams store large quantities of floodwater, providing flood protection to areas downstream. However, the reservoirs may submerge large tracts of productive agricultural land and environmentally sensitive areas. Forests and wildlife may be lost. The soil, geologic stability, and ecosystem of the area may be altered. People living in these areas may be permanently displaced. Varying the amount of water released for flood control from the reservoirs may change the natural flow pattern downstream. The amount of silt and sediment transported by the water may change. The riverine ecology may be destroyed. Also, some of the worst floods in history have been caused by failures of flood-control structures.

A thorough review of the potential environmental impact is necessary before flood-control projects are launched. Throughout the world, many large dams have been built for flood control. The environmental implications of those projects were not adequately investigated, perhaps because of the difficulty in forecasting ecosystem changes that might occur. However, increasing awareness of the adverse impact that dams and floodplain developments may have on the environment has caused public opposition to the construction of flood-control structures to increase. The Narmada Valley project in India and the Three Gorges Dam project in China are two widely known examples.

Levees have been constructed in many places along the Missouri and Mississippi Rivers in the United States. The floods of 1993, 1995, and 1997 saw a number of these levees fail, resulting in huge amounts of damage. Some people believe that levees, dams, and dikes that increase flood heights by straight-jacketing rivers should not have been built. Instead, the floodplains should have been left alone so that they can store and slowly release flood waters. Natural floodplains filter runoff, provide fish and wildlife habitat, and contribute organic nutrients for the aquatic food chain. Levees eliminate these environmental benefits. Also, levees provide a false sense of security for people encroaching into the floodplains.

Another approach to flood control involves the management of the land in the drainage basins and the construction of flood-retarding reservoirs in the upstream areas of the rivers. Grass and trees on the slopes of the land that drains to rivers help the soil absorb and retain more water. Ditches and terraces along the slopes slow the flow of water toward the rivers. Typically, the upstream approach includes land management methods such as wood- and rangeland management, strip cropping, and contour cultivation on farm lands. Land management alone may be insufficient against even local floods of small magnitude. Therefore, check dams, gully plugs, water spreaders, and other minor engineering structures are included in this approach. Small dams about 15 meters (50 feet) high are also used. This approach has become

The Los Angeles River, which was paved during the 1930's in an attempt to reduce the effects of the flooding that frequently accompanied winter rains in Southern California. (McCrea Adams)

synonymous with conservation and environmental sustainability. Many conservation groups have rallied around this approach.

Without the significant flood-control advances that occurred during the twentieth century, some of the most productive agricultural areas in the United States would be vast swamplands. Flood-control projects are justified on the basis of a comparison of benefits to cost. However, some argue that environmental and sociological damages, if included, will lower the estimated benefits. All benefits and damages cannot be reduced to monetary terms. Flood-control projects must also be justified on the basis of their impact on the environment. Although the intent of flood control is to reduce flood losses, without effective local floodplain management, the structural control tends to encourage the development of flood-prone areas. Floodplain and swamp channelization for drainage and other developments in the floodplain often result in the loss of valuable vegetation and sensitive wetlands.

Many communities find that upstream land-use decisions to build levees or engineer farm fields to drain quickly make their flooding worse. Rather than build flood-control levees, relocation of homes and businesses to higher grounds may be a better solution. Many people believe that the best way to reduce flood damage is to implement a strictly enforced floodplain zoning plan that would prevent growth of towns and cities in flood-prone areas and encourage the use of land close to the river for low-damage uses such as public parks. Federal flood-control policies provide for building and repairing levees, paying relief, and subsidizing flood insurance. These measures discourage state and local officials from directing new development away from flood-prone areas, resulting in poor floodplain management and environmental degradation.

G. Padmanabhan

SUGGESTED READINGS: An excellent review of the history, development, and genesis of structural versus nonstructural methods of flood control is provided in Luna B. Leopold and Thomas Maddock, Jr., in *The Flood Control Controversy* (1954). Mark Reisner's *Cadillac Desert* (1986) is a moving account of the efforts to control rivers in the American West. *The Social and Environmental*

Effects of Large Dams (1984), by Edward Goldsmith and Nicholas Hildyard, provides a good discussion on the structural versus ecological approach to flood control. *Channelized Rivers: Perspectives for Environmental Management* (1988), by Andrew Brookes, discusses environmental consequences of channelization of rivers and improved design procedures to accommodate environmental concerns.

SEE ALSO: Dams and reservoirs.

Fluoridation

CATEGORY: Water and water pollution

Fluoridation is the treatment of water with chemicals that release fluoride into a community's water-supply system. It is an example of preventative medicine and was first introduced into the United States in the 1940's in an attempt to reduce tooth decay. Since then, many cities have added fluoride to their public water-supply systems.

Proponents of fluoridation have claimed that it has dramatically reduced tooth decay, which was a serious and widespread problem in the early twentieth century. Opponents of fluoridation have not been entirely convinced of its effectiveness and have been concerned about possible health risks that many be associated with fluoridation. The decision to fluoridate drinking water has generally rested with local governments and communities and has always been a controversial issue.

Fluoride is the water soluble, ionic form of the element fluorine. It is present naturally in most water supplies at low levels, generally less than 0.2 parts per million (ppm), and nearly all food contains traces of fluoride. Tea contains more fluoride than most foods, while fish and vegetables also have relatively high levels. Most scientific studies have suggested that water containing a concentration of about 1 ppm fluoride dramatically reduces the incidence of tooth decay.

Tooth decay occurs when food acids dissolve the protective enamel surrounding teeth and create holes, or cavities, in the teeth. These acids are present in food and can also be formed by acid-produced bacteria that convert sugars into acids. Americans have always consumed large quantities of sugar, which is a significant factor in the high incidence of tooth decay. By contrast, studies of people in primitive cultures reveal tooth decay to be less common, which has been attributed to their more natural diets.

Early fluoridation studies between 1930 and 1950 demonstrated that fluoridation of public water systems produced a 50 to 60 percent reduction in tooth decay and that there were no immediate health risks associated with increased fluoride consumption. Consequently, many communities quickly moved to fluoridate their water, and fluoridation was endorsed by most major health organizations in the United States.

Strong opposition to fluoridation began to emerge in the 1950's as opponents claimed that the possible side effects of fluoride had not been adequately investigated. This concern was not unreasonable, since high levels of ingested fluoride can be lethal. However, it is not unusual for a substance that is lethal at high concentration to be safe at low levels, as is the case with most vitamins and trace elements. Opponents of fluoridation were also concerned on moral grounds because fluoridation represented compulsory mass medication.

Since the 1960's, controversy and heated debate have surrounded the issue of fluoridation across the country. Critics have pointed to the harmful effects of large doses of fluoride, including bone damage and special risks for some people with kidney disease or those who are particularly sensitive to toxic substances. Between the 1950's and 1980's, some scientists suggested that fluoride may have a mutagenic effect—that is, it may be associated with human birth defects, including Down syndrome.

Controversial claims that fluoride can cause cancer were also raised in the 1970's, most notably by biochemist John Yiamouyiannis, who claimed that U.S. cities with fluoridated water had greater cancer death rates than cities with unfluoridated water. Fluoridation proponents were quick to discredit his work by pointing out that he had failed to take other factors into consideration, such as the levels of known environmental carcinogens. Most scientific opinion suggests that the link between cancer and fluoride is a tenuous one. Nevertheless, it is a link that cannot be completely ignored, and a number of respected scientists continue to argue that the benefits of fluoridation are not without potential health risks. A 1988 article in the American Chemical Society publication *Chemical and Engineering News* created considerable attention by suggesting that scientists opposing fluoridation were more credible than had been previously acknowledged. By the 1990's even some fluoridation proponents began suggesting that observed tooth decay reduction as a result of water fluoridation may have been at levels of only around 25 percent. Other factors—such as education, better dental hygiene, and the addition of fluoride to some foods, salt, toothpastes, and mouthwashes—may also contribute to the overall reduction in tooth decay levels.

The development of the fluoridation issue in the United States was closely observed by other countries. Dental and medical authorities in Australia, Canada, New Zealand, and Ireland endorsed fluoridation, although not without considerable opposition from various groups. Fluoridation in Western Europe was greeted less enthusiastically, and scientific opinion in some countries, such as France, Germany, and Denmark, concluded that it was unsafe. As a result, few Europeans drink fluoridated water.

While there is little doubt that fluoride does reduce tooth decay, the exact degree to which fluoridated water contributes to the reduction remains unanswered. It also remains unclear what, if any, side effects are involved in ingesting 1 ppm levels of fluoride in water over many years. Although it has been argued that any risks associated with fluoridation are small, these risks may not necessarily be acceptable to everyone.

Since the 1960's and 1970's, concerns over environmental and health issues have been growing, and it has often been difficult, if not impossible, for science to resolve completely the potential hazard of small amounts of chemical substances in the environment. The fact that only about 50 percent of U.S. communities have elected to adopt fluoridation is indicative of people's cautious approach to the issue. In 1993 the National Research Council published a report on the health effects of ingested fluoride and attempted to determine if the Environmental Protection Agency's maximum recommended level of 4 ppm for fluoride in drinking water should be modified. The report concluded that this level was appropriate but that further research may indicate a need for revision. The report also found inconsistencies in the scientific studies of fluoride toxicity and recommended further research in this area.

Nicholas C. Thomas

SUGGESTED READINGS: Brian Martin provides a detailed look at the fluoride debate in *Scientific Knowledge in Controversy: The Social Dynamic of the Fluoridation Debate* (1991). An easy-to-read article is Bette Hileman's "Fluoridation of Water," in the readily accessible journal *Chemistry and Engineering News* 66 (August 1, 1988). John De Zuane's *Handbook of Drinking Water Quality* (1990) and John Cary Stewart's *Drinking Water Hazards* (1989) both contain good summaries of fluoridation. The report by the National Research Council can be found in *Health Effects of Ingested Fluoride* (1993).

SEE ALSO: Drinking water; Water treatment.

Food and Agriculture Organization

DATE: established October 24, 1945
CATEGORY: Agriculture and food

The fundamental objective of the United Nations Food and Agriculture Organization (FAO) is to eliminate hunger and improve human nutrition by enhancing efficiency in the production and distribution of food and agricultural products around the world. Increasingly, it does so by promoting sustainable practices that will minimize deleterious environmental consequences.

The FAO is the oldest and largest permanent specialized agency of the United Nations (U.N.). Established at the end of World War II, the FAO was initially concerned with feeding people whose countries had been devastated by war. It also sought to minimize the boom and bust cycles that long characterized agricultural markets. The FAO has operated in the context of the rapid rise in human population that distinguished the second half of the twentieth century and has concentrated on working in the developing countries of the world.

In conjunction with the World Health Organization (WHO), the FAO establishes international food standards, providing an objective basis on which to evaluate food consumption patterns. It convenes conferences to address problems of hunger, malnutrition, and food security, seeking viable solutions and stimulating their adoption. In 1960, for example, the FAO launched the Freedom from Hunger Campaign to mobilize nongovernmental support for the issue. In 1981 it initiated the first World Food Day, observed by 150 countries, to promote awareness of the continuing problem.

The agency coordinates efforts by governments and technical agencies to develop programs in agriculture, nutrition, forestry, and fisheries, as well as related economic and social policy, including rural development. In so doing, the FAO attempts to provide impartial recommendations for such programs. To that end, it conducts research into areas requiring further knowledge and documentation. It also provides direct technical assistance and develops materials for education in food production, processing, transportation, and consumption. In addition, the FAO maintains worldwide statistics on the production, trade, and consumption of agricultural commodities. Specifically, in 1986 it started AGROSTAT, a highly comprehensive source of agricultural information and statistics for all countries. It also publishes numerous periodicals and yearbooks.

Long-term solutions to food security must be sustainable; that is, they must be able to be maintained indefinitely. Examples of unsustainable practices that degrade the environment and make it less likely that future generations will be able to feed themselves include overfishing, overgrazing, overexploitation of water supplies, overdependence on exhaustible fossil fuels, and pollution of air and water. Sustainable development, which has long undergirded the activities of the FAO, was formally recognized as one of its eight departments when it was reorganized in 1994. The organization plays a crucial role in the intertwined triumvirate of population growth, food security, and the environment.

James L. Robinson

SEE ALSO: Population growth; Sustainable agriculture; Sustainable development; Sustainable forestry.

Food chains

CATEGORY: Ecology and ecosystems

A food chain is a linear depiction of food items and the organisms that consume them. In a lake, for example, algae are eaten by microscopic zooplankton, which are, in turn, eaten by zooplanktivorous fishes, which are eaten by piscivorous fishes, which may ultimately be consumed by humans.

The original source of energy for almost every food chain is the sun. At the base of a food chain is a primary producer, such as a terrestrial plant or aquatic algae. Organisms that consume plants are the primary consumers (or herbivores), which are themselves consumed by secondary consumers (predators), which are consumed by tertiary consumers, and so on. Organisms that are the same number of feeding levels away from the original source of energy are said to be at the same trophic level. Also important to the concept of a food chain is the energy transfer between adjacent trophic levels, which is not very efficient: Only about 10 percent of the energy at one trophic level is actually transferred to the next one. This makes much less energy available for animals than for the primary producers in a system.

A number of factors make natural systems more complex than suggested by the basic concept of a food chain. First, most organisms consume more than one type of food and are consumed by more than one type of predator. Such complexity leads to the concept of a food web, in which a series of food chains are interrelated in a weblike arrangement. Second, some organisms can feed on more than one trophic level at the same time. For example, bluegill sunfish can feed on zooplankton (primary consumers), predatory invertebrates (secondary consumers), and even young fishes (secondary or tertiary consumers). Third, changes occur in the foods that an organism consumes and the predators that can consume an organism as it grows, a concept termed the "ontogenetic niche." For example, a frog begins life by hatching from an aquatic egg and living for a period of time as an aquatic larvae. After a metamorphosis (change in form), it becomes a terrestrial frog, clearly consuming different food types and facing consumption from a different array of predators than the aquatic larval stage. All of these factors combine to make the concept of food chains and food webs much more complex than is apparent from the initial simple definition.

An important issue surrounding the concept of a food chain or food web from an environmental perspective relates to bioaccumulation and biomagnification, specifically as related to some pesticides. The group of pesticides known as organochlorides includes dichloro-diphenyltrichloroethane (DDT), which was widely used during the twentieth century to control pests such as mosquitoes. While DDT helped eliminate small pest organisms at low concentrations, these concentrations were not harmful to larger organisms. Because DDT is not biodegradable, however, it is not broken down or eliminated by organisms; rather, it is stored in fat within those organisms to which it is not lethal. As animals continue to eat food containing DDT, they accumulate more and more of it, increasing the concentration of DDT in their bodies above that in the surrounding environment. This process is known as bioaccumulation.

Those organisms in which DDT bioaccumulates are eventually consumed by their predators, which further increases the concentration of the pesticide as it proceeds up the food chain. This process is called biomagnification. For some organisms at lower or intermediate trophic levels, the pesticide was not present at concentrations that were harmful. However, as its concentration increased up the food chain, harmful levels were eventually reached. For mammals, this did not always lead to direct harm to adults but sometimes caused the death of their young. In addition, many predatory birds that fed on fish in which DDT had become concentrated (for example, ospreys, bald eagles, peregrine falcons, and brown pelicans) produced thin-shelled eggs, resulting in the death of unborn chicks during incubation.

Another related consequence for the food web was that the target pest was not the only organism affected by the pesticide. Nontarget organisms could also be affected, and predator-prey balances that existed in the food web could be altered, releasing prey from predatory control (if a predator was reduced), potentially leading to another pest problem. As a consequence, DDT was banned in many developed nations, including the United States, in the early 1970's. Following this ban, scientists have documented that populations of affected predatory birds have displayed dramatic recoveries.

Food chains are also important in the area of biodiversity. Two aspects of biodiversity are important from the perspective of food chains and food webs—species extinction and introduction of exotic species. All living organisms occupy a niche, which is simply a definition of what they eat, where they live, what eats them, and so on, a portion of which is portrayed in a food web. Because organisms interact with other organisms, as implied by the concepts of food chains and food webs, the loss of a species through extinction opens a gap in the food web that might eliminate food for a predator or eliminate predatory control on their own food resources. Similarly, the introduction of an exotic species that did not previously occur in that food chain or food web would add competitive or predatory interactions that did not occur previously. Both of these situations alter the food web such that new interactions must be developed and a new system equilibrium reached. As an example, the introduction of the sea lamprey to Lake Michigan led to the decline and eventual disappearance of many native fish species because of the lamprey's predatory influence. Based on an understanding of the food web that had occurred in Lake Michigan, biologists were able to reintroduce a suite of predators, many of which were nonnative, that reestablished a food web in Lake Michigan that was similar to the one that existed prior to lamprey introduction.

Dennis R. DeVries

SUGGESTED READINGS: A nice scientific overview of the concept of the ontogenetic niche is given in E. E. Werner and J. F. Gilliam, "The Ontogenetic Niche and Species Interactions in Size-Structured Populations," *Annual Review of Ecology and Systematics* 15 (1984). For the history and situation regarding the use and eventual ban of DDT, see J. Turk and A. Turk, *Environmental Science* (1988). For a summary of the food web studies that have been conducted in Lake Michigan, see J. F. Kitchell and L. B. Crowder, "Predator-Prey Interactions in Lake Michigan: Model Predictions and Recent Dynamics," *Environmental Biology of Fishes* 16 (1986).

SEE ALSO: Biomagnification; Dichlorodiphenyl-trichloroethane (DDT); Extinctions and species loss; Introduced and exotic species.

Food irradiation

CATEGORY: Agriculture and food

Food irradiation is a process that uses nuclear radiation to sterilize foods in order to reduce spoilage and decrease the incidence of illness from contaminated food.

The U.S. Food and Drug Administration (FDA) has certified that irradiation is safe for many foods, including spices, fresh fruit, fish, poultry, and hamburger meat. Opponents of food irradiation focus on the inherent hazards of nuclear technology—especially the production, transportation, and disposal of radioactive materials—and criticize the possible creation of harmful radiation products in foods.

The most common radioactive source used for food processing is cobalt 60, which is produced by irradiating ordinary cobalt metal in a nuclear reactor. The shape of the source is typically a bundle of many thin tubes mounted in a rack. The source is kept in a building with thick walls and under 4.6 meters (15 feet) of water for shielding. The food to be irradiated is packaged and put on a conveyor belt. The operator raises the cobalt-60 source out of the water by remote control, while the food packages slowly travel past the source on the moving belt. The typical exposure time is three to thirty minutes, depending on the type of food, the required dose, and the source intensity. Radiation dose is measured in a unit called the kilogray (kGy), where one kGy equals 1,000 Grays, and one Gray equals 100 radiation absorbed doses (rads). The Gray is named for a British radiation biologist. The rad is an older unit still commonly used in the medical profession. When radiation passes through the food, it interacts with atomic electrons and breaks chemical bonds. Microorganisms that cause food spoilage or illnesses are inactivated so they cannot reproduce.

Fresh strawberries, sweet cherries, and tomatoes have a normal shelf life of only seven to ten days. Research has shown that a radiation dose of 2 kGy can double their shelf life without affecting the flavor. Trichinosis parasites in fresh pork can be controlled with a dose of 1 kGy. Doses up to 3 kGy are used to destroy 99.9 percent of salmonella in chicken meat and *Escherichia coli* in ground hamburger. A dose of less than 0.1 kGy is sufficient to interrupt cell growth in onions and potatoes to prevent undesirable sprouting in the spring. Larger doses, up to 30 kGy, are used to eliminate insects, mites, and other pests in spices, herbs, and tea. The American Dietetic Association supports food irradiation as an effective technique to reduce outbreaks of food-borne illnesses, which cause several million cases of sickness and more than nine thousand fatalities annually in the United States.

Irradiated food does not become radioactive. It does not "glow in the dark" as some opponents have claimed. Hundreds of animal feeding studies with irradiated foods have been done since the late 1950's. Irradiated chicken, wheat, oranges, and other foods were fed to four generations of mice, three generations of beagles, and thousands of rats and monkeys. No increase in cancer or other inherited diseases was detected in comparison to a control group eating nonirradiated food. In one experiment, several thousand mice were fed nothing but irradiated food. After sixty generations—about ten years—the cancer rate for the experimental group was no greater than for the control group.

Chemical analyses of irradiated foods have looked for potentially harmful by-products from radiation. Small quantities of benzene and formaldehyde were found. However, canning, cooking, and baking have been shown to create these same by-products even more abundantly. Based on the accumulated evidence, food irradiation has been endorsed by an impressive list of organizations: the American Medical Association, the U.S. Department of Agriculture, the American Diabetic Association, the World Health Organization, the United Nations Food and Agriculture Organization, and the FDA. Astronauts on space missions have eaten irradiated foods since 1972, and many hospitals use irradiated foods for patients with an impaired immune system.

The antinuclear movement was born after World War II when the United States and the Soviet Union were conducting nuclear weapons tests that released large amounts of radioactive fallout into the atmosphere. Public pressure eventually led to a Limited Test Ban Treaty in 1963. Two serious accidents at nuclear power plants, at Three Mile Island (1979) in the United States and Chernobyl (1986) in the Soviet Union, dramatized the hazards of nuclear technology for the general public. A consumer activist organization called Food and Water, based in Walden, Vermont, connects fear of the atomic bomb with food irradiation by showing a picture of a mushroom cloud hovering over a plate filled with food. The caption says, "The Department of Energy has a solution to the problem of radioactive waste. You're going to eat it." The most serious safety issues relating to food irradiation are in

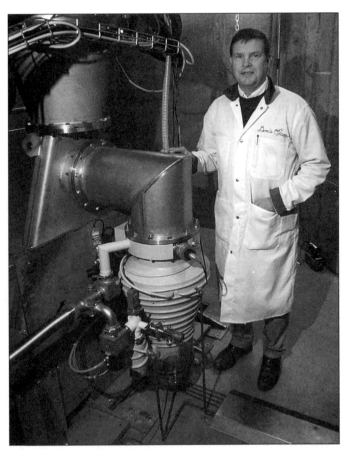

Dennis Olsen, director of Iowa State University's irradiation research program, stands next to a linear accelerator. The device is used to apply doses of radiation to food for the purpose of eliminating harmful microorganisms. (AP/Wide World Photos)

the production of radioactive cobalt; the safety of workers while transporting, installing, and using the source; and the eventual disposal of radioactive waste. Such issues, however, do not have the same impact on consumers as the implied claim that foods may become tainted by irradiation. Commercial food processors have been deterred from installing irradiation facilities by fear of negative publicity and a potential consumer backlash.

The FDA requires that irradiated foods be labeled with a special radura symbol and the statement "treated with radiation." It will be up to consumers to decide if the benefits of a safer food supply outweigh the potential hazards of an expanded nuclear industry.

Hans G. Graetzer

SUGGESTED READINGS: A good introduction to the uses and hazards of radioactivity is given in *Living with Radiation: The Risk, the Promise* (1989), by Henry Wagner and Linda Ketchum. The application of radiation to food preservation is described in Morton Satin's *Food Irradiation: A Guidebook* (1996). For a critical assessment of this technology, see *The Food That Would Last Forever: Understanding the Dangers of Food Irradiation* (1993), by Gary Gibbs. The American Dietetic Association issued a position paper entitled "Food Irradiation" in 1995, which is available from their national headquarters. An informative article quoting both supporters and opponents of food irradiation appeared in the Massachusetts Institute of Technology's *Technology Review* (November-December, 1997).

SEE ALSO: Antinuclear movement; Environmental illnesses.

Foreman, Dave

BORN: October 18, 1946; Albuquerque, New Mexico
CATEGORY: Preservation and wilderness issues

Author and activist Dave Foreman was one of the cofounders of the radical environmental group Earth First!

Dave Foreman, the son of Benjamin and Lorane Foreman, was born in 1946. Benjamin was an Air Force pilot, and the family moved often. They were living in Blythe, California, in 1964 when Foreman graduated from high school. He attended junior college for one year and enrolled the following fall at the University of New Mexico in Albuquerque. His involvement in politics began in high school when he did volunteer work for republican senator Barry Goldwater. In 1966 he was the state chairman of the conservative Young Americans for Freedom and a vocal supporter of the country's involvement in the Vietnam War. In 1967 Foreman enlisted in the Marines Corps but found the military unsatisfactory. He fled into the mountains of New Mexico but eventually turned himself in. He served time in prison and was dishonorably discharged.

In 1970 Foreman took to the woods again, supporting himself with odd jobs while spending time backpacking and rafting. He joined an environmental organization called the Black Mesa Defense Fund, and his knack for waging effective campaigns against developers was apparent. The Wilderness Society employed Foreman in 1973 as their principal consultant in New Mexico, and he continued to lead campaigns against polluters and land developers. By 1980 he was one of the Wilderness Society's most visible members.

Foreman, however, felt the need to become more confrontational. In 1980 he resigned from the Wilderness Society and cofounded Earth First!, a radical environmental movement that used direct action to combat environmental destruction. The founders were inspired in part by Edward Abbey's novel *The Monkey Wrench Gang* (1975), which detailed the exploits of a small group of people who destroyed construction equipment to stop environmental destruction in the southwestern United States. One of Earth First!'s early actions was the unfurling of a 300-foot piece of black plastic down the front of Glen Canyon Dam to represent a crack. In 1985 Foreman published *Ecodefense: A Field Guide to Monkeywrenching*, which provided detailed instructions on how to destroy heavy machinery, deface billboards, and spike trees.

Earth First!, however, developed problems keeping its aims directed to environmental protection, and in 1989 Foreman left the movement. Around the same time he was arrested on charges of conspiring to topple power lines in Arizona. The case was heard in 1991, and Foreman pled guilty to a reduced conspiracy charge with delayed sentencing and an agreement that the charge would be reduced after five years of good behavior. In 1995 Foreman accepted a three-year term as a director of the Sierra Club but resigned to devote more time to the Wildlands Project, which he initiated in 1993. Through this project Foreman is "trying to articulate a vision of restoring wilderness across North America."

Kenneth H. Brown

SEE ALSO: Earth First!; Ecotage; Ecoterrorism; Glen Canyon Dam; Monkeywrenching.

Forest and range policy

CATEGORY: Forests and plants

Many national governments have established legal policies for protecting forests and rangelands. Such legislation usually seeks to sustain and protect biodiversity while setting guidelines for the sustainable use of natural resources.

Rangeland—land that supplies forage for grazing and browsing animals—covers almost one-half of the ice-free land on earth. More than three billion cattle, sheep, goats, camels, buffalo, and other domestic animals graze on rangelands. These animals are important in converting forages into milk and meat, which provide nourishment for people around world. Forests cover almost 30 percent of the earth and provide humans with lumber, fuel woods, spices, chocolate, tropical fruits, nuts, latex rubber, and valuable chemicals that constitute prescription and nonprescription drugs, as well as cancer-fighting drugs. Rangelands and forests also function as important ecosystems that play a vital role in providing food and shelter for wildlife, controlling erosion, and purifying the atmosphere. Forests and rangelands have been facing alarming rates of destruction and degradation at the hands of humans.

PROTECTING FORESTS AND RANGELANDS

The 14.8 billion acres of forest that originally existed has been reduced to 11.4 billion acres by human conversion of land to cropland, pastureland, cities, and nonproductive land. Forests, if properly maintained or left alone, are the most productive and self-sustaining ecosystem that land can support. Tropical rain forests are the natural habitat for at least 50 percent, and possibly up to 90 percent, of the species on Earth. Harvard biologist Edward O. Wilson believes that 25 percent of the earth's species could become extinct by the year 2050 if the current rate of tropical forest destruction is not stopped.

Many national governments have established legal policies for protecting forest habitats and the important biological diversity found within them. National parks and reserves provide protection for both forests and rangelands. Some countries have local laws that protect particular forests or prohibit clearing, burning, or logging of forests. China, which suffered from erosion and terrible floods as a result of centuries of deforestation, began an impressive reforestation campaign. Almost 11 million acres of new tress were planted in China during the 1990's. Korea attained 70 percent reforestation after losing almost all of its forested land in a civil war during the 1950's. Japan has put strict environmental laws into effect, which have allowed it to reforest 68 percent of its land area. Japan has relied upon imported timbers in order to allow its new forest projects to flourish. Even with such worldwide success stories in reforestation, it is estimated that protection and sustainable management of forests and rangelands still need to be increased by a factor of three if forests are to be saved.

MULTIPLE USE

Protecting forestland involves an interdisciplinary approach. In the United States, 191 million acres of forestland are managed by the U.S. Forest Service. The Forest and Rangeland Renewable Resource Planning Act (RPA) of 1974 and the National Forest Management Act (NFMA) of 1976 direct that management plans must be developed for forests and rangelands to ensure that resources will be available on a sustained basis. Management policies must be made to sustain and protect biodiversity; old-growth forests; riparian areas; threatened, endangered, and sensitive species; rangeland; water and air quality; access to forests; and wildlife and fisheries habitat. The Forest Service provides inexpensive grazing lands for more than three million cattle and sheep every year, supports multimillion-dollar mining operations, maintains a network of roads eight times longer that the U.S. interstate highway system, and allows access to almost one-half of all national forest land for commercial logging. The Forest Service is responsible

328 • Forest and range policy

for producing plans for the multiple use of national lands. Management plans are to be reviewed and updated every ten years.

Sustainability policies require that the net productive capacity of the forest or rangeland does not decrease with multiple use. This involves making sure that soil productivity is maintained by keeping erosion, compaction, or displacement by mining or logging equipment or other motorized vehicles within tolerable limits. It further requires that a large percentage of the forest remains undeveloped so that soils and habitats, as well as tree cover, will remain undisturbed and in their natural state.

The RPA and NFMA, along with the Endangered Species Act (ESA) of 1973, mandate policies that encourage the proliferation of species native to and currently living in the forest. Natural ecosystem processes are followed to ensure their survival. Even though forests and rangelands are required to be multiple-use areas, policy maintains that there can be no adverse impact to threatened, endangered, or sensitive species. Species habitats within the forest are to remain well distributed and free of barriers that can cause fragmentation of animal populations and ultimately species loss. If a forest contains fragmented areas created by human activity, corridors that connect the forest patches are constructed. In this way species are not isolated from one another, and viable populations can exist.

In the case of natural disasters, the Forest Service creates artificial habitats to encourage the survival of species. When Hurricane Hugo devastated the Francis Marion National Forest in South Carolina in 1989, winds snapped 90 percent of the trees with active woodpecker cavities in some areas of the forest. The habitat destruction caused 70 percent of the red-cockaded woodpecker population to disappear. The Forest Service and university researchers created nesting and roosting cavities to save the woodpeckers. Within a four-year period, the population dramatically recovered.

TIMBER, OIL, AND MINERAL LEASING

Logging activities in forests are covered by the Resource Planning Act of 1990. Forested land must be evaluated for its ability to produce com-

mercially usable timber without negative environmental impact. There must be reasonable assurance that stands managed for timber production can be adequately restocked within five years of the final harvest. Further, no irreversible resource damage is allowed to occur. Policy further requires that the silviculture practices that are best suited to the land management objectives of the area be used. Cutting practices are then monitored. The 1990's were characterized by a trend toward restricting logging methods in order to protect habitats and preserve older stands of trees. In the 1993 Renewable Resource Assessment update, the Forest Service found that timber mortality was 24.3 percent and was still interfering with biological diversity. Some forested areas have been withdrawn from timber production because of their fragility.

Multiple use under the NFMA also allows forests to be available for oil and gas leasing. Certain lands have been exempted from mineral exploration by acts of Congress or executive authority. However, the search for and production of mineral and energy sources remain under the jurisdiction of the Forest Service, which must provide access to national forests for mineral resources activities. The Federal On-Shore Oil and Gas Leasing Reform Act of 1987 gave the Forest Service more authority in making lease decisions.

PEST AND WEED CONTROL

Insecticides are sometimes used during attempts to ensure the health of forestland. Policy in the United States requires the use of safe pesticides and encourages the development of an integrated pest management (IPM) plan. Any decision to use a particular pesticide must be based on an analysis of its effectiveness, specificity, environmental impact, economic efficiency, and effects on humans. The application and use of pesticides must be coordinated with federal and state fish and wildlife management agencies to ensure that no harm occurs to either fish or wildlife. Pesticides can only be applied to areas that are designated as wilderness when their use is necessary to protect or restore resources in the area. Other methods of controlling disease include removing diseased trees and vegetation from the forest, cutting infected areas from plants and removing the debris, treating trees with antibiotics, and developing disease-resistant plant varieties.

Forest Service policy on integrated pest management was revised in 1995 to emphasize the importance of integrating noxious weed management into the forest plan for ecosystem analysis and assessment. Noxious weed management must be coordinated in cooperation with state and local government agencies, as well as private landowners. Noxious weeds include invasive, aggressive, or harmful nonindigenous or exotic plant species. They are generally poisonous, toxic, or parasitic, or may carry insects or disease. The Forest Service is responsible for the prevention, control, and eradication of noxious weeds in national forests and grasslands.

One strategy for promoting weed-free forests that has been implemented in Wyoming, Colorado, Idaho, Utah, and Montana requires pack animals on national forest land to eat state-certified weed-free forage. In North Dakota, goats have been used to help control leafy spurge. The goats graze on designated spurge patches during the day and return to portable corrals during the night. A five-year study found that the goats effectively reduced stem densities of spurge patches to the extent that livestock forage plants were able to reestablish themselves. Another strategy involves the use of certified weed-free straw and gravel in construction and rehabilitation efforts within national forests. Biocontrols, herbicides, and controlled burning are also commonly used during IPM operations in forests.

OTHER PROTECTION ISSUES

Natural watercourses and their banks are referred to as riparian areas. The plant communities that grow in these areas often serve as important habitats for a large variety of animals and birds, and also provide shade, bank stability, and filtration of pollution sources. It is therefore important that these areas remain in good ecological condition. Riparian areas and streams are managed according to legal policies for wetlands, floodplains, water quality, endangered species, and wild and scenic rivers.

Dirt roads in national forests are often closed when road sediment pollutes riparian areas and harms fish populations. Forest and rangeland roads are also closed to prevent disruption of breeding or nesting colonies of various species of animals. Seemingly harmless human endeavors—such as seeking mushrooms, picking berries, or hiking in the forest—can cause problems for calving elk and nesting eagles. Therefore, the amount of open roads in the forest is being reduced in order to preserve habitat and return land to a more natural state.

Studies have shown that the vast number of roads, and the accompanying vegetation loss, on Bureau of Land Management (BLM) and Forest Service lands has made it too easy to kill elk. It was noted that in Oregon there were few mature bulls in the elk population. Lack of mature bulls can cause many problems, including disruption of breeding seasons and conception dates, and decreases in calf survival rates. Younger bulls tend to breed later and over a longer period of time than mature bulls. This results in a calving season that lasts longer—calves that are born late in the spring do not have enough time to feed on high-quality forage before winter. Long calving seasons also make the calves more vulnerable to predators, such as coyotes, bears, and mountain lions.

Fire management is important to healthy forests. In many cases fires are prevented or suppressed, but prescribed fires are used to protect and maintain ecosystem characteristics. Some conifers, such as the giant sequoia and the jack pine, will release their seeds for germination only after being exposed to intense heat. Lodgepole pines will not release their seeds until they have been scorched by fire. Ecosystems that depend on the recurrence of fire for regeneration and balance are called fire climax ecosystems. Prescribed fires are used as a management tool in these areas, which include some grasslands and pine habitats.

In 1964 the U.S. Congress passed the Wilderness Act, which mandates that certain federal lands be designated as wilderness areas. These lands must remain in their natural condition, provide solitude or primitive types of recreation, and be at least 5,000 acres in area. They usually

New trees grow in an area that has been clear-cut. Forest policy in many nations includes provisions for reforestation of regions devastated by logging operations. (Ben Klaffke)

wide is three times that of tropical forests, and the area lost is six times that of tropical forests. There are more threatened plant species in North American rangelands than any other major biome.

Rangeland grasses are known for their deep, complex root systems, which makes the grasses hard to uproot. When the tip of the leaf is eaten, the plant quickly regrows. Each leaf of grass on the rangeland grows from its base, and the lower half of the plant must remain for the plant to thrive and survive. As long as only the top half of the grass is eaten, grasses serve as renewable resources that can provide many years of grazing. Each type of grassland is evaluated based on grass species, soil type, growing season, range condition, past use, and climatic conditions. These conditions determine the herbivore carrying capacity or the maximum number of grazing animals a rangeland can sustain and remain renewable.

Overgrazing occurs when too many animals are allowed to graze in one area for too long or herbivore numbers exceed the carrying capacity. Grazing animals tend to eat their favorite grasses first and leave the tougher, less palatable plants. If animals are allowed to do this, the vegetation begins to grow in patches, allowing cacti and woody bushes to move into vacant areas. As native plants disappear from the range, weeds also begin to grow. As the nutritional level of the forage declines, hungry animals pull the grasses out by their roots, leaving the ground bare and susceptible to damage from hooves. This process initiates the desertification cycle. With no vegetation present, rain quickly drains off the land and does not replenish the groundwater. This makes the soil vulnerable to erosion. Almost one-third of rangeland in the world is degraded by overgrazing. Among the countries suffering severe range degradation are Pakistan, Sudan, Zambia, Somalia, Iraq, and Bolivia.

contain ecological or geological systems of scenic, scientific, or historical value. No roads, motorized vehicles, or structures are allowed in these areas. Furthermore, no commercial activities are allowed in wilderness areas except livestock grazing and limited mining endeavors that began before the area received wilderness designation.

GRAZING PRACTICES AND PROBLEMS

Approximately 42 percent of the world's rangeland is used for grazing livestock; much of the rest is too dry, cold, or remote to serve such purposes. It is common for these rangelands to be converted into croplands or urban developments. The rate of loss for grazing lands worldwide

The United States has approximately 788 million acres of rangeland. This represents almost 34 percent of the land area in the nation. More than one-half of the rangeland is privately owned, while approximately 43 percent is publicly owned and managed by the Forest Service and the BLM. State and local governments manage the remaining 5 percent. Efforts to preserve rangelands include close monitoring of herbivore carrying capacity and removal of substandard ranges from the grazing cycle until they recover. New grazing practices, such as cattle and sheep rotation, help to preserve the renewable quality of rangelands. Grazing is also managed with consideration to season, moisture, and plant growth conditions. Noxious weed encroachment is controlled, and native forages and grasses are allowed to grow.

Most rangelands in the United States are short-grass prairies located in the western part of the nation. These lands are further characterized by thin soils and low annual precipitation. They undergo numerous environmental stresses. Woody shrubs, such as mesquite and prickly cactus, often invade and take over these rangelands as overgrazing or other degradation occurs. Such areas are especially susceptible to desertification. Recreational vehicles, such as motorcycles, dune buggies, and four-wheel-drive trucks, can also damage the vegetation on ranges. According to the 1993 Renewable Resource Assessment update, many of the rangelands in the United States are in unsatisfactory condition.

Steps to restore healthy rangelands include restoring and maintaining riparian areas and priority watersheds. These areas are monitored on a regular basis, and adjustments are made if their health is jeopardized by sediment from road use or degradation of important habitats caused by human activity. The Natural Resources Conservation program is teaching private landowners how to burn unwanted woody plants on rangelands, reseed with perennial grasses that help hold water in the soil, and rotate grazing of cattle and sheep on rangelands so that the land is able to recover and thrive. Such methods have proven to be successful.

Toby Stewart and Dion Stewart

SUGGESTED READINGS: *Conserving the World's Biological Diversity* (1990), by Jeffrey A. McNeely, provides an interesting look at deforestation around the world. A good history of the National Forest Service and its cooperation with state and private landowners is *American Forestry* (1985), by William G. Robbins. *Endangered Species: Opposing Viewpoints* (1996), edited by Brenda Stalcup, contains interesting articles on forest priorities as well as other biodiversity issues. *Forest Management* (1966), by Kenneth P. Davis, contains an interesting discussion of management techniques. Annual reports and local management plans from the Bureau of Land Management and the National Forest Service are excellent sources of information.

SEE ALSO: Deforestation; Forest management; Logging and clear-cutting; National forests; Rain forests and rain forest destruction; Range management; Sustainable forestry.

Forest management

CATEGORY: Forests and plants

Forests provide lumber for homes, fuelwood for cooking and heating, and raw materials for making paper, latex rubber, resin, dyes, and essential oils. Forests are also home to millions of plants and animal species and are vital in regulating climate, purifying the air, and controlling water runoff. A 1993 global assessment by the United Nations Food and Agriculture Organization (FAO) found that three-fourths of the forests in the world still have some tree cover, but less than one-half of these have intact forest ecosystems. Deforestation is occurring at an alarming rate, and management practices are being sought to try to halt this destruction.

Thousands of years ago, before humans began clearing the forests for croplands and settlements, forests and woodlands covered almost 15 billion acres of the earth. Approximately 16 percent of the forests have been cleared and converted to pasture, agricultural land, cities, and nonproductive land. The remaining 11.4 billion acres of forests cover approximately 30 percent of the earth's land surface.

Clearing forests has severe environmental consequences. It reduces the overall productivity of the land, and nutrients and biomass stored in trees and leaf litter are lost. Soil once covered with plants, leaves, and snags becomes prone to erosion and drying. When forests are cleared, habitats are destroyed and biodiversity is greatly diminished. Destruction of forests causes water to drain off the land instead of being released into the atmosphere by transpiration or percolation into groundwater. This can cause major changes in the hydraulic cycle and ultimately in the earth's climate. Forests also remove a large amount of carbon dioxide from the air; the clearing of forests causes more carbon dioxide to remain in the air, thus upsetting the delicate balance of atmospheric gases.

RAIN FOREST DESTRUCTION

The destruction of tropical rain forests is of great concern. These forests provide habitats for at least 50 percent (some estimates are as high as 90 percent) of the total stock of animal, plant, and insect species on earth. They supply one-half of the world's annual harvest of hardwood and hundreds of food products, such as chocolate, spices, nuts, coffee, and tropical fruits. Tropical rain forests also provide the main ingredients in 25 percent of the world's prescription and nonprescription drugs, as well as 75 percent of the three thousand plants identified as containing chemicals that fight cancer. Industrial materials, such as natural latex rubber, resins, dyes, and essential oils, are also harvested from tropical forests.

Tropical forests are usually cleared to produce pastureland for large cattle ranches, establish logging operations, construct large plantations, grow marijuana and cocaine plants, develop mining

Trees are grown in a reforestation nursery. The seedlings will be planted in areas that have been logged. (Ben Klaffke)

Environmental Effects of Select Silvicultural Methods

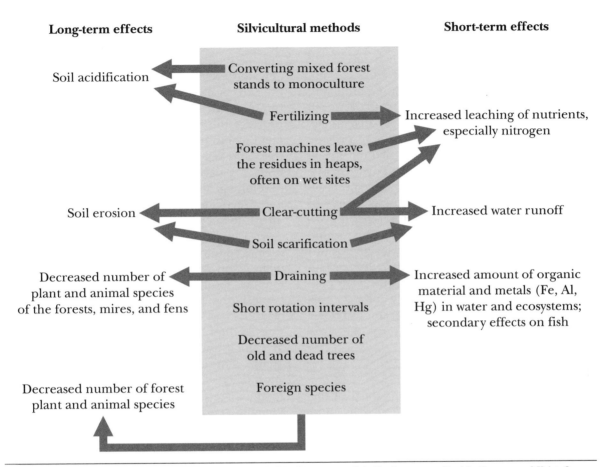

Source: Adapted from I. Stjernquist, "Modern Wood Fuels," in *Bioenergy and the Environment,* edited by Pasztor and Kristoferson, 1990.

operations, or build dams to provide power for mining and smelting operations. In 1985 the FAO's Committee on Forest Development in the Tropics developed the Tropical Forestry Action Plan to combat these practices. Fifty nations in Asia, Africa, and Latin America have adopted the plan, which seeks to develop sustainable forest methods and protect precious ecosystems.

Several management techniques have been successfully applied to tropical forests. Sustainable logging practices and reforestation programs have been established on lands that allow timber cutting, with complete bans of logging on virgin lands. Certain regions have set up extractive reserves to protect land for the native

people who live in the forest and gather latex rubber and nuts from mature trees. Sections of some tropical forests have been preserved as national reserves, which attract tourists while preserving trees and biodiversity. Developing countries have also been encouraged to protect their tropical forests by using a combination of debt-for-nature swaps and conservation easements. In debt-for-nature swaps, tropical countries act as custodians of the tropical forest in exchange for foreign aid or relief from debt. Conservation easement involves having another country, private organization, or consortium of countries compensate a tropical country for protecting a specific habitat.

Another management technique involves putting large areas of the forest under the control of indigenous people who use swidden or milpa agriculture. This traditional, productive form of agriculture follows a multiple-year cycle. Each year the native people clear a forest plot approximately 2.5 acres in size to allow the sun to penetrate to the ground. Leaf litter, branches, and fallen trunks are burned and leave a rich layer of ashes. Fast-growing crops such as bananas and papayas are planted and provide shade for root crops, which are planted to anchor the soil. Finally, crops such as corn and rice are planted. Crops mature in a staggered sequence, thus providing a continuous supply of food. The natives' use of mixed perennial polyculture helps prevent insect infestations, which can destroy monoculture crops. After one or two years the forest begins to take over the agricultural plot. The native farmers continue to pick the perennial crops but essentially allow the forest to reclaim the plot for the next ten to fifteen years before clearing and planting the area again.

U.S. FORESTS AND MANAGEMENT

Forests cover approximately one-third of the land area of the continental United States and comprise 10 percent of the forests in the world. Only about 22 percent of the commercial forest area in the United States lies within national forests. The rest is primarily managed by private companies that grow trees for use in commercial logging. The land managed by the U.S. Forest Service provides inexpensive grazing lands for more that three million cattle and sheep every year, supports multimillion-dollar mining operations, and consists of a network of roads eight times longer than the U.S. interstate highway system. Almost 50 percent of national forest land is open for commercial logging, and nearly 14 percent of the timber harvested in the United States each year comes from national forest lands. Total wood production in the United States has caused the loss of more than 95 percent of the old-growth forests in the lower forty-eight states. This loss includes not only high-quality wood but also a rich diversity of species not found in early-growth forests.

National forests in the United States are required by law to be managed in accordance with principles of sustainable yield. Congress has mandated that forests be managed for a combination of uses, including grazing, logging, mining, recreation, and protection of watersheds and wildlife. Healthy forests also require protection from pathogens and insects. Sustainable forestry, which emphasizes biological diversity, provides the best management. Other management techniques include removing only infected trees and vegetation, cutting infected areas and removing debris, treating trees with antibiotics, developing disease-resistant species of trees, using insecticides and fungicides, and developing integrated pest management plans.

Two basic systems are used to manage trees: even-aged and uneven-aged. Even-aged management involves maintaining trees in a given stand that are about the same age and size. Trees are harvested at the same time, then seeds are replanted to provide for a new even-aged stand. This method, which tends toward the cultivation of a single species or monoculture of trees, emphasizes the mass production of fast-growing, low-quality wood (such as pine) to give a faster economic return on investment. Even-aged management requires close supervision and the application of both fertilizer and pesticides to protect the monoculture species from disease and insects.

Uneven-aged management maintains trees at many ages and sizes to permit a natural regeneration process. This method helps sustain biological diversity, provides for long-term production of high-quality timber, allows for an adequate economic return, and promotes a multiple-use approach to forest management. Uneven-aged management also relies on selective cutting of mature trees and reserves clear-cutting for small patches of tree species that respond favorably to such logging methods.

HARVESTING METHODS

The use of a particular tree-harvesting method depends on the tree species involved, the site, and whether even-aged or uneven-aged management is being applied. Selective cutting is used on intermediate-aged or mature trees in uneven-aged forests. Carefully selected trees are cut in a prescribed stand to provide for a continuous and attractive forest cover that preserves the forest ecology.

Shelterwood cutting involves removing all the mature trees in an area over a period of ten years. The first harvest removes dying, defective, or diseased trees. This allows more sunlight to reach the healthiest trees in the forest, which will then cast seeds and shelter new seedlings. When the seedlings have turned into young trees, a second cutting removes many of the mature trees. However, enough mature trees are left to provide protection for the younger trees. When the young trees become well established, a third cutting harvests the remaining mature trees, leaving an even-aged stand of young trees from the best seed trees to mature. When done correctly, this method leaves a natural-looking forest and helps reduce soil erosion and preserve wildlife habitat.

Seed-tree cutting harvests almost every tree at one site with the exception of a few high-quality, seed-producing, and wind-resistant trees, which will function as a seed source to regenerate new crops. This method allows a variety of species to grow at one time and aids in erosion control and wildlife conservation. Clear-cutting removes all the trees in a single cutting. The clear-cut may involve a strip, an entire stand, or patches of trees. The area is then replanted with seeds to grow even-aged or tree-farm varieties. More than two-thirds of the timber produced in the United States, and almost one-third of the timber in national forests, is harvested by clear-cutting. A clear-cut reduces biological diversity by destroying habitat, can make trees in bordering areas more vulnerable to winds, and may take decades to regenerate.

FOREST FIRES

Forest fires can be divided into three types: surface, crown, and ground fires. Surface fires tend to burn only the undergrowth and leaf litter on the forest floor. Most mature trees easily survive, as does wildlife. These fires occur every five years or so in forests with an abundance of ground litter and help prevent more destructive crown and ground fires. Such fires can even re-

lease and recycle valuable mineral nutrients, stimulate certain tree seeds, and help eliminate insects and pathogens. Crown fires are very hot fires that burn both ground cover and tree tops. They normally occur in forests that have not experienced fires for several decades. Strong winds allow these fires to spread from deadwood and ground litter to treetops. They are capable of killing all vegetation and wildlife, leaving the land prone to erosion. Ground fires are more common in northern bogs. They can begin as surface fires but burn peat or partially decayed leaves below the ground surface. They can smolder for days or weeks before anyone notices them, and they are difficult to douse.

Natural forest fires can be beneficial to some species of trees, such as the giant sequoia and the jack pine, which release seeds for germination only after being exposed to intense heat. Grassland and pine forest ecosystems that depend on fires to regenerate are called fire climax ecosystems. They are managed for opti-

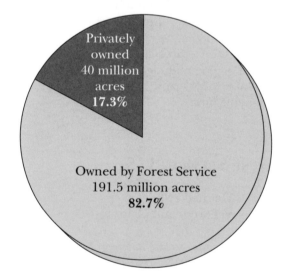

Ownership of Land Within National Forest and Grassland Boundaries, 1992

Privately owned 40 million acres 17.3%

Owned by Forest Service 191.5 million acres 82.7%

Total acres: 231.5 million

Source: U.S. Department of Commerce, *Statistical Abstract of the United States, 1996,* 1996.

mum productivity with prescribed fires.

The Society of American Foresters has begun advocating a concept called new forestry, in which ecological health and biodiversity, rather than timber production, are the main objectives of forestry. Advocates of new forestry propose that any given site should be logged only every 350 years, wider buffer zones should be left beside streams to reduce erosion and protect habitat, and logs and snags should be left in forests to help replenish soil fertility. Proponents also wish to involve private landowners in the cooperative management of lands.

Toby Stewart and Dion Stewart

SUGGESTED READINGS: *Conserving the World's Biological Diversity* (1990), by Jeffrey A. McNeely, is an excellent book on forest management problems throughout the world. *American Forestry* (1985), by William G. Robbins, provides a good history of cooperation among national, state, and privately owned forests. *Endangered Species: Opposing Viewpoints* (1996), edited by Brenda Stalcup, contains interesting articles on forests and other environmental issues. *Forest Management* (1996), by Kenneth P. Davis, deals with forest management techniques. Annual reports from the Bureau of Land Management and the U.S. Forest Service supply extensive information on policies and management of forests.

SEE ALSO: Deforestation; Forest and range policy; Logging and clear-cutting; National forests; Rain forests and rain forest destruction; Sustainable forestry.

Forest Service, U.S.

DATE: established 1905
CATEGORY: Forests and plants

The U.S. Forest Service, a federal agency housed within the United States Department of Agriculture and mandated by Congress to manage the national forests, is involved in environmental issues ranging from local flood control to attempting to prevent global warming through reforestation.

The mission of the Forest Service is to conserve national forests for multiple uses. It has the difficult task of balancing the needs of the nation's commercial forest products industries with the desire of the general public to pursue a variety of recreational activities. In all cases the Forest Service attempts to pursue what its professional foresters and research scientists believe are the most environmentally sustainable management policies.

Under the administration of President Theodore Roosevelt, Congress created the U.S. Forest Service in 1905 in response to a perceived impending timber famine. Nineteenth century lumbermen left a swath of barren, eroding land in their wake as they clear-cut their way across the North American continent from New England to the Pacific Northwest. Forest resources that had appeared inexhaustible at the start of the nineteenth century seemed about to vanish forever at the opening of the twentieth century. Navigable streams became choked with silt and debris when rain poured across deforested hillsides, and flash floods devastated riverside towns.

In a comparatively short time, the Forest Service successfully reversed the loss of forest habitat within the United States as it met the goals of its original mandate to ensure a sustainable forest reserve for future generations. As public awareness of other environmental issues—such as the importance of preserving old-growth forests for the continued viability of certain wildlife species—increased during the 1970's and 1980's, the Forest Service found itself facing growing criticism for policies that apparently favored logging interests over ecosystem preservation. For example, environmental activists argued that since the commercial forest industry now has access to millions of acres of privately owned plantation forests, logging should cease in the national forests. Criticisms of the Forest Service persisted into the 1990's, often overshadowing significant gains made in maintaining forest health and promoting global environmental stability.

In addition to managing national forests and undertaking research projects, the Forest Service provides advice and training to private landowners and state and local government in areas such as reforestation, community forestry, and agroforestry. Through its research stations, the Forest Service has developed environmentally sound harvesting methods, or best management practices (BMPs), for loggers to follow when removing timber from the forest. BMPs include directions on building haul roads in ways that discourage erosion, maintaining buffer zones between logged areas and nearby rivers and streams, and the optimum size and shape for clear-cuts to minimize loss of wildlife habitat. The Forest Service's technical experts also work with the U.S. Agency for International Development on forest management projects in other countries and have been active in trying to prevent the loss of forest habitat in environmentally sensitive regions around the world.

Nancy Farm Männikkö

SEE ALSO: Forest and range policy; Forest management; National forests; Sustainable forestry.

Forest Summit

DATE: April 3, 1993
CATEGORY: Forests and plants

Representatives from the timber industry and environmental groups met with government officials in an attempt to resolve controversies relating to the management of federally owned forests in the Pacific Northwest.

The Forest Summit, officially known as the Portland Timber Summit, brought together disputing parties in the conflict between loggers and supporters of the northern spotted owl to present their perspectives and to identify mutually acceptable new directions in forest management. In June, 1990, the northern spotted owl was declared an endangered species. In May, 1991, U.S. District Judge William Dwyer ruled that the George Bush administration was deliberately violating the Endangered Species Act by failing to develop a plan that adequately protected the owl from extinction. The judge

placed an injunction against logging on approximately three million acres of federally owned old-growth forest land in Northern California, Oregon, and Washington until an acceptable plan was developed.

President Bill Clinton held the day-long summit with Vice President Al Gore, Jr., cabinet members, Environmental Protection Agency (EPA) administrators, and other federal officials. Prior to the summit, environmental, timber, industry, and logger groups made a number of efforts to "sell" their perspectives on the logging situation to administrative officials. On the day before the conference, helicopter, airplane, and four-wheel-drive vehicle tours of the forests were given to administration officials. Employees of timber mills were given the day off and bused to Portland to picket the conference hall with signs proclaiming the need to save jobs and protect families.

The conference was expressly designed as a forum for participants to present their perspectives on the conflict. Loggers tended to focus on the negative effects of the logging ban on logging communities. In these presentations, loggers were often depicted as environmentalists with a vested interest in a healthy forest ecology and as modern-day Paul Bunyans unable to pass on their culture to the next generation. Logging company officials often focused on the economic difficulties they faced and on the country's need for housing lumber and other wood products. Small logging companies described how they had successfully adapted to changes in timber supplies and markets. Native Americans described the traditional and continuing significance of the forest in their lives. Environmentalists suggested that economic problems of logging communities were not a result of the restriction of cutting on federal land but rather were a consequence of general economic conditions, such as the low number of housing starts and a general decline in the logging industry. Environmentalists claimed that the loggers' desire for federal old-growth timber was a result of mismanagement of private lands. Environmentalists also asserted that logging-related jobs in the United States could be saved if log exports to foreign countries were reduced.

Following the summit, the task of developing a management plan was turned over to thirty-seven physical scientists and economists. The teams developed ten options, ranging from option 1, the "save it all" option, which allowed 190 million board feet to be cut and would have saved all the federally owned old-growth forests, to option 10, which permitted the cutting of 1.84 billion board feet. In June, 1994, a committee of senior federal officials began considering the options and prepared decision memoranda for the president. Clinton selected option 9, which set annual harvests at 1.2 billion board feet, a level about one-fourth of 1980's harvests but more than twice the level permitted since the northern spotted owl had been declared an endangered species. This option was known as the "efficiency option" because it focused on watersheds as the basic building blocks of the ecosystem and included measures designed to protect dwindling salmon stocks.

The management plan also called for the provision of $1.2 billion over five years to offset economic losses from logging. The plan eliminated 6,000 jobs in 1994 but created more than 8,000, retaining an additional 5,400 jobs. The option also established ten adaptive management areas in which local community and government groups would work together to allow logging and protect wildlife. Salvage logging, the cutting of fire- and insect-damaged trees, and thinning would be permitted in some sections of old-growth forest if an interagency team determined that these practices would not be detrimental to the northern spotted owl's habitat.

Clinton introduced his timber-management plan with the prophetic words, "Not everyone is going to like this plan. . . . Maybe no one will." Although the plan was judged to be a fair solution by many observers, none of the parties directly involved publicly expressed satisfaction with it. Loggers did not believe the promise of economic aid, especially at a time when the federal government was attempting to balance the budget through spending cuts. Loggers further concluded that the plan was unfair because it contained few provisions for meeting their main concern of preserving logging jobs and the logging-based economy of their small towns. Re-

training funds were offered for the purpose of equipping loggers with the skills to be employed in other jobs. Timber industry leaders believed that the plan was based on faulty assumptions about forest productivity and that it unduly restricted logging.

Environmentalists believed that the plan was unfair because it provided for more cutting than was currently permitted under the court injunction and it allowed loopholes that would be used to permit even more logging. They believed that the provision to allow salvage logging would be used to cut green trees. Definitions of salvage logging had been stretched in the past. Whistleblowers within the National Forest Service had leaked a memorandum that explicitly directed employees to allow green cutting to happen. Workers in the Agriculture Department Inspector General's Office had also found documents indicating that Forest Service officials may have in the past made questionable agreements with logging company officials prior to timber sales. Expected large cuts in the number of Forest Service staff were also thought to make policing of the plan more difficult.

Representatives of all sides in the dispute stated that they were considering bringing lawsuits against the plan. One basis for a lawsuit against the plan was the lack of public participation in and review of the plan's development. The Portland Timber Summit, however, was a putative effort to bring the sides together and to discover mutually agreeable solutions. Nevertheless, once the summit had concluded, the alternative management options were developed by experts working in seclusion. The summit gave participants voice by allowing them to articulate their ideas. The participants did not, however, have the power to determine the specifics of the selected management plan.

On June 6, 1994, Judge Dwyer ruled that the option 9 management plan satisfactorily addressed concerns about protection of the northern spotted owl that had prompted the original injunction. Despite the continued threat of lawsuits, U.S. Forest Service officials proceeded with plans to sell timber in the disputed areas.

George Cvetkovich
Timothy C. Earle

SUGGESTED READINGS: William Dietrich's *The Final Forest: The Battle for the Last Great Trees of the Pacific Northwest* (1992) focuses on how one Washington logging community was affected by the debate. Keith Ervin's *Fragile Majesty: The Battle for North America's Last Great Forest* (1989) provides a history of the area's forest-management controversies. The Forest Ecosystem Management Team's *Forest Ecosystem Management: An Ecological, Economic, and Social Assessment* (1993) presents the options developed following the Forest Summit.

SEE ALSO: Endagered species; Northern spotted owl; Old-growth forests.

Fossey, Dian

BORN: January 16, 1932; San Francisco, California
DIED: December 26-27 (?), 1985; Virunga Mountains, Rwanda
CATEGORY: Animals and endangered species

Dian Fossey was an American anthropologist known for her passionate attempt to save the mountain gorillas from Central African poachers.

Dian Fossey was born on January 16, 1932, in San Francisco, California. She was the daughter of George Fossey III, an insurance agent, and Kitty Fossey, a homemaker. When Fossey was six her parents divorced, and she grew up with her mother and her stepfather, Richard Price, a building contractor. After high school, Fossey enrolled in a veterinary medicine program at the University of California at Davis while supporting herself with low-paying jobs. Her academic difficulties with chemistry and physics courses made her transfer to San Jose State College, where she earned her bachelor's degree in occupational therapy in 1954. After her postcollege clinical training, she became the director of the occupational therapy department at the Kosair Crippled Children's Hospital in Louisville, Kentucky.

Fossey's love for Africa was inspired by a book on mountain gorillas written by American zoologist George Schaller. In 1963 she took a bank loan to finance a seven-week safari trip to Africa. At Olduvai Gorge in Tanzania, she met with Mary and Louis S. B. Leakey who were involved with the search for hominid fossils. Fossey's first encounter with a mountain gorilla had a tremendous impact on her. After the end of her African trip, she returned to Kentucky and resumed her work with handicapped children. Three years later Leakey went to Louisville and tried to convince Fossey to study the gorillas in the wild as part of a long-term expedition. She agreed and, after paying a short visit to Jane Goodall in Tanzania, where she learned her methods of data collection, Fossey set up her first campsite and work station at Kabara, Zaire.

Fossey managed to approach and study the gorillas in the remote mountain areas for about seven months, until political unrest took over Zaire. On July 10, 1967, she was arrested by armed guards and kept in custody for two weeks, while subjected to repeated raping. She eventually managed to escape and found refuge in Uganda. In 1970 she enrolled at Cambridge University in England, where she earned her doctorate in zoology six years later. Immediately thereafter Fossey traveled to Rwanda, where she stayed until 1980, at which time she accepted a visiting associate professorship at Cornell University. She continued to act as the project coordinator at the Karisoke Research Center in Rwanda.

While at Cornell Fossey became aware of increased poaching and the rapid deterioration of the research center. She decided to improve the situation and returned to Karisoke in June, 1983. She actively took over the management of the center but failed to keep the support of the National Geographic Society. Her incessant, passionate antipoaching activities created many enemies. On December 27, 1985, Fossey was found dead from machete wounds in her camp in the Virunga Mountains. She was buried in the gorilla cemetery she had built near the camp. Wayne Richard McGuire, an American wildlife researcher, was prosecuted as the primary suspect in her death but fled Rwanda.

Fossey's life was driven by the fighting of poachers and continuation of the mountain gorilla conservation efforts that covered a period of almost twenty years in the mountains of Zaire, Uganda, and Rwanda. Poachers, slaughters, and revolutions, as well as loneliness, were part of her daily routine. At the Karisoke Research Center she studied more than fifty gorillas that she described as rather peaceful, charging at humans only when threatened. Fossey was acknowledged as the world's leading authority on the mountain gorillas after the 1983 publication of her book *Gorillas in the Mist*, which dramatically enlarged the contemporary knowledge of gorilla habits, communication, and social structure. Fossey believed that gorillas are altruistic and regal animals whose family structure is unbelievably strong.

Soraya Ghayourmanesh

SEE ALSO: Animal rights; Endangered species; Mountain gorillas; Poaching.

Fossil fuels

CATEGORY: Energy

Fossil fuels such as coal, oil, and natural gas are the primary fuel sources for the industrial nations of the world. Developing nations are trying to raise their standard of living by becoming more industrialized, so their utilization of fossil fuels is increasing. As consumption has increased, however, people have become more aware that fossil fuels are nonrenewable resources that produce atmospheric pollution when they are burned.

Transportation by car, truck, train, ship, or airplane depends on oil and its by-products. Winter heating for homes, factories, and public buildings is mostly supplied by burning natural gas or coal. In the United States more than 70 percent of the electricity is generated at plants that burn fossil fuels. Modern agricultural technology enables less than 5 percent of the workforce to provide food for the whole population, but substantial fuel supplies are needed for farm machinery, irrigation, and agricultural chemicals.

The world's resources of coal, oil, and natural gas have come from plants and animals that lived on the earth's surface millions of years ago. Therefore, the present supply cannot be replaced once it is used. Known reserves of oil and natural gas are declining in spite of extensive exploration efforts to find new deposits. Coal is still abundant, with an estimated supply that would last for many centuries.

Mining coal and shipping it to a factory or power plant creates several environmental problems. Surface mining defaces the landscape and causes erosion. Mine operators can be required to replace soil and replant vegetation, but it takes time to regrow. Deep mining is a hazardous occupation for workers because of long-term lung damage and the possibility of an underground explosion. Coal is normally shipped by train. A typical power plant burns about two hundred fully loaded freight cars of coal per day, which equals over seven million tons per year. Problems encountered along a train track include noise and the hazards of derailment or collisions at track crossings.

Oil is the preferred fuel for the various methods of transportation. To a lesser extent, it is used to heat buildings, generate electricity, and make plastics. Oil is transported from wells to refineries by means of pipelines and ocean tankers. The Trans-Alaskan Pipeline, built during the 1970's, had to be designed to protect wildlife migration routes and avoid damaging the permafrost environment. A pipeline can break because of natural causes or sabotage. Tanker accidents that result in oil spills have caused harm to fish and wildlife. The United States imports more than 50 percent of its oil. Overdependence on foreign sources can generate the need for military intervention to assure a reliable oil supply.

Natural gas is found in association with oil and coal deposits. It burns cleaner than coal and oil, so its use for electric power plants and home heating has rapidly increased. New gas pipeline construction has brought environmental objections when a route is chosen that will cross a Native American reservation or valuable farmland.

Some coal contains as much as 6 percent sulfur, which is converted to gaseous sulfur dioxide during combustion. When sulfur dioxide is released into the atmosphere, it eventually combines with water to form sulfuric acid and returns to earth as acid rain. Also, nitrogen from the air combines with oxygen during combustion, leading to the formation of nitric acid. The harmful effects of acid rain have been documented by the death of fish populations in acidified lakes, deterioration of forests, and surface erosion of marble monuments, cathedrals, and public buildings. To reduce sulfur dioxide emission from coal-burning plants, several technological procedures are available. Washing coal at the mine before it is shipped can remove as much as one-third of its sulfur content. Coal is pulverized before entering the combustion chamber, making the sulfur accessible for chemical removal. Finally, sulfur dioxide can be captured after combustion in so-called scrubbers, where a spray of fine limestone combines with the gas to form a thick slurry. These procedures all add to the cost, and scrubbers cannot be installed economically at older plants. For nitrogen oxides produced during combustion, no chemical procedure for large-scale removal is available, so emission-control regulations apply only to sulfur.

When coal, oil, or natural gas are burned, carbon combines with oxygen from the air to form carbon dioxide gas. Each ton of coal that is burned releases almost four tons of carbon dioxide to the atmosphere, which acts like the glass roof of a greenhouse, allowing solar radiation to enter but preventing infrared heat from escaping. Whether the extra carbon dioxide actually causes global warming is still uncertain because summer and winter temperatures have large natural fluctuations from year to year. If global warming does take place, serious consequences would follow. A small rise in the level of the oceans would cause disastrous flooding in

Oil, a fossil fuel, is used to power millions of vehicles around the world. The combustion of oil and other fossil fuels creates harmful emissions that pollute the air. (Archive Photos)

coastal communities. Also, a hotter, drier climate would create worldwide problems in food production. The only way to reduce production of carbon dioxide is to burn less coal, oil, and natural gas. At the 1997 conference held in Kyoto, Japan, the industrialized nations pledged to reduce pollution output 5 percent below their 1990 levels by the year 2012. The United States signed the Kyoto Accords in 1998, but the U.S. Senate did not immediately ratify them.

Hans G. Graetzer

SUGGESTED READINGS: *Global Warming: Assessing the Greenhouse Threat* (1990), by Laurence Pringle, presents an elementary overview of global warming, with photographs on almost every page. Robert Orstmann, *Acid Rain: A Plague upon the Waters* (1982), provides good historical background. Richard Balzhiser and Kurt Yeager, "Coal-Fired Power Plants for the Future," *Scientific American* (September 1987), provides information on pollution-reduction strategies for power plants that use coal. George J. Mitchell, U.S. senator from 1980 to 1994, makes an urgent appeal to control global warming in *World on Fire: Saving an Endangered Earth* (1991).

SEE ALSO: Acid deposition and acid rain; Air pollution; Climate change and global warming; Greenhouse effect; Kyoto Accords.

Freon

CATEGORY: Atmosphere and air pollution

Du Pont introduced Freon in 1930 as a non-toxic, nonflammable refrigerant. It was later shown to harm the earth's ozone layer.

Freon is an example of a class of gases known as chlorofluorocarbons (CFCs), carbon compounds that contain fluorine and chlorine. According to some scientists, CFCs appear to erode the ozone layer in the stratosphere. They are derivatives of simple alkanes—such as methane, ethane, propane, and butane—by direct or selective ultraviolet halogenation using chlorine or fluorine gas. Freon, a Du Pont trade name, has found extensive uses in industry. CFCs have served as dispersing gases in aerosol cans, in the preparation of foamed plastics, and, primarily, as refrigerants. Their manufacture, together with that of the closely related halons, rose in the mid-1970's and peaked in 1986 with the production of almost 1.25 million tons. At that time these compounds were universally used in aerosol products that ranged from insecticides to shaving foams and hair sprays, as well as in the insulation of buildings and cleaning of circuit boards and other electronic parts. One of them, bromotrifluoromethane (Freon 13B1), has been used as fire extinguisher in situations where the use of water is avoided, such as electrical fires.

Common members of this family include tri-chlorofluoromethane (Freon or CFC 11), dichlorodifluoromethane (CFC 12), and 1,2-dichloro 1,1,2,2-tetrafluoroethane (CFC 114). They are all either gases or low-boiling liquids at room temperature and are virtually insoluble in water. They are generally dense, easily liquefied, not flammable, thermally stable, virtually odorless, and inexpensive to manufacture. They do not undergo decomposition via the ordinary chemical reactions that take place in the troposphere. As a result, they are used as ideal propellants in aerosol cans of deodorants, hair sprays, and other commercially available food products. Their relative inertness toward other chemicals allows them to persist in the atmosphere and is responsible for their great environmental problems.

Since they are water insoluble, rain cannot dissolve them and wash them down to the ground. As a result, they drift upward into the stratosphere and the ozone layer, which they reach after approximately seven to ten days. They may stay in the stratosphere for several decades, absorb the sun's ultraviolet light, and yield free radicals, which appear to undergo chemical reactions that lead to the depletion of the ozone layer. Although toxic to human lungs, the ozone's presence in the stratosphere is critical in protecting the earth from the harmful ultraviolet part of the electromagnetic radiation associated with sunlight. If the ozone layer gets thin, exposure to ultraviolet radiation will exponentially increase the cases of skin cancer and other diseases, while at the same time destroying crops and other plants.

CFCs have been found to escape into the atmosphere from old refrigerators and air conditioning units. Most industrialized countries have banned their use and have replaced them with methylene chloride or nonhalogenated hydrocarbons, such as isobutane. The flammability of those hydrocarbons and the suspected carcinogenicity of methylene chloride have created an incentive for the development of CFC substitutes. In 1970 the United States Congress passed legislation aimed at curbing the sources of air pollution by setting standards for air quality. In 1987 more than twenty nations signed an agreement to downscale the production of CFCs with the intent eventually to eradicate their use.

Soraya Ghayourmanesh

SEE ALSO: Aerosols; Air pollution; Chlorofluorocarbons; Clean Air Act and amendments; Ozone layer and ozone depletion.

G

Gaia hypothesis

CATEGORY: Philosophy and ethics

The Gaia hypothesis suggests that the earth's troposphere was actively formed by and is presently being modulated and regulated by life itself.

While working as a consultant to the National Aeronautics and Space Administration (NASA), James Lovelock suggested that the key to ascertaining whether life existed on other planets was to look at their atmospheric gases. Lovelock claimed that the levels of carbon dioxide and oxygen held the answers. Mars, a carbon-dioxide-based planet, is extraordinarily dry with temperatures that seldom reach 0 degrees Celsius (32 degrees Fahrenheit). This means that water vaporizes at a low pressure. Consequently, life cannot exist on Mars. Venus, on the other hand, is very hot with trace amounts of water. Its atmosphere is 95 percent carbon dioxide with clouds of sulfuric acid. This atmosphere is extremely reactive and unstable. Earth is neither too hot nor too cold. Moreover, its atmosphere is uniquely high in oxygen and low in carbon dioxide.

Physiochemical theories of the earth's evolution strongly suggest that its atmosphere originally was not always this way. It is assumed that the troposphere contained large amounts of ammonia, methane, and sulfurous gases. As the earth alternately cooled and warmed under the sun's influence, the gases cooled and mixed. With further help from volcanic action and lightning, the amounts of water increased. This provided a unique reaction medium wherein early life forms could evolve, a so-called primordial soup.

Over time, single-celled prokaryotic microorganisms became the prominent life form. As the atmosphere cooled and settled, more sunlight was able to penetrate. Occasionally, some of those microorganisms mutated to utilize carbon dioxide and water in the presence of sunlight, and thus photosynthesis began. As a resultant by-product, oxygen gas was emitted. The Gaia hypothesis suggests that this was the crucial step. Oxygen, now a toxic gas to some of the early bacteria, forced them to eventually be sequestered in areas such as bacterial mats or in the large masses of stromatolites in mud flats. Other organisms found ways to utilize the glucose energy provided by photosynthesis in the presence of the oxygen. As time progressed and the number of green plants continued to thrive, oxygen levels gradually increased, while carbon dioxide was continually consumed. Further, organisms that utilized carbon in the form of carbonates continued to store carbon in their shells. As they died, layers of this carbon became trapped in rock as reefs or cliffs.

The Gaia hypothesis suggests that it is the sum of the biota, all things that photosynthesize and respire, that continues to modulate and maintain the atmosphere. Since it is the plant life that continuously processes respired carbon dioxide and the animal life that utilizes the photosynthetic by-product of oxygen, life itself keeps the earth's atmosphere in check. Lovelock, as advised by his friend William Golding, chose the Greek Goddess of the earth, Gaia, to name this theory.

Kathleen Rath Marr
SEE ALSO: Biosphere concept; Lovelock, James.

Garbology

CATEGORY: Waste and waste management

Garbology is the study of human discards. The term is most commonly used for the archeological study of refuse, but garbology can include any activity in which refuse is closely studied.

Garbology is an informal term. No university has an official garbology program, but many people call themselves garbologists. Despite the formal definition of garbage as moist refuse, in practice garbology includes the study of all types of household and commercial refuse. Professor William Rathje is recognized as the foremost expert in garbology. Combining his anthropological and archeological expertise, Rathje directed the Garbage Project at the University of Arizona. He and his team of researchers analyzed more than fifteen thousand samples of household refuse from around the world. Rathje and Culleen Murphy coauthored *Rubbish: The Archaeology of Garbage* (1992), which describes the more interesting findings from these Garbage Project investigations.

A typical Garbage Project study uses recently collected refuse from a particular neighborhood, keeping the names of refuse generators anonymous. Garbage Project researchers then comb through this garbage, sorting it into 150 categories and recording information about its size, weight, and other characteristics. These garbage sorts provide a wide range of consumer information, ranging from the quantity of recyclable materials thrown away to trends in liquor consumption over time.

Other people have popularized garbology by examining the discards of famous people. In the 1970's A. J. Weberman sifted through musician Bob Dylan's garbage for many months and wrote magazine articles about his findings. He and his associates at the National Institute of Garbology promote their garbological investigations of luminaries such as Jacqueline Kennedy Onassis, Spiro Agnew, and Dustin Hoffman. A 1975 *National Enquirer* article featuring journalist Jay Gourley's analysis of Secretary of State Henry Kissinger's garbage sparked widespread controversy about journalistic standards and invasion of privacy.

Likewise, law enforcement and intelligence agents analyze the discards of suspected criminals and political figures to build evidence about their activities. Elevating garbology to interna-tional intrigue, the Central Intelligence Agency (CIA) reportedly paid $1,000 in the 1950's for the refuse from a single Soviet airliner. Although the U.S. Constitution protects against illegal search and seizure of private property, the Supreme Court ruled in 1988 that garbage left for collection can be searched without a warrant. Some states and municipalities have laws to prevent refuse scavenging, although such statutes may not be rigidly enforced.

The opportunity to look for useful items in refuse has led to the growing popularity of "dumpster divers," who may or may not consider themselves garbologists. Some of these dumpster divers are desperate for food and necessities, but others are environmentalists who dislike seeing usable items go to waste.

The term garbology is sometimes used to designate other refuse-related activities. Refuse collection and recycling professionals may refer to themselves as master garbologists. Antilitter campaigns use garbology to analyze the types of litter strewn on highways and other public places. Litter prevention programs can then target key sources of litter. Even artist garbologists use items from the refuse to make "trash art."

Andrew P. Duncan

SEE ALSO: Landfills; Recycling; Resource recovery; Waste management.

General Motors solar-powered car race

CATEGORY: Energy

General Motors (GM) initiated the GM Sunrayce USA in 1990 to encourage the development of solar-powered transportation and technology.

In 1987 Hughes Aircraft, a subsidiary of GM, and Aero Vermont, a company in which GM held stock, built a solar-powered car to compete in the Pentex World Solar Challenge in Australia. The car—the GM Sunraycer—won the 3,000-kilometer (1,864-mile) race with an average speed of 66.9 kilometers per hour (41.6 miles per hour).

The University of Michigan's Sunrunner, which won the inaugural General Motors Sunrayce USA in 1990. (Jim West)

GM decided to use the Sunraycer to promote public interest in scientific and environmental education. The corporation distributed information about solar technology and initiated a solar-powered car race to challenge students to pursue scientific innovation. Sponsored by GM, the United States Department of Energy (DOE), and Electronic Data Systems (EDS), the GM Sunrayce USA has become the largest race for solar-powered transportation in the United States. The race's goal is to educate Americans about energy options that are efficient, economical, and nonpolluting.

In the spring of 1989, GM invited university students to prepare proposals explaining their designs for solar-powered race cars. Participants assembled their solar cars, developed prototypes through computer modeling, and assessed their vehicles' aerodynamic qualities in wind tunnels. The DOE Office of Energy Efficiency and Renewable Energy provided technical assistance, tested photovoltaic arrays, and offered information from the National Renewable Energy Labo-

ratory. Rules stated that sunlight was the only permissible power source and that contestants had to use commercially available solar cells manufactured in North America. All cars had to incorporate safety devices for the driver. Each team fashioned a car covered with photovoltaic solar cells to generate electricity, which powered a motor. Excess electricity was stored in batteries. Competitors attempted to create lightweight, durable, and aerodynamic vehicles that were low to the ground to prevent drag.

Thirty-two college teams qualified to race in the first GM Sunrayce USA, July 9-19, 1990. The back-road race route—from Lake Buena Vista, Florida, to the GM Technical Center in Warren, Michigan—totaled 1,625 miles (2,600 kilometers). The teams completed one leg of the course each day, with a recharge day midway through the contest. Pit stops and overnight rests provided opportunities to service cars and contemplate strategy. The team that accumulated the lowest time while covering the mileage was champion. The University of Michigan's

Sunrunner won the race with a time of seventy-two hours, fifty minutes, and forty-seven seconds and an average speed of 39.7 kilometers per hour (24.7 miles per hour). Western Washington University's Viking (74:10:06) and the University of Maryland's Pride of Maryland (80:10:55) finished second and third. Awards were presented that recognized engineering design, technical innovation, sportsmanship, and teamwork. The top three teams earned slots to compete in the November, 1990, World Solar Challenge in Australia, where the Michigan team placed third.

After the success of the first GM Sunrayce USA, the DOE agreed to organize the solar car race every two years. Teams submit proposals, participate in technical workshops, and raise and budget thousands of dollars for vehicle components and expenses, often paid for by corporate sponsors. Every solar car is inspected for safety and is required to log 160 kilometers (100 miles) to enter preliminary trials at GM's Milford Proving Grounds in Michigan and other regional sites. A maximum of forty qualifying teams may compete. The race is usually scheduled for late June to coincide with summer solstice for optimum sunlight. In 1993, the Sunrayce covered a 1,770-kilometer (1,100-mile) route from Arlington, Texas, to Apple Valley, Minnesota, in eight days. Thirty-four teams competed, and the University of Michigan repeated its victory. Despite rain and cloudy conditions, the average winning speed was 43.9 kilometers per hour (27.29 miles per hour).

The theme of GM Sunrayce 1995 was "Education, Energy, and the Environment," and cosponsors included the Environmental Protection Agency (EPA). Teams raced from Indiana to Colorado, and the average speed of the winner, the Massachusetts Institute of Technology, was 59.9 kilometers per hour (37.2 miles per hour). California State University at Los Angeles's Solar Eagle III won GM Sunrayce 1997 from the Indianapolis Motor Speedway to Colorado Springs, Colorado, at an average speed of 69.7 kilometers per hour (43.29 miles per hour). The GM Sunrayce 1999 was scheduled to begin in Washington, D.C., and end in Orlando, Florida.

Every year teams refine designs, introduce new materials, and produce more sophisticated, energy-efficient cars in an attempt to beat their personal best times and secure the national championship. Winning cars have been exhibited in the Henry Ford Museum and the Chicago Museum of Science and Industry. Race veterans have formed professional companies to further develop the application of renewable energy resources to transportation. The GM Sunrayce USA has inspired related races for solar boats, bicycles, and model cars. By showcasing solar power, the GM Sunrayce USA encourages awareness of how transportation energy utilizes and affects the environment.

Elizabeth D. Schafer

SUGGESTED READINGS: James Neal Blake, *Run Your Car on Sunshine: Using Solar Energy for a Solar Powered Car* (1980), is a how-to manual for the construction of solar-powered vehicles. Ken Butti and John Perlin, *A Golden Thread: 2500 Years of Solar Architecture and Technology* (1980), provides a narrative history of solar energy and its applications. Jan F. Kreider, Charles J. Hoogendoorn, and Frank Keith, *Solar Design: Components, Systems, Economics* (1989), contains a technical exploration of solar technologies. Maryann Keller, "GM Lets Its People Go," *Motor Trend* 40 (April, 1988), contains an account of the GM Sunraycer's victory in Australia. Brian Nadel, "Sun Racers," *Popular Science* 237 (August, 1990), describes preparations for the first GM Sunrayce.

SEE ALSO: Alternative energy sources; Alternative fuels; Alternatively fueled vehicles; Energy conservation; Renewable resources; Solar energy.

Genetically altered bacteria

CATEGORY: Biotechnology and genetic engineering

Bacteria may be genetically altered through the introduction of recombinant deoxyribonucleic acid (DNA) molecules into the cells. Such bacteria may be used to produce human insulin or introduce disease-resistant genes into plants.

Types of Bacteria

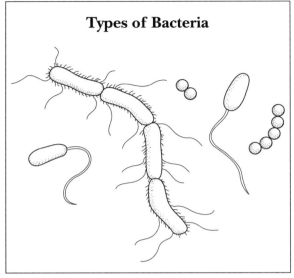

Bacteria occur in a variety of shapes and sizes.

The ability to genetically alter bacteria is the outcome of several independent discoveries. In 1944 Oswald Avery and his coworkers demonstrated gene transfer among bacteria using purified DNA, a process called "transformation." In the 1960's the discovery of restriction enzymes permitted the creation of hybrid molecules of DNA. Such enzymes cut DNA molecules at specific sites, allowing fragments from different sources to be joined within the same piece of genetic machinery. Restriction enzymes are not species-specific in choosing their targets. Therefore, DNA from any source, when treated with the same restriction enzyme, will generate identical cuts. The treated DNA molecules are allowed to bind with each other, while a second set of enzymes called "ligases" are used to fuse the hybrids. The recombinant molecules may then be introduced into bacteria cells through transformation. In this manner, the cell has acquired whatever genetic information is found in the DNA. Descendants of the transformed cells will be genetically identical, forming clones of the original.

The most common forms of genetically altered DNA are bacterial plasmids, small circular molecules separate from the cell chromosome. Plasmids may be altered to serve as appropriate vectors for genetic engineering, usually containing an antibiotic resistance gene for selec-tion of only those cells that have incorporated the DNA. Once the cell has incorporated the plasmid, it acquires the ability to produce any gene product encoded on the molecule. The first such genetically altered bacterium used for medical purposes, *Escherichia coli*, contained the gene for the production of human insulin. Prior to creation of the insulin-producing bacterium, diabetics were dependent upon insulin purified from animals. In addition to being relatively expensive, insulin obtained from animals produced allergic reactions among some individuals. Insulin obtained from genetically altered bacteria is identical to that of human insulin. Subsequent experiments also produced bacteria able to produce a variety of human proteins, including human growth hormone, interferon, and granulocyte colony-stimulating factor.

Genetically altered bacteria may also serve as vectors for the introduction of genes into plants. The bacterium *Agrobacterium tumefaciens*, the etiological agent for a plant disease called crown gall, contains a plasmid called Ti. Following infection of the plant cell by the bacterium, the plasmid is integrated into the host chromosome, becoming part of the plant's genetic material. Any genes that were part of the plasmid are integrated as well. Desired genes can be introduced into the plasmid, promoting pest or disease resistance within plants infected by the bacterium.

In April of 1987 scientists in California sprayed strawberry plants with genetically altered bacteria to improve the plant's freeze resistance, marking the first deliberate release of genetically altered organisms in the United States to be sanctioned by the Environmental Protection Agency (EPA). The release of the bacteria climaxed more than a decade of public debate over what would happen when the first products of biotechnology became commercially available. Fears centered on the creation of bacteria that might radically alter the environment through elaboration of gene products not normally found in such cells. Others feared that super bacteria might be created with unusual resistance to conventional medical treatment.

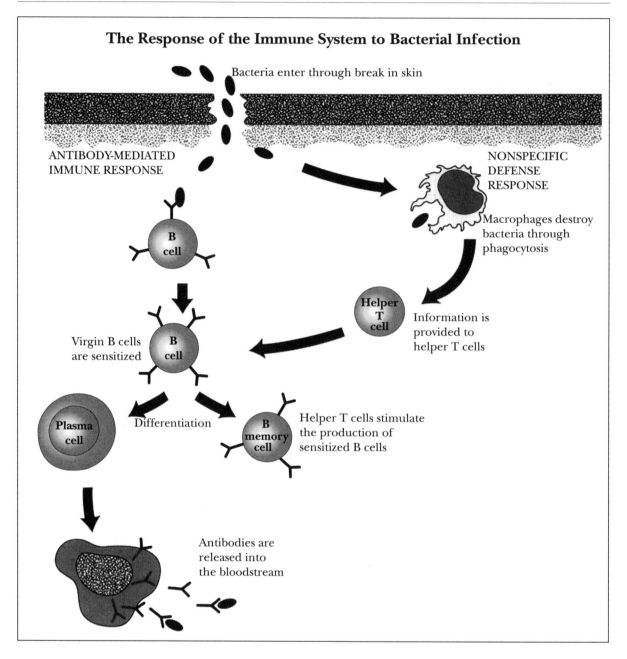

The Response of the Immune System to Bacterial Infection

Bacteria enter through break in skin

ANTIBODY-MEDIATED
IMMUNE RESPONSE

NONSPECIFIC
DEFENSE
RESPONSE

B cell

Macrophages destroy
bacteria through
phagocytosis

Helper
T cell

Information is
provided to
helper T cells

Virgin B cells
are sensitized

B cell

Plasma
cell

Differentiation

B memory
cell

Helper T cells stimulate
the production of
sensitized B cells

Antibodies are
released into
the bloodstream

Despite these concerns, approval for further releases of genetically altered bacteria soon followed, and the restrictions on release were greatly relaxed. By 1991 permits for field tests of more than 180 genetically altered plants and microorganisms had been granted.

Richard Adler

SEE ALSO: Biotechnology and genetic engineering; Genetically engineered organisms.

Genetically engineered foods

CATEGORY: Biotechnology and genetic engineering

Genetically engineered foods are derived from living organisms that have been modified by gene-transfer technology. Applications of genetic engineering in agriculture and the food industry could increase world food supplies, reduce environmental problems associated with food production, and enhance the nutritional value of certain foods. However, these benefits are countered by food-safety concerns, the potential for ecosystem disruption, and fears of unforeseen consequences resulting from altering natural selection.

Humans rely on plants and animals as food sources and have long used microbes to produce foods such as cheese, bread, and fermented beverages. Conventional techniques such as cross-hybridization, production of mutants, and selective breeding have resulted in new varieties of crop plants or improved livestock with altered genetics. However, these methods are relatively slow and labor intensive, are generally limited to intraspecies crosses, and involve a great deal of trial and error.

Recombinant deoxyribonucleic acid (DNA) techniques developed in the 1970's enable researchers to rapidly make specific, predetermined genetic changes. Since the technology

Genetically Engineered Crop Plants No Longer Regulated by the U.S. Department of Agriculture

CROP	PATENT HOLDER	GENETICALLY ENGINEERED TRAIT(S)
Canola	AgrEvo	herbicide tolerance
Corn	AgrEvo	herbicide tolerance
	Ciba-Geigy	insect resistance
	DeKalb	herbicide tolerance; insect resistance
	Monsanto	herbicide tolerance; insect resistance
	Northrup King	insect resistance
Cotton	Calgene	herbicide tolerance; insect resistance
	Du Pont	herbicide tolerance
	Monsanto	herbicide tolerance; insect resistance
Papaya	Cornell	virus resistance
Potato	Monsanto	insect resistance
Squash	Asgrow	virus resistance
	Upjohn	virus resistance
Soybean	AgrEvo	herbicide tolerance
	DuPont	altered oil profile
	Monsanto	herbicide tolerance
Tomato	Agritope	altered fruit ripening
	Calgene	altered fruit ripening
	Monsanto	altered fruit ripening
	Zeneca	altered chemical content in fruit

Source: U.S. Department of Agriculture Animal and Plant Health Inspection Service (APHIS).

also allows for the transfer of genes across species and kingdom barriers, an infinite number of novel genetic combinations are possible. The first animals and plants containing genetic material from other organisms (transgenics) were developed in the early 1980's. By 1985 the first field trials of plants engineered to be pest resistant were conducted. In 1990 the U.S. Food and Drug Administration (FDA) approved chymosin as the first substance produced by engineered organisms to be used in the food industry for dairy products such as cheese. That same year the first transgenic cow was developed to produce human milk proteins for infant formula. The well-publicized Flavr Savr tomato obtained FDA approval in 1994.

By the mid-1990's, more than one thousand genetically engineered crop plants were approved for field trials. The goals for altering food crop plants by genetic engineering fall into three main categories: to create plants that can adapt to specific environmental conditions to make better use of agricultural land, increase yields, or reduce losses; to increase quality, nutritional value, and flavor; and to alter transport, storage, and processing properties for the food industry. Many genetically engineered crops are also sources of ingredients for processed foods and animal feed.

Herbicide-resistant plants such as the Roundup Ready soybean can be grown in the presence of glyphosphate, a herbicide that normally destroys all plants with which it comes in contact. Beans from these plants have been approved for food-industry use in several countries, but there has been widespread protest by activists such as Jeremy Rifkin and organizations such as Greenpeace. Frost-resistant fruit containing a fish antifreeze gene, insect-resistant plants with a bacterial gene that encodes for a pesticidal protein (*Bacillus thuringiensis*), and a viral disease-resistant squash are examples of other genetically engineered food crops that have undergone field trials. Scientists have also created plants that produce healthier unsaturated fats and oils rather than saturated ones, coffee plants whose beans are caffeine-free without processing, and tomatoes with altered pulp content for improved canned products.

Animals can also be genetically engineered food sources. Transgenic research in animals is technically more difficult than with plants, although the technology used to clone Dolly the sheep in 1997 was a significant advancement. Animal rights issues, vegetarian and religious objections to animal-based components in food, and concerns over infectious agents that could be transferred to humans have all hindered developments in this field. The most notorious application of genetic engineering in animals involves the bovine growth hormone (BGH; also known as BST) synthesized by bacteria containing the bovine BGH gene. When BGH is given to cows as a supplement, milk production can increase up to 20 percent, but concerns over the health of treated cows and the safety of the milk have made this practice controversial.

Genetically engineered microbes are used for the production of food additives such as amino acid supplements, sweeteners, flavors, vitamins, and thickening agents. In some cases, these substances previously had to be obtained from slaughtered animals. Altered organisms are also used for improving fermentation processes in the food industry.

Food safety and quality are at the center of the genetically engineered food controversy. Concerns include the possible introduction of new toxins or allergens into the diet and changes in the nutrient composition of foods. Proponents argue that food sources could be designed to have enhanced nutritional value.

A large percentage of crops worldwide are lost each year to drought, temperature extremes, and pests. Plants have already been engineered to exhibit frost, insect, disease, and drought resistance. Such alterations would increase yields, allow food to be grown in areas that are currently too dry or infertile, and positively impact the world food supply.

Environmental problems such as deforestation, erosion, pollution, and loss of biodiversity have all resulted, in part, from conventional agricultural practices. Use of genetically engineered crops could allow better use of existing farmland and lead to a decreased reliance on pesticides and fertilizers. Critics fear the creation of "superweeds"—either the engineered plants or new plant varieties formed by the transfer of recombinant genes conferring various types of resistance to wild species. These weeds, in turn, would compete with valuable plants and have the potential to destroy ecosystems and farmland unless stronger poisons were used for eradication. The transfer of genetic material to wild relatives (outcrossing, or "genetic pollution") might also lead to the development of new plant diseases. As with any new technology, there may be other unpredictable environmental consequences.

Despite the risks, genetic engineering may be required to develop food sources that can survive rapidly changing environmental conditions. Pollution, global climate change, and increased ultraviolet irradiation result in stress conditions for living organisms, and all impact agriculture. The processes of natural selection and adaptation may be too slow to maintain required food supplies.

Diane White Husic

SUGGESTED READINGS: A thorough overview of the technology, applications, and risks associated with genetically engineered foods can be found in *Genetically Modified Foods: Safety Issues* (1995), published by the American Chemical Society. Jane Rissler discusses the potential environmental impact of this technology in *The Ecological Risks of Engineered Crops* (1996). Developments in agricultural genetic engineering are reported in *The Gene Exchange*, a public newsletter published by the Union of Concerned Scientists.

SEE ALSO: *Bacillus thuringiensis*; Biotechnology and genetic engineering; Flavr Savr tomato; Genetically engineered organisms; Sustainable agriculture.

Genetically engineered organisms

CATEGORY: Biotechnology and genetic engineering

Genetically engineered organisms are living organisms whose genetic compositions have been altered by technology. Organisms can be engineered for use in scientific research, human and veterinary medicine, industry, agriculture, and environmental remediation. Despite the beneficial applications, potential risks and ethical issues associated with the technology have led to controversy.

With the advent of recombinant deoxyribonucleic acid (DNA) technology in the 1970's came the ability to modify and create genes and to transfer genetic material between unrelated species in a rapid and specific manner. The development of new traits is no longer limited to mutation or natural selection from a limited pool of genes; to alter an organism, scientists can introduce genetic traits from any species. Transgenic animals and plants were first developed in the early 1980's using genetic engineering in combination with techniques such as cell fusion, tissue culture, in vitro fertilization, and embryo transplantation. Clonal propagation, possible in cell and tissue culture for years, was successfully applied to mammals with the creation of Dolly the sheep in 1997.

Industry has made widespread use of genetically engineered organisms. Modified microbes are used in fermentation processes and to produce food ingredients. In the chemical industry, organisms are engineered to produce reagents and novel catalysts and to convert hazardous waste into harmless or useful substances. The result has been increased efficiency of certain industrial processes and decreased waste. In 1982 the U.S. Food and Drug Administration (FDA) approved the use of human insulin derived from genetically engineered bacteria. Subsequently, the pharmaceutical industry has made significant use of genetically engineered organisms to develop new drugs, vaccines, and diagnostic tests.

Genetic engineering has also been used extensively in agriculture. Products of engineered organisms are used to protect plants from frost and insects, manipulate lactation and growth processes in livestock, and improve animal health. In 1986 the Environmental Protection Agency (EPA) approved the release of the first genetically engineered crop plant; by the end of the 1990's, more than one thousand others had

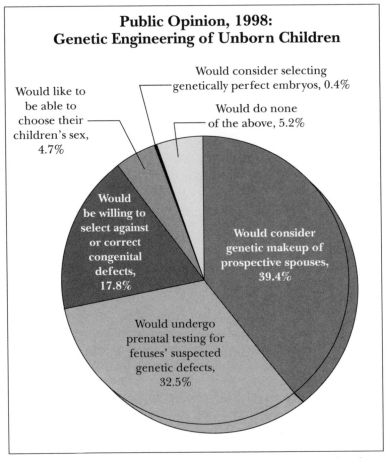

Public Opinion, 1998:
Genetic Engineering of Unborn Children

Would consider selecting genetically perfect embryos, 0.4%

Would like to be able to choose their children's sex, 4.7%

Would do none of the above, 5.2%

Would be willing to select against or correct congenital defects, 17.8%

Would consider genetic makeup of prospective spouses, 39.4%

Would undergo prenatal testing for fetuses' suspected genetic defects, 32.5%

An informal 1998 survey conducted on the Internet elicited the above responses to the question, "How far would you go in genetically manipulating your unborn child?"

Source: Moms Online

The use of living organisms to remove toxic chemicals from the environment is known as bioremediation. Genetically engineered microorganisms (GEMs) are essential to this process since many pollutants are human-made chemicals that cannot normally be degraded by living organisms. GEMs contain new or modified enzymes that enable them to digest the pollutants and convert them to harmless compounds such as carbon dioxide and water. GEMs are also advantageous in that they can be used for in situ (at site) treatments including soil decontamination, detoxification of wetlands and streams, and groundwater purification, eliminating some of the technical problems associated with other remediation methods. GEMs are also used to recover minerals from mining and industrial waste.

Genetically modified plants can also remove environmental pollutants. Field trials demonstrated that engineered plants absorbed lead from contaminated soil into their roots. Harvested plants can be incinerated, and the toxic metal waste is then confined to a small volume of ash. In the future, similar phytoremediation technologies may be used at sites contaminated with hydrocarbons and radioactivity.

Modification of natural selection and disruption of ecosystems are the major concerns associated with the introduction of genetically engineered organisms into the environment. In addition, not all of the consequences of this technology are predictable because of a lack of data on the stability of artificial genetic changes, the tendency of DNA manipulation to induce mutations in organisms, and the natural complexity of organisms. Unforeseen environmental problems posed by genetically engineered organisms are intractable, since once introduced to the environment, it may be impossible to re-

been field-tested. Plants have been designed to resist disease, drought, frost, insects, and herbicides, and to improve the nutritional value or flavor of foods. Plants have even been engineered to produce synthetic rubber, plastics, vaccines, and renewable fuels.

One of the first examples of an organism genetically engineered to address environmental problems was a bacterium created to degrade oil. These oil-eating bacteria were at the center of an important U.S. Supreme Court case (*Diamond v. Chakrabarty*, 1980) in which it was ruled that the bacteria were living inventions and patentable. Such bacteria have proven effective in cleaning up oil spills both in the ocean and on land.

call them. Critics of the technology make reference to "genetic wastes" that are able to propagate, mutate, and migrate.

The possibility of gene flow or escape—the transfer of genes from the modified organisms to related species in the wild—is of concern. Antibiotic resistance genes used as markers in the development of transgenic organisms might be transferred to bacteria, leading to new treatment-resistant strains. Modified viruses used in many recombinant DNA techniques might escape to create new disease-causing agents. Transgenic organisms designed to be more vigorous or any new species created by the gene transfer may have selective advantages over native species, leading to the disruption of natural balance in ecosystems and possibly exacerbating biodiversity restriction. However, even without human intervention, the exchange of genes can occur in nature.

Underlying the controversy over genetic engineering are ethical concerns related to the fair treatment of animals and the capability to selectively modify humans. Furthermore, there is potential for the misuse of the technology in human experimentation, the development of biological weapons, and ecoterrorism. Several agencies in the United States and throughout the world have the authority to regulate and restrict the use of genetic engineering but have been reluctant to do so because of the potential benefits of the applications of this technology.

Diane White Husic and H. David Husic

SUGGESTED READINGS: For a perspective on this topic from an outspoken opponent of genetic engineering, see *The Biotech Century: Harnessing the Gene and Remaking the World* (1998), by Jeremy Rifkin. Ecologist Peter Kareiva discusses the environmental impact of this technology in *Environmental Risks of Genetically Engineered Organisms and Key Regulatory Issues* (1995). *Molecular Biotechnology: Principles and Applications of Recombinant DNA* (1994), by Bernard R. Glick and Jack J. Pasternak, provides a technical but comprehensible discussion of the techniques, applications, and regulatory issues related to genetic engineering. A perspective on the environmental im-

pact of genetic engineering is found in the position statement of the Ecological Society of America, "The Planned Introduction of Genetically Engineered Organisms: Ecological Considerations and Recommendations," by J. M. Tiedje et al., *Ecology* 70 (1989).

SEE ALSO: Bioremediation; Biotechnology and genetic engineering; *Diamond v. Chakrabarty*; Dolly the sheep; Flavr Savr tomato; Genetically engineered bacteria; Genetically engineered foods; Genetically engineered pharmaceuticals.

Genetically engineered pharmaceuticals

CATEGORY: Biotechnology and genetic engineering

Genetical technology is being used to develop state-of-the-art drugs to treat human diseases and health conditions.

Genetics is the scientific study of heredity—the biological factors that determine the characteristics of all living things. All reproductive life forms develop under the laws of genetics. The basis of genetics is the gene, a tiny unit of matter that determines some identifiable characteristic of an individual. Genes are located in fixed positions on chromosomes (molecular chains) that reside in the center of cells. A major part of the chromosome is deoxyribonucleic acid (DNA), which is responsible for transmitting genetic information, in the form of genes, when new life is created. This transfer of traits applies to organisms of all sizes, from microscopic to larger and more complex systems, such as humans. Inherited traits include color of hair, eyes, and skin, as well as susceptibility to various ailments.

As the complex chemical-biological process of reproduction takes place at the gene-chromosome level, it is possible for a random processing error called a mutation to be introduced. Mutant cells may cause many human defects and diseases. Genetic abnormalities, also known as birth defects, include such ailments as hemophilia (resistance to blood clotting), color blindness, anatomical defects, speech disorders, hormonal disorders, brain disorders, and psychiatric illness. Besides birth defects, genetic cellular mutations can occur anytime during a lifetime. Normally the body contains certain controlling genes that destroy mutant genes that spontaneously appear. If these controlling genes become defective, the mutant genes can take over the body, as occurs with cancerous tumors.

Pharmaceuticals are drugs used to treat human diseases and conditions. An "engineered drug" implies that scientific principles and manufacturing processes are applied in creating the drug. A genetically engineered pharmaceutical is a specialized drug made by the application of specific genetic principles. Gene-based technology is being used to investigate, test, and apply state-of-the-art pharmaceuticals to invasive and widespread diseases, such as cancer and acquired immunodeficiency syndrome (AIDS).

As an example, scientists are working on ways to control cancer cells using genetic medicines instead of killing the cancer with radiation, conventional drugs, or surgical removal. Large numbers of mice are used as test subjects since their physical systems are somewhat similar to humans. Herceptin, a genetically engineered drug approved by the Food and Drug Administration (FDA), is used in treating certain breast cancers. Herceptin is an antibody engineered to attack specific cancer cells, which helps to reduce the cancer tumor by keeping a particular protein from reproducing.

Production of natural body chemicals in the past could only be done by harvesting the needed chemicals from human or animal materials. Supply sources are sometimes minimal, and concentrating chemicals from human and animal tissue can also multiply the chances of carrying diseases from those sources. Genetic engineering avoids many traditional problems. The genes for producing the desired chemicals can be implanted in the genetic code of plants or microorganisms, especially the benign bacteria *Escherichia coli*. These sources can enable high-volume production at high levels of concentration. Because plants and microorganisms are very different from people, the chance of spreading disease is minimal.

Genetically engineered pharmaceuticals can increase production of natural body chemicals and supply toxins to attack targeted microorganisms. Vaccines work by triggering the body's immune system, which then defends itself. Genetic engineering allows faster development of safer vaccines. Prominent techniques include recombinant vaccines, which are being used against malaria. Disease-producing genes are placed in contact with cells specifically prepared to promote growth. This produces antigens, a substance that triggers the body's immune system. Individuals are vaccinated with these antigens instead of the virus itself. Naked DNA vaccines are being tested on some cancers. This method injects parts of DNA from a pathogen (a virus, bacterium, fungus, or parasite) into a person. The body's cells accept this added DNA as instructions to make antigens. This activates the immune system.

A reassortant virus technique may conquer the diarrhea virus. Genes from different forms of this disease are combined in a way that makes a harmless simulator virus. While the mixture is harmless to the individual, it looks like an invader at the cellular or molecular level, causing the body's immune system to develop antibodies to attack the active microorganisms. Live vector vaccines have been used to treat AIDS. This procedure inserts genes from a dangerous virus into a harmless (to humans) virus, then injects this altered virus into a person. The immune system develops resistance to the dangerous virus.

Conjugate vaccines are used experimentally to treat ear infections. Bacterial antigens for which one desires an immune response that are not easily recognized by the body are obtained. They are then joined with easily recognizable antigens located on a harmless bacterial shell and injected into the body to trigger the immune response system. Subunit vaccines have been used to treat strep throat. For this method, a basic protein is identified as common to all known strains of a disease (strep A has 120 identified strains). Portions of its genetic code are inserted into harmless bacteria. This prepared substance is sprayed into the nose of a person, where the foreign protein triggers the body's immune system.

The use of genetically engineered pharmaceuticals has raised some concerns about their possible effects on the environment. In 1989 the Virginia Department of Health approved the field testing of baits spiked with genetically engineered rabies vaccine to control the spread of rabies in raccoons. However, an island location was chosen to prevent the possible spread of vaccinated animals to larger, mainland populations because health officials were concerned about the possible danger of the vaccinia virus to humans. Several researchers questioned the long-term effects of releasing a non-native virus into the environment. Although vaccinia was used for many years to prevent smallpox, little was known about its host range or its ability to cause disease. The United States Department of Agriculture concluded in a 1991 report, however, that laboratory and field tests had shown the genetically engineered rabies vaccine to have had no adverse effects on any species. In the same report, the department approved field testing on the grounds that such tests were safe and posed no significant environmental risk.

Robert J. Wells

SUGGESTED READINGS: The frivolous title of Larry Gonick's and Mark Wheelis's heavily illustrated *The Cartoon Guide to Genetics* (1983) belies the usefulness of the to-the-point, hand-lettered introduction to genetics. *The Thread of Life: The Story of Genes and Genetic Engineering* (1996), by Susan Aldridge, is a guide to the world of DNA and biotechnology applications. Another useful reader is *Genetic Engineering Technology in Industrial Pharmacy* (1989), edited by John M. Tabor.

SEE ALSO: Biotechnology and genetic engineering; Birth defects, environmental.

Geothermal energy

CATEGORY: Energy

Natural radioactive isotopes occur throughout the earth and release heat with each radioactive decay. This internally generated heat results in a temperature gradient of about 1 degree Fahrenheit per 100 feet of depth on average. In areas where hot, molten rock (magma) is moving toward the surface or has recently done so, the geothermal gradient may be much higher than the average. Geothermal gradients two to three times higher than normal have the potential to provide useful heat.

Geothermal resources encompass a range that includes hot water, hot water mixed with steam, and superheated water that flashes into hot, dry steam when the pressure on the water is reduced. As a general rule, the higher the temperature of the resource, the higher its economic value. Steam and liquids are extracted from below the earth's surface through wells, or hot water is taken from surface springs.

While people have been enjoying the pleasures and benefits of natural hot springs for many centuries, the first use of geothermal energy for electric power generation was at Larderello, in northern Italy, in 1904. The first geothermal electric power generation outside Italy was in New Zealand in 1958. The first geothermal electric power plant in the United states, at the Geysers in California, began operation in 1960.

During the 1970's the world price of petroleum more than doubled. This and an expanded concern for the environment greatly stimulated interest in geothermal energy resources, and many new facilities were constructed all over the world. In 1992 the leading countries in geothermal electric power generation were the United States, Mexico, the Philippines, and Italy. None of these countries produce more than 1 percent of its electric power from geothermal energy.

The geothermal energy use that attracts the most interest is electric power generation. However, in many places where hot water or high-temperature geothermal resources are available, they are used to heat homes, schools, and other types of buildings and for a wide variety of industrial and agricultural processes that are heat dependent. In Iceland, for example, hot geothermal water has long been used to heat homes, public buildings, and greenhouses and to dry seaweed, hay, and diatomite.

In the early years of intense concern over the environment, geothermal energy was seen as absolutely nonpolluting because no fuel was burned. In actuality, some early projects produced considerable air pollution because geothermal steam is normally heavily charged with carbon dioxide, hydrogen sulfide, and other gases regarded as pollutants. In addition, the hot water, or water from condensed steam, commonly contained dissolved solids such as arsenic, mercury, and boron that were released into local rivers or evaporation ponds. Where hot water or condensate was released into streams, thermal pollution also occurred. These problems can largely be solved by returning the water to the geothermal reservoir, which isolates the pollutants. The residual heat tends to extend the useful life of the reservoir. In some cases the hydrogen sulfide can be reduced to elemental sulfur, which can be safely stored or used in industry.

In geothermal electric power plants and heating systems, dissolved solids may precipitate in production wells and piping, a process called pipe scaling. The scale, usually calcium carbonate or silica, may contain arsenic, lead, mercury, and other toxic metals, and disposal of the scale may constitute an environmental problem.

A great advantage of geothermal energy is that it can replace petroleum as an energy source, reducing pollution. A 100-megawatt geothermal electric power plant will replace a petroleum-based plant that would use about one million barrels of fuel oil per year. The principal advantage of geothermal energy over nuclear energy is that it produces no radioactive waste.

The initial cost of a geothermal power plant, because of the cost of supply wells, may be higher than the cost of an oil-fired plant of equal capacity. If scaling is a problem, the operating cost may be considerably higher, but in the geothermal plant there is no fuel cost. A problem is that most geothermal resources have a rather limited energy output and a limited economic life that cannot be determined in advance of development. In addition, the number of high-energy geothermal prospects worldwide is rather low, although for heating purposes the number of geothermal prospects is much higher than for electric power generation. However, transmission of hot water from the source to the point of use normally requires relatively large, insulated pipes and is invariably subject to heat losses, which limit transmission distances. In contrast, natural gas or liquefied petroleum gases such as pressure-liquefied propane and butane can be supplied through small-diameter pipes that do not require insulation. Also, the transmission of natural gas involves minimal energy loss.

From highs in the 1970's, petroleum and natural gas prices sharply declined in the 1980's and 1990's. Correcting for inflation, the prices of these resources were as low in the late 1990's as they had ever been. That fact greatly reduced the incentive for geothermal energy development in the last decades of the twentieth century.

Robert E. Carver

SUGGESTED READINGS: Excellent general treatments of geothermal energy are given in Robert Bowen's *Geothermal Resources* (1989) and Harsh K. Gupta's *Geothermal Resources: An Energy Alternative* (1981). A more rigorous approach is available in John Elder's *Geothermal Systems* (1981).

SEE ALSO: Alternative energy sources; Energy policy; Fossil fuels; Power plants.

Gibbons, Euell

BORN: September 8, 1911; Clarksville, Texas
DIED: December 29, 1975; Sunbury, Pennsylvania
CATEGORY: Forests and plants

A popular ethnobotanist and nature writer of the 1960's and 1970's, Euell Gibbons improved the public image of wild food foraging and thus environmentalism in general. His staid, avuncular image made his environmentalism acceptable to an American culture that sometimes perceived environmental activism as subversive.

By becoming a best-seller, Euell Gibbons's first book, *Stalking the Wild Asparagus* (1962), took the art of collecting and preparing wild foods out of the realm of the eccentric and into the mainstream of popular culture. Gibbons's writing, as well as his lifetime of research both in the field and at the Pendle Hill Quaker Study Center in

Wallingford, Pennsylvania, offered original arguments for the necessity of maintaining the balance of nature. Whereas many ecological critics blamed a fundamentalist reading of the biblical mandate to "subdue the world" for much of the ecological problems of the twentieth century, Gibbons placed equal blame on a misunderstanding of Darwinism. Much of the notion of conquering nature that Gibbons, like most ecologists, despised stemmed, he thought, from the mistaken notion of natural selection. "Survival of the fittest," Gibbons taught, does not always mean "survival of the strongest" but rather "survival of the most cooperative." A forager who uproots a plant is not a destroyer from a larger ecological point of view, since such an action scatters seeds and invariably creates more plants than it destroys. By seeing the human forager as a vital part of the plant's reproductive cycle, Gibbons offered an ideological framework for understanding natural selection without the insistence on predator and prey.

Gibbons's impact on ecological thought went beyond the influence of his books, however. Both the Boy Scouts of America and the United States Navy hired him to teach foraging as a survival skill, though Gibbons disliked that context for his discipline. His advocacy of wild foods was culinary rather than survivalist, and he resented the romanticized notion of going into the woods with no provisions and "living off the land." Besides, as a Quaker and pacifist, he resented the military uses to which his skills might be put by the Navy. Nevertheless, he welcomed the opportunity to spread the gospel of cooperation with nature. Another forum for his ideas was a series of seminars he ran at Ithaca College and Bucknell University in the 1970's. These seminars assured a wide distribution of his methods by training college students in the first half of the seminar and then guiding them in training high school students in the second half.

Though not formally educated in botany—his college studies were in anthropology and creative writing—Gibbons read virtually every article and book, scholarly and popular, that dealt with the gathering, preparation, and nutritive value of wild plants. He coined the term "ethnobotony" because he saw his field as recovering the botanical and culinary knowledge of Native American culture.

John R. Holmes

SEE ALSO: Balance of nature; Environmental ethics.

Gibbs, Lois

BORN: June 25, 1951; Buffalo, New York
CATEGORY: Human health and the environment

In 1977 Lois Gibbs discovered that her neighborhood school was built on top of Love Canal, a toxic chemical dump containing 20,000 tons of waste. Gibbs united the community by forming the Love Canal Homeowners Association and leading efforts to compel state and federal officials to relocate residents whose homes were compromised.

Lois Gibbs was alerted to the presence of the Love Canal dump in December, 1977, when her son, Michael, began to experience asthma and seizures just four months after he entered kindergarten at the school. She went door to door questioning residents about their health in an attempt to determine the full extent of contamination. Two years later, Gibbs traveled to Washington, D.C., where she testified on behalf of the Love Canal homeowners at the hearings on hazardous waste disposal held before the House Subcommittee on Oversight and Investigations.

Although homes immediately adjacent to the dump were evacuated, residents living outside the "inner ring" remained until January, 1980, when the federal government's Environmental Protection Agency (EPA) financed a controversial scientific investigation by Biogenics Corporation of Houston, Texas. On May 17, 1980, the EPA released a report claiming that residents were believed to have damaged chromosomes and consequently had an increased risk of cancer, miscarriages, birth defects, and seizures. Gibbs described the report as "one more frightening, scary thing, and we couldn't take it any more. . . . People were very, very frightened, almost panicked." Shortly afterward, U.S. president Jimmy Carter declared a state of emergency at Love Canal and authorized relocation of residents at a cost of $35 million to the federal government.

Lois Gibbs's fight to evacuate Love Canal captured the public's imagination and was depicted in *Lois Gibbs and the Love Canal*, a 1980 Columbia Broadcasting System (CBS) film starring Marsha Mason. The most significant development following Love Canal was the Superfund Act of 1980. This legislation created an industry-funded program for cleaning toxic chemical dumps across the nation. In 1981 Gibbs formed the Citizen's Clearinghouse for Hazardous Waste. This nonprofit organization began by helping community groups suffering from the effects of toxic dumps similar to Love Canal. The clearinghouse eventually expanded its pro-

grams to match the growing needs of grassroots environmental organizations, working with over eight thousand community-based groups.

Peter Neushul

SEE ALSO: Citizen's Clearinghouse for Hazardouse Waste; Love Canal; Superfund.

Gila Wilderness Area

CATEGORY: Preservation and wilderness issues

The Gila Wilderness Area, a wilderness of more than 500,000 acres within the Gila National Forest in southwestern New Mexico, was the first designated wilderness area in the world.

After Lois Gibbs discovered that houses in her neighborhood had been built on top of a toxic waste dump, irate homeowners demanded action by government officials. Gibbs went on to found the Citizen's Clearinghouse for Hazardous Waste in 1981. (AP/Wide World Photos)

The Gila Wilderness Area is a rugged area of highly diverse landforms, plants, and animals. It includes the termini of both the Rocky Mountains to the north and the Sierra Madre to the south. The Chihuahuan and Sonoran Deserts also reach into the Gila Wilderness, contributing to its high biodiversity.

The designation of this area as the world's first official wilderness area grew out of the nineteenth century conservation movement in the United States. As the United States became more settled, some people grew concerned that areas of natural beauty needed to be preserved. This led to the establishment of the first national park, Yellowstone, in 1872. At the same time, however, plans were being formulated to make it and similar areas readily accessible to visitors. About thirty years later, people began realizing that some wildlands needed to be preserved in a manner that was devoid of roads and other amenities for visitors.

In 1899 President William McKinley set aside the Gila River Forest Reserve. Seven years later, the 2.7-million-acre Gila National Forest was established. In the early 1920's threats to the area by developers led Aldo Leopold, then a young forester in the Southwest District of the U.S. Forest Service, to awaken the American public to the need to save wildlands. An avid hunter, Leopold observed that wildlands and wildlife habitat were shrinking. He blamed the loss on road building. Upon his arrival in the Southwest in 1909, he had identified six roadless backcountry areas—each more than a 500,000 acres in size—in the region's national forests. By 1921, only one, at the headwaters of the Gila River, remained.

Passage of the Federal Highway Act in 1922 threatened to bring more road building and tourism, leading Leopold to fear the loss of this last roadless area. As assistant district forester, he led a movement to establish the Gila headwaters area as the first officially designated wilderness, to keep it roadless and without buildings or artificial trails. Persuaded by his arguments, the district forester designated 755,000 acres as the Gila Wilderness Area in June, 1924—a harbinger of the national wilderness system later established under the Wilderness Act of 1964.

Under the 1924 U.S. Forest Service wilderness designation, hunting was encouraged in the Gila Wilderness Area. Predator control then led to deer overpopulation. In 1933 the Forest Service responded by rebuilding an abandoned wagon road through the wilderness to provide easier access for hunters. This resulted in division of the wilderness. After some modifications in the laws and name changes in the intervening years, the New Mexico Wilderness Act of 1980 designated 557,873 acres west of the road as the Gila National Wilderness and 202,016 acres east of the road as the Aldo Leopold National Wilderness. The combined acreage of these areas exceeds the size of the original Gila Wilderness Area.

Jane F. Hill

SEE ALSO: Forest Service, U.S.; Hunting; Leopold, Aldo; National forests; Predator management; Wilderness Act; Wilderness areas.

Glen Canyon Dam

DATE: completed 1983
CATEGORY: Preservation and wilderness issues

The Glen Canyon Dam is a large hydroelectric dam on the Colorado River in northeastern Arizona built and operated by the U.S. Bureau of Reclamation. The controversy surrounding the dam's construction is often cited as the beginning of the environmental movement.

Congress authorized the Glen Canyon Dam in 1956 to provide water storage in the second-largest human-made reservoir in the United States, Lake Powell. Construction of the dam began in 1956, and the lake reached its full mark in 1983. The dam provides hydroelectrical power generation and recreation, and decreases siltation in downstream Lake Mead reservoir, which is formed by Hoover (Boulder) Dam. The combined storage behind the Glen Canyon and downstream Hoover Dams manages the flow of water in the Colorado River to California, Arizona, Nevada, and Mexico. Agriculture uses approximately 85 percent of the water, with most of the remaining water going to urban areas in Southern California; Phoenix, Arizona; and Las Vegas, Nevada. The dam, with power plants, cost $272 million to build. Although designed to generate up to 1,300 megawatts of electric power annually, environmental damage to the riparian (shoreline) ecosystem downstream in the Grand Canyon proved too large, and the government limited electricity production to fewer than 800 megawatts annually.

When construction began, Glen Canyon was extremely remote and rarely visited. The Sierra Club initially fought the project but agreed to end their resistance in a compromise agreement with the Bureau of Reclamation that stopped construction of two other dams in Utah. They soon regretted their decision when they learned of the spectacular beauty of Glen Canyon. Sierra Club executive director David Brower wrote a beautifully illustrated book called *The Place No One Knew* (1963), which featured photographs of Glen Canyon before the dam. This book launched a Sierra Club policy of publishing illustrated books on outstanding natural areas to try to ensure that Americans sacrificed no other spectacular areas because they did not know of their beauty. In 1996 the Sierra Club called on the federal government to drain Lake Powell and abandon the dam.

To many people, the Glen Canyon Dam is an icon of environmental destruction. The central theme of Edward Abbey's landmark novel on ecotage, *The Monkey Wrench Gang* (1975), involves plotting to destroy the dam. The radical preservationist group Earth First! leaped into national prominence in the early 1980's when members unfurled a 300-foot length of black plastic from the top of the dam, giving the appearance of a giant crack descending down the front. Brower suggested that the lake be drained but the dam left "as a tourist attraction, like the Pyramids, with passers-by wondering how humanity ever built it, and why."

The most obvious postdam change was the flooding of a spectacular canyon and its tributaries, along with the accompanying loss of hiking and river-running experiences. The entire ecosystem of the inundated area, 653 square kilometers (252 square miles), was destroyed and replaced with a lake community. The reservoir water alters the color of the adjacent red sandstone rocks, resulting in a prominent, white "bathtub" mark along the shore most of the year.

The reservoir behind the Glen Canyon Dam has caused several other changes. Since the reservoir is located in the desert, an estimated 600,000 acre-feet (one acre of water, one foot deep) of water are lost to evaporation each year. Annually, more than 350,000 acre-feet of water seep into the ground. The river loses approximately 8 percent of its volume in Lake Powell. An additional concern is the rate of siltation. Silt and sand carried by the Colorado River settle onto the reservoir floor. Approximately two million acre-feet of water storage has been loss to siltation.

Dramatic changes to the riparian environment in the Grand Canyon are less obvious. Prior to the dam, muddy river temperatures seasonally ranged from 26 degrees Celsius (80 degrees Fahrenheit) to nearly freezing. The dam now releases water from the depths of the reservoir, and the river below the dam is approximately 7 degrees Celsius (45 degrees Fahrenheit). In addition, the river below the dam is now nearly sediment-free. These changes have impacted the aquatic fauna and flora of the Grand Canyon's Colorado River. Several species have become extinct, and many nonnative or previously sparse species have flourished.

Releasing water from the dam to meet hydroelectric needs eliminates large floods but causes a daily tide. More water is released during afternoon, high-power demand times, while less flows through the dam during the low-demand nights. The resulting tide stresses aquatic organisms downstream and causes rapid beach erosion. Lack of natural flooding prevents rebuilding of the beaches. The rapid loss of beaches along the river in the Grand Canyon brought about ecosystem changes and decreased recreational experiences for customers of the multimillion dollar rafting industry.

During the 1980's and 1990's, the Department of the Interior undertook an extensive study of the impacts of the Glen Canyon Dam, resulting in the Glen Canyon Environmental Impact Study. The major outcome of the $70 million study was changes to the way water is released from the dam. Extreme and rapid daily fluctuations in river level to meet power needs were banned. In the spring of 1996, the dam released an experimental flood with a slow rise to modest flood levels, followed by a gradual lowering of the river. The release was designed to scour the river channels and rebuild the beaches, as happens in a natural flood. The results were modest but promising. Environmental groups hope that similar experiments will continue along the Colorado and other dammed rivers and that more environmentally friendly management plans will develop for large dams in the future.

Louise D. Hose

SUGGESTED READINGS: Steven W. Carothers and Bryan T. Brown's *The Colorado River Through Grand Canyon* (1991) provides a comprehensive discussion of changes that occurred after the construction of the Glen Canyon Dam. *The Place*

No One Knew (1963), by David Brower, is a beautiful coffee-table book showing the canyons that were lost when water backed up behind the Glen Canyon Dam. John McPhee's *Encounters with the Archdruid* (1971) entertainingly documents informal debates between a prominent environmentalist and the Bureau of Reclamation commissioner about the impact of the Glen Canyon Dam on the Grand Canyon. The *U.S. Department of the Interior, Bureau of Reclamation, March 1995, Operation of Glen Canyon Dam—Final Environmental Impact Statement—Summary* (1995) provides a wealth of information about the environmental impact of the Glen Canyon Dam.

SEE ALSO: Abbey, Edward; Brower, David; Dams and reservoirs; Grand Canyon; Monkeywrenching; Sedimentation; Sierra Club.

Global Biodiversity Assessment

DATE: 1995
CATEGORY: Ecology and ecosystems

The Global Biodiversity Assessment is an independent scientific analysis of all the issues, theories, and views regarding biodiversity from a global perspective. The report concludes that the earth's biological resources are under serious threat from rapidly expanding human populations that are degrading the environment at an accelerating rate.

Between 1993 and 1995 a group of scientific experts helped develop, write, and review contributions to the Global Biodiversity Assessment. Funded by the Global Environment Facility (GEF) and the United Nations Environment Programme (UNEP), this independent, peer-reviewed assessment is the work of more than fifteen hundred scientists and experts from all parts of the world. The final report focuses on assessing the scientific understanding of biodiversity's various components—namely ecosystems, species, and genes—and on identifying the gaps in the knowledge base that should be targeted for future research. Preservation of biodiversity must include a blend of strategies, including pro-

grams to save species by creating controlled environments and policies to manage natural environments in ways that minimize adverse impacts on ecosystems and species.

While great advances in understanding the earth's biological diversity were made during the 1980's and 1990's, the assessment demonstrates that this understanding is incomplete. There is a great range of opinion on many basic theoretical issues, and gaps in data are enormous, with estimates sometimes differing by several orders of magnitude. Ecosystem dynamics, such as the optimal size of a nature reserve necessary to preserve species diversity, suffer greatly from a lack of scientific understanding, as does the understanding of how genetic diversity is distributed within populations and how species evolve and function.

The assessment concludes that ecosystems of all kinds around the world are under great pressure. Lowland and coastal areas, native grasslands, wetlands, and many types of woodlands and forests have been adversely affected or destroyed. During the 1980's humid tropical forests were being lost at an annual rate of nearly 25 million acres, dry tropical forests were being destroyed at an even faster rate, and 10 percent of the coral reefs of the world were eroded beyond recovery.

The report estimates that only about 13 percent of the total number of species on earth have been scientifically described. In addition, the number of species reported as being threatened by extinction is far from the actual total. Most of the damage is being caused by human activities. In fact, the root causes of the loss of biodiversity in general are closely tied to the way that human societies use resources.

The assessment not only evaluates the existing problems but also analyzes various options for ensuring that biodiversity is conserved and used wisely in the future. The report concludes that biodiversity management must go far beyond just establishing nature reserves or setting up agricultural seed banks. It must be fully integrated into all aspects of landscape management. Specific actions of the report include the characterization of biodiversity and the assessment of the magnitude and distribution of biodiversity, basic principles of the functioning of biodiversity and ecosystems, human influences on biodiversity, economic values of biodiversity, and measures for conserving and sustaining biodiversity.

Alvin K. Benson

SEE ALSO: Biodiversity; Ecology; Ecosystems; Global Environment Facility.

Global Environment Facility

DATE: established 1991
CATEGORY: Ecology and ecosystems

Initiated in 1991, the Global Environment Facility (GEF) provides grants and concessional funds to recipient countries for projects and activities that protect the global environment. GEF programs focus on environmental problems dealing with climate change, biological diversity, international waters, and depletion of the ozone layer.

Jointly implemented by the United Nations Development Program (UNDP), the United Nations Environment Programme (UNEP), and the World Bank, the GEF was launched as a pilot program in 1991. The GEF secretariat, who is functionally independent from the three implementing agencies, reports to the Council and Assembly of the GEF. The Council is the main governing body and consists of representatives from thirty-two countries, eighteen from countries receiving GEF funds and fourteen from nonrecipient countries. Any country that is a member of the United Nations or one of its specialized agencies may become a GEF participant by filing a notification of participation with the GEF secretariat. By 1998 160 countries were participating in the GEF program. Countries may be eligible for GEF project funds in one of two ways: if they are eligible to borrow from the World Bank or receive technical assistance funds from the UNDP, or if they are eligible for financial assistance through either the Convention on Biological Diversity or the Climate Change Convention. In either case, the country must be a participant in the Convention on Biological Diversity or the Climate Change Convention.

The organizing principle of the GEF is that no new bureaucracy will be created. The UNDP is responsible for technical assistance activities and helps identify projects and activities consistent with the Small Grants Program for nongovernmental organizations (NGOs) and community groups around the world. Responsibility for initiating the development of scientific and technical analysis and for advancing environmental management of GEF-financed activities rests with the UNEP, as does the management of the Scientific and Technical Advisory Panel that provides scientific and technical guidance to the GEF. The World Bank serves as the repository of the GEF trust fund and is responsible for investment projects. It also seeks resources from the private sector in accordance with GEF objectives.

GEF resources aim to facilitate projects with global environmental benefits. Funds are available for proposals dealing with four focal areas: climate change and global warming, biodiversity, international waters, and depletion of the ozone layer. In addition, funding is available for problems dealing with land degradation, particularly desertification and deforestation, as they relate to the four focal areas. Projects and activities must be approved by the GEF Council. If at least four Council members request a review of the final project documents, the Council will conduct the review prior to granting approval of GEF funds.

The GEF and the Montreal Protocol (1987) provide a complementary relationship in funding ozone projects. Although the Montreal Protocol will provide funds for ozone-related projects to countries where ozone-depleting substance (ODS) production or consumption is low, the GEF provides funding for countries where production or consumption of ODSs is high. This applies mainly to countries in Central and Eastern Europe, as well as the former Soviet Union.

Alvin K. Benson

SEE ALSO: Biodiversity; Climate change and global warming; Ecosystems; Global Biodiversity Assessment; Ozone layer and ozone depletion.

Global ReLeaf Program

CATEGORY: Forests and plants

The Global ReLeaf Program is a conservation campaign that focuses on the protection of forestland from overdevelopment and pollution, as well as on the replanting of trees to replace those that have already been eliminated by a variety of causes.

In the United States, Global ReLeaf Program efforts have been led by the American Forestry Association, which was founded in 1875 and remains the nation's oldest national nonprofit conservation organization. The Global ReLeaf Program sponsors educational programs to show the benefits of trees and forests for the environment and for the enhancement of people's lives. These programs highlight the value of trees for filtering air and water, sheltering and feeding wildlife, absorbing greenhouse gases, and reducing the runoff of polluted soil into rivers and streams. The organization also provides funding, collected from private and corporate donations, for tree-planting projects across the United States through its Global ReLeaf Forest Ecosystem Restoration Program.

Global ReLeaf Forests are typically planted on public or private land that was once forested but has been cleared by wildfires, hurricanes, tornadoes, insects, or other natural occurrences; by developers; or by unintentional human interference, such as the spread of accidentally introduced exotic species. The American Forestry Association works with local groups to ensure that the new trees are native to the area and that they are properly planted and maintained.

Between 1990 and 1998, the Global ReLeaf Program in the United States was responsible for planting seven million trees in ninety-three forests, but the numbers dramatically increased toward the end of the decade. In the first half of 1998 more than two million trees were planted, and the goal for 1999 was to plant several million trees in watershed restoration projects in Chesapeake Bay, Puget Sound, the Mississippi River Delta, and other areas. The American Forestry Association also declared a special Global ReLeaf 2000 program, dedicated to planting twenty million trees around the world by the year 2000.

Global ReLeaf has been successful in part because it works with governmental agencies, local organizations, and large corporations, making it easy for individual citizens to participate. Through extensive advertising and publicity, and through a colorful presence on the Internet, the American Forestry Association has encouraged donations of as little as ten dollars, with one tree being planted for each dollar received. In partnership with a major breakfast cereal, Global ReLeaf Kids supported a project to plant trees in rain forests in the Philippines and Hawaii.

Other organizations throughout the world have also used the Global ReLeaf Program name. One prominent group based in Slovakia in Eastern Europe was the former Slovak Union of Nature and Landscape Conservation, now called the Global ReLeaf Foundation. Like the American Forestry Association, this organization sponsors educational programs and prepares school curriculum materials, especially about the dangers of pollution and overdevelopment.

Cynthia A. Bily

SEE ALSO: Deforestation; Forest management; Logging and clear-cutting.

Global 2000 Report, The

DATE: 1980
CATEGORY: Ecology and ecosystems

The Global 2000 Report (1980) is a published discussion of the first attempt by the United States government to use a computer to create a model for global human activities.

As computers became more numerous in the early 1970's, the world was shocked by the publication of two books whose computer models projected the collapse of the current world system if population growth and resource consumption were not curtailed: *The Limits to Growth* (1972), by Donella H. Meadows et al., and *Mankind at the Turning Point* (1974), by Mihajlo Mesarovic and Eduard Pestel. The U.S. government was unprepared for the conclusions presented in these books. In 1977 President Jimmy Carter focused the attention of the federal government on producing a computer model to serve as a planning tool for human activity. The simulation was completed in 1979 and discussed in the Council on Environmental Quality's *Global 2000 Report.*

The Global 2000 Report concluded that the population would increase by 55 percent and reach 6.35 billion in the year 2000 and that the number of people being fed on 2.5 acres of land would increase from 2.6 in 1970 to 4 by 2000, requiring the increased use of biocides, fertilizers, and irrigation. The report also predicted that between 1970 and 2000, the per capita consumption of food would increase by 15 percent, but the increase would be confined to the well-fed, industrialized nations. Finally, the report claimed that the number of malnourished people in the world would increase from 500,000 in 1970 to 1.3 billion in 2000. The report concluded that "the needed changes go far beyond the capability and responsibility of this or any other single nation. An era of unprecedented cooperation and commitment is essential."

A number of serious flaws existed in *The Global 2000 Report.* Unlike the simulations developed for *The Limits to Growth* and *Mankind at the Turning Point, The Global 2000 Report* was not a single unified model. Every government agency produced its own simulation, and their models were not combined into an integrated whole. As a result, various agencies assumed no interruption in the flow of necessary goods and services in contrast to what the agency overseeing those goods and services was predicting. In addition, while other groups were extending their models to the year 2100, the government's simulation stopped at the year 2000, severely limiting its usefulness and longevity. Nevertheless, the projections of *The Global 2000 Report* were not significantly different from what other groups were reporting.

In spite of the pessimistic projections in *The Global 2000 Report,* it was revised and reissued in 1988 because its original projections were too optimistic. In 1992 Senator Albert Gore, Jr., ech-

oed the difficulties noted in *The Global 2000 Report* when he said in his book *Earth in the Balance,* "We find it difficult to imagine a realistic basis for hope that the environment can be saved, not only because we still lack widespread agreement on the need for this task, but also because we have never worked together globally on any problem even approaching this one in degree of difficulty."

Gary E. Dolph

SEE ALSO: Carter, Jimmy; Gore, Albert, Jr.; Population-control movement; Population Growth.

Gore, Albert, Jr.

BORN: March 31, 1948; Washington, D.C.
CATEGORY: Ecology and ecosystems

U.S. politician Albert Gore, Jr., has devoted a large part of his political career to environmental issues. Born into a political family—his father, Albert Gore, Sr., was a senator from Tennessee—Gore became aware of environmental concerns as a child when he recognized the potential problems soil erosion could cause on his family farm.

Gore earned a bachelor's degree in government from Harvard in 1969 and was subsequently drafted into the military, where he served in Vietnam as an army reporter and witnessed firsthand the effects of Agent Orange. After returning to Tennessee, he worked as a reporter with the Nashville *Tennessean* and attended divinity school and, later, law school at Vanderbilt University. In 1976 Gore mounted a successful campaign to become the representative from Tennessee's Fourth Congressional District. In his first of four terms in the House, Gore led the first congressional hearing on the dangers of hazardous waste. These hearings focused on the damage done in a small town east of Memphis, Tennessee, and along the Love Canal in upstate New York.

What Gore learned in these hearings enhanced his already keen interest in the environment. Putting his reporter's investigative skills to

Al Gore, Jr., whose political career has been characterized by a concern for the environment. In 1992 he published Earth in the Balance *and was elected vice president of the United States.* (Library of Congress)

work, he began a serious study of global warming and other related environmental issues. In 1980 he and other lawmakers who shared his concerns succeeded in passing the Superfund bill, which focused on cleaning up chemical and other hazardous dump sites. Gore was elected to the Senate in 1984, where he continued his work on behalf of the environment. In 1990 he cosponsored the first Interparliamentary Conference on the Global Environment.

In March, 1987, Gore declared his intentions to seek the Democratic presidential nomination and promised to focus his campaign on nuclear arms control and several environmental issues, including global warming and ozone depletion. His campaign failed, but he maintained his support for these issues. In 1992 he published *Earth in the Balance: Ecology and the Human Spirit*, in which he proposed a Global Marshall Plan that would focus on distributing technology equitably throughout both the developed and develop-

ing world, stabilizing world populations, and requiring industrial nations to accelerate their transition toward environmental responsibility.

The same year his book was published, Gore was named Bill Clinton's running mate in Clinton's successful bid for the U.S. presidency. As vice president, Gore continued his work for the ailing global environment. In 1997, for example, he was instrumental in turning the tide at the Kyoto Climate Change Summit when his last-minute plea to U.S. negotiators to be more flexible helped bring about the Kyoto Accords, in which more than 150 nations agreed to reduce emissions of greenhouse gases.

Jane Marie Smith
SEE ALSO: Environmental policy and lobbying; Kyoto Accords; Superfund.

Grand Canyon

CATEGORY: Preservation and wilderness issues

The Grand Canyon is a deep, 450-kilometer-long (280-mile) segment of the Colorado River and its tributary canyons in northern Arizona. Grand Canyon National Park, one of the most heavily visited national parks, was established by President Theodore Roosevelt in 1919 and has been designated a World Heritage Site. The Hualaipai and Navajo Tribal Councils manage the other parts of the Grand Canyon that lie outside of the park's boundaries.

The arid climate of the Grand Canyon influences every resource in the region. Well-exposed rock layers reveal more than 1.8 billion years of the earth's history. The arid climate preserves ancient human and animal remains, including many extinct animals that lived more than ten thousand years ago. Cliff dwellings, human artifacts, and old adobe buildings represent habitation dating back more than four thousand years. Grand Canyon tourists, however, report that the grandeur and beautiful views most dramatically mark their visits. People who hike into the canyon commonly feel overwhelmed by the immensity of the chasm.

Grand Canyon National Park, with more than 2,000 meters (6,500 feet) of relief, contains many ecosystems. The South Rim, a flat plateau, has an elevation of 2,100 meters (7,000 feet). The north side is an even higher plateau, 2,400 meters (8,000 feet) above sea level. Conifer forests cover both rims and provide homes to deer, squirrels, and mountain lions. No streams cross the plateaus, as rain and snowmelt immediately flows underground in the karst terrains.

The conifer forests extend down into the canyon, transitioning into an arid environment at lower elevations. Desert plants—such as cacti, acacia, mesquite, brittle bush, ocotillo, rabbitbrush, and agave—grow on the walls of the Grand Canyon. Desert bighorn sheep, lizards, snakes, skunks, and mice populate the slopes and side canyons. Water is scarce, particularly on the south side of the Colorado River. Cottonwood, ash, and redbud trees—as well as ferns, columbine, and other water-loving plants and animals—cluster around small seeps throughout the canyon and larger karst springs in some tributary canyons on the north side.

Another distinct ecological zone in the Grand Canyon is the riparian (riverside) habitat along the Colorado River at the bottom. Willow, arrowweed, and exotic tamarisk line the riverbanks. Otters, beavers, muskrats, fish, and other aquatic organisms call the Colorado River home.

After visiting part of the Grand Canyon in 1858, Lieutenant Joseph C. Ives wrote, "The region is, of course, altogether valueless . . . It seems intended by nature that the Colorado River . . . shall be forever unvisited and undisturbed." If his prediction had come true, the Grand Canyon would not be facing the many environmental problems that now threaten it.

More than five million people visit Grand Canyon National Park each year. Support facilities (campgrounds, hotels, visitor centers, shops, and toilets) require water and sewage treatment. Yet no water is available on either rim. Drilling to the closest aquifer, several thousands of feet underground, would be extremely costly. Thus all water used in the park comes from a cave about halfway up the north side of the canyon. A transcanyon pipeline and associated pumphouses

The Colorado River flows though the Grand Canyon beneath a bridge in Kaibab National Park. Upstream dams have drastically changed the ecology of riparian zones in the canyon. (Lynn Abigail)

lift the water to two places on the South Rim and one location on the North Rim. The purity and quantity of this modest stream is critical to keeping the Grand Canyon open to visitors. Occasionally, the water exceeds state water quality limits, causing the park to truck water at tremendous expense. The number of visitors and further development are strictly limited unless other sources of water are developed.

The popularity of the Grand Canyon brings other problems. In an attempt to provide quality wilderness experiences, the park requires overnight campers to register for permits and limits the number of visitors. In popular backcountry areas, campers are restricted to designated campgrounds with solar-powered compost toilets. The desert recovers slowly from erosion, and the park rangers enforce strict rules concerning cutting switchbacks and vandalism. They warn hikers to treat archaeological sites with respect. As in all national parks, visitors may not take archaeological or historical materials, rocks, animals, or plants.

Many visitors choose to see the park from the air. Helicopter and low-flying airplane flights have created a volatile issue. Many tourists wish to experience the canyon from these spectacular aerial views, but hikers commonly express dismay, and even anger, at the accompanying noise and disruption of solitude. The government struggles to balance the demands of these two different user groups by strictly controlling the routes and heights of overflights.

Once known for its spectacularly clear air, pollution from nearby coal-burning power plants occasionally mar views of the Grand Canyon. Since the 1970's, environmental battles have raged between people interested in preserving air quality in the area and those interested in developing the region's abundant coal supplies.

The most dramatic and contentious environmental issues in the Grand Canyon involve the Colorado River. The greatest change to the riparian zone resulted from construction of the Glen Canyon Dam, which has controlled the Colorado River flow through the Grand Canyon since 1966. The dam eliminated large floods. The maintenance of more consistent flows throughout the year has greatly impacted the riparian ecosystem, including elimination of many beaches and a general increase in vegetation (including exotic plants) and wildlife. The dam also alters water temperature and clarity. Predam water temperatures in the river fluctuated from 26 degrees Celsius (80 degrees Fahrenheit) in summer to nearly freezing in winter. Today, the river temperature remains constant at 7 degrees Celsius (46 degrees Fahrenheit). Once noted for its load of sediment, the river below the dam is now clear. The combined changes in temperature and clarity dramatically impacted the aquatic ecology, resulting in the loss of many plant and fish species and the proliferation of nonnative carp and trout.

Louise D. Hose

SUGGESTED READINGS: John McPhee's *Encounters with the Archdruid* (1971) entertainingly documents informal debates between a prominent environmentalist and the Bureau of Reclamation commissioner about the impact of the Glen Canyon Dam on the Grand Canyon. The *U.S. Department of the Interior, Bureau of Reclamation, March 1995, Operation of Glen Canyon Dam—Final Environmental Impact Statement—Summary* (1995) provides information about the Grand Canyon ecosystem and changes since the Glen Canyon Dam. *A Century of Change: Rephotography of the 1889-90 Stanton Expedition* (1996), by Robert Webb, provides fascinating comparative photographs of Grand Canyon scenes taken one century apart.

SEE ALSO: Kaibab plateau deer disaster; National parks; Powell, John Wesley; Roosevelt, Theodore.

Grand Coulee Dam

DATE: completed 1941
CATEGORY: Preservation and wilderness issues

The Grand Coulee Dam, located on the Columbia River west of Spokane, Washington, is the largest concrete structure ever built. Construction of the dam blocked the largest salmon spawning run in the world.

The multipurpose, 168-meter-high (550-foot-high) Grand Coulee Dam provides downstream flood control, irrigation, and hydroelectricity. The facility delivers irrigation water to more than 550,000 acres of agricultural lands. Associated electrical power production facilities are the largest in North America. President Franklin D. Roosevelt initiated construction of the dam in 1933 under the Public Works Administration of the National Industrial Act, which Congress authorized for emergency projects to relieve Depression-era unemployment. Electricity was first generated in 1941. Placing a dam across the fourth-largest river in North America created a 243-kilometer-long (151-mile-long) reservoir, Franklin D. Roosevelt Lake, which affords fishing and water sports to visitors.

While salmon ladders were built adjacent to downstream dams to allow the annual Columbia River salmon spawning migration to continue, the Grand Coulee Dam is too high to accommodate this dam-passing technique. At the time of construction, an elevator to lift the fish up and into the reservoir was considered but rejected because too many fish would die in the process. Instead, more than 1,600 kilometers (1,000 miles) of the world's most prolific salmon breeding river was eliminated upstream of the dam. Many miles of the impacted river flow through Canada, but the Canadian government expressed a lack of concern about the potential loss, stating that no commercial salmon fisheries existed along the river in Canada.

In the spring of 1939 the U.S. Bureau of Reclamation captured chinook and blueback salmon downstream from the Grand Coulee Dam site and transported them to spawning grounds in both tributaries and the main river upstream from the dam. The fish that survived the truck ride and transplant spawned. They and their offspring returned over the dam spillways and through the turbines, causing high mortality rates. Captured fish the following year were transferred to a hatchery for spawning. The young fish were released in downstream tributaries with the hope that they would return to the new spawning sites the following year. In 1943 the Bureau of Reclamation declared the salmon relocation program successful and turned the efforts over to the U.S. Fish and Wildlife Service.

The salmon population in the Columbia River, once the largest natural salmon hatchery in the world, dropped dramatically with construction of the Grand Coulee and other dams. At the downstream Rock Island Dam, 51,879 salmon passed through ladders to bypass the dam in 1933. A 1942 census at the same site counted only 7,086 salmon. Improved understanding and accommodation of salmon diet and environmental needs has helped the Columbia River populace grow, but it remains far smaller than the predam population.

Louise D. Hose

SEE ALSO: Dams and reservoirs; Flood control; Hydroelectricity; Irrigation.

Grazing

CATEGORY: Land and land use

Grazing is the consumption of any plant species by any animal species. While grazing is of mutual benefit to plants and animals, overgrazing is ultimately detrimental to both the plant and animal populations, as well as the environment. Sustaining a balance between grazing animals and the plants on which they feed prevents deleterious consequences.

Grasslands are characterized by the presence of low plants, mostly grasses, and are distinguished from woodlands, tundra, and deserts. Grasslands occupied vast areas of the world more than ten thousand years ago, before the development of agriculture, industrialization, and the subsequent explosive growth of the human population. They experience sparse to moderate rainfall and are found in both temperate and tropical zones. Grassland plants coevolved over millions of years with the grazing animals that depended on them. Wild ancestors of cattle and horses, as well as antelope and deer, were found in Eurasian grasslands. On the North American prairie, bison and antelope prospered. Wildebeest, gazelle, zebra, and buffalo dominated African savannas, whereas the kangaroo was the preponderant grazer in Australia.

Grazing is a symbiotic relationship whereby animals gain their nourishment from plants, which in turn benefit from the activity. Grazing removes the vegetative matter required for grasses to grow, facilitates seed dispersal, and disrupts mature plants, permitting young plants to take hold. Urine and feces from grazing animals recycle nutrients to the plants. The grassland ecosystem also attracts other animals, including invertebrates, birds, rodents, and predators. The grasses, grazing animals, and grassland carnivores, such as wolves or cat species, constitute a food chain.

Grasses are generally well-suited to periods of low rainfall because of extensive root systems and can go dormant during periods of drought. Humans have been an increasing presence in grassland areas, where more than 90 percent of contemporary crop production occurs and much urbanization and industrialization has also taken place. Because of inadequate rainfall or unsuitable terrain, remaining grasslands are used for grazing domesticated or wild herbivores. In addition, many woodland areas around the world have been cleared and converted to grasslands, where animals can graze.

Continued heavy grazing leads to deleterious environmental consequences. Even repeated removal of leaf tips will not adversely affect regeneration of grasses provided that the basal zone of the plant remains intact. While the upper half of the grass shoot can generally be safely eaten, ingesting the lower half, which sustains the roots and fuels regrowth, will eventually kill the plants. Overgrazing leads to denuding of the land, invasion by less nutritious plant species, erosion caused by decreased absorption of rainwater, and starvation of animal species. Because the loss of plant cover changes the reflectance of the land, climate changes can follow that make it virtually impossible for plants to return, with desertification an ultimate consequence.

The number of animals is not the only factor in overgrazing; the timing of the grazing that can also be detrimental. Grasses require time to regenerate, and continuous grazing will inevitably kill them. Consumption too early in the spring can stunt their development. Semiarid regions are particularly prone to overgrazing because of low and often unpredictable rainfall; regrettably, these are the areas of the world to which much livestock grazing has been relegated, because the moister grassland areas have been converted to cropland.

Overgrazing has contributed to environmental devastation worldwide. Excessive grazing by cattle, sheep, goats, and camels is partly responsible for the desert of the Middle East. Uncontrolled livestock grazing during the late nineteenth century and early twentieth century negatively affected many areas in the American West, where sagebrush and juniper trees have invaded the grasslands. Livestock overgrazing has similarly devastated areas of Africa and Asia. Feral horses in the American West and the Australian outback continue to damage those environments. Overgrazing by wildlife can also be deleterious. The Kaibab Plateau deer disaster in Arizona is one such example, where removal of the natural predators led to overpopulation, overgrazing, starvation, and large die-offs. Protection of elk and bison in the Yellowstone National Park has similarly led to high populations, excessive grazing, and changes to the environment; only the provision of winter feed prevents the die-offs that would naturally ensue.

Grassland areas need not deteriorate if properly managed, whether for livestock, wild animals, or both. Carrying capacity, which is the number of healthy animals that can be grazed indefinitely on a given unit of land, must not be exceeded. Because of year-to-year changes in weather conditions and hence food availability, determining carrying capacity is not simple; worst-case estimates are preferred to minimize exceeding it. The goal should be a healthy grassland by optimizing, not maximizing, the number of animals. For private land, optimizing livestock numbers is in the long-term self-interest of the landowner. For publicly held land, managed in common or with unclear or disputed ownership, restricting animals to the optimum level is particularly difficult to achieve. Personal short-term benefit often leads to long-term disaster, as described in Garrett Hardin's 1968 essay "The Tragedy of the Commons."

Grazing areas will remain healthy as long as their carrying capacities are not exceeded. (Ben Klaffke)

Managing grasslands involves controlling the number of animals and enhancing the habitat. Cattle and sheep may be physically restricted through the use of herding and fencing, although such restrictions can be difficult to achieve through political means. Much more problematic is controlling wildlife when natural predators have been eliminated and hunting is severely restricted. As for habitat improvement, the use of chemical, fire, mechanical, and biological approaches can increase carrying capacity for either domesticated or wild herbivores. Removing woody vegetation by burning or mechanical means will increase grass cover, fertilizing can stimulate grass growth, and reseeding with desirable species can enhance the habitat. Plants native to a particular region are best for preserving that environment. Effective grassland management requires matching animals with the grasses on which they graze.

James L. Robinson

SUGGESTED READINGS: Balanced treatments of grazing can be found in chapter 13 of *Natural Resource Conservation* (7th edition, 1998), by Oliver S. Owen, Daniel D. Chiras, and John P. Reganold, and chapter 7 of *Contemporary Issues in Animal Agriculture* (2d edition, 1999), by Peter R. Cheeke.

SEE ALSO: Desertification; Erosion and erosion control; Hardin, Garrett; Kaibab Plateau deer disaster; Range management.

Great Swamp Wildlife Refuge

DATE: established May, 1964
CATEGORY: Animals and endangered species

The Great Swamp Wildlife Refuge in New Jersey provides an important nesting and feeding habitat for migratory birds. The existence of the refuge is primarily the result of grassroots efforts to stop the construction of a new airport in the area.

The Great Swamp Wildlife Refuge is a 7,400-acre region of bottomland hardwood swamps and mixed hardwood forests containing cattail marshes, grasslands, ponds, and streams located in Morris County, New Jersey. It was designated a Registered National Natural Landmark in 1966. An estimated 300,000 people visit the refuge each year. The refuge supports approximately 220 species of birds, 39 species of reptiles and amphibians, 29 species of fish, 33 species of mammals, and 600 species of plants, 215 of which are wildflowers. Of those species, 26 are designated as threatened or endangered by the state of New Jersey, including the bog turtle, the wood turtle, and the blue-spotted salamander.

In 1959 the New York-New Jersey Port Authority identified a 10,000-acre area in rural New Jersey as the site of a new airport serving the New York metropolitan area. The proposed site would cover twice the area of Kennedy Airport (then known as Idlewild). When *The Newark Evening News* broke the story, citizens, politicians, and conservationists banded together to fight the proposed project. Citizens objected to the Port Authority's expansion plans for a multitude of reasons: destruction of homes and business, unacceptable noise levels from the new airport, traffic, and contamination of underground water supplies.

Fourteen volunteer groups joined forces as the Jersey Jetport Site Association (JJSA). The JJSA fought the expansion on a variety of fronts: political, legal, economic, and, most important, public opinion. Some of the same people also became involved with the North American Wildlife Foundation (NAWF), which purchases threatened lands and holds them for future government purchase or donates the property outright. Through the efforts of the NAWF and the Bureau of Sport Fisheries and Wildlife, the United States Department of the Interior agreed to grant the area wildlife refuge status if the NAWF could raise the funds to purchase 3,000 acres. The NAWF acquired the first 1,000 acres in 1960 and turned them over to the U.S. Fish and Wildlife Service later that same year.

In 1962 public hearings began in New Jersey on a bill that would prohibit airport construction in seven northern counties—including the Port Authority site in Morris County. The bill passed by a wide margin but was vetoed by New Jersey governor Robert Meyner, who declared it unconstitutional. Governor Meyner lost his bid for reelection and was replaced by Richard Hughes, who supported the bill and the refuge. By 1963 the state had even provided $25,000 to purchase additional acreage, and the refuge was opened in May of 1964. Additional parcels of land were added throughout the following years, until by 1990 the Great Swamp Wildlife Refuge consisted of more than 7,000 acres of land.

P. S. Ramsey

SEE ALSO: Environmental policy and lobbying; Fish and Wildlife Service, U.S.; Nature preservation policy; Wildlife refuges.

Green marketing

CATEGORY: Philosophy and ethics

Green marketing includes a consideration of the environmental impact associated with the production or consumption of goods. It is significant because it represents a new way of thinking about the relationship between business and the environment.

The term "green marketing" first arose in the 1970's during the period of rising environmental awareness among consumers. New companies that emphasized environmentally friendly products formed. Existing corporations began to give more consideration to environmental issues, partly because of government regulations, but also because suddenly there were economic benefits from being "green." Corporations discovered that they could create a loyal base of customers by exceeding regulatory compliance and becoming environmentally proactive.

There is some confusion in discussing green marketing because the expression is applied in two different circumstances. The phrase sometimes refers to a company making a conscious decision to change its production process to reduce its environmental impact and then using the change as a basis for a public relations campaign. At other times the term refers to the retail marketing of a product based upon its biodegradability, recycling potential, energy efficiency, ozone impact, or other environmental attributes. Adding more confusion, in some urban areas the term green market is used to refer to small-scale farmers' markets held in parking lots on weekends during the local fruit and vegetable season.

Green marketing represents an emerging challenge to the traditional view that there is an inherent conflict between economic profits and environmental quality. By "going green," companies can make profits while producing environmentally friendly products or increase profits by changing the production process to use less energy or material, produce less pollution, or find new uses for one-time waste products. Corporate consideration of environmental consequences can increase efficiency and profitability by forcing executives to rethink normal operating procedures. For example, at the Atlantic Richfield Company's Los Angeles refinery, a series of relatively low-cost changes during the 1980's reduced waste volume by 8,600 tons per year. Because disposal costs were about $300 per ton, the company saves more than $2 million each year in disposal costs alone. Atlantic Richfield also found markets for much of its former waste and began selling spent alumina catalyst to chemical companies, spent silica catalyst to cement makers, and alkaline carbonate sludge to a nearby sulfuric acid plant.

There are numerous other examples of corporate changes benefitting both profitability and environmental quality. The Pollution Prevention Pays program created by the giant 3M company is the project most often cited. Initiated in 1975 by two environmental engineers and a communications specialist, the program cut the company's generated pollution by 50 percent. According to 3M, the program has saved the company more than $500 million since its inception. In addition to the savings, these programs benefit the public image of the corporation.

Many major corporations now consider their environmental efforts to be an important public relations tool. Market research indicates that roughly one-half of the consumers in the United States and other industrialized nations are willing to pay a premium for environmentally friendly products. Even among those unwilling to pay a higher price, environmental attributes serve as a tie-breaker between competing brands, thus serving as a potential source of brand loyalty and product differentiation. During the 1980's many companies began using product labels that contained information touting the environmental friendliness of the contents and packaging. Many of these claims were dubious at best. For example, soon after dolphin-safe tuna became available, Greenpeace reported that some of the tuna being sold under this claim had been taken in the traditional manner, with no attempt to prevent dolphins from dying in the nets.

In 1992 the Federal Trade Commission announced its guidelines governing the circumstances in which terms such as "biodegradable," "recyclable," and "ozone safe" could be used in advertising. Firms that cannot substantiate their environmental claims are now required to drop these terms from their advertising. Moving beyond the regulation of marketing phrases, the actual environmental impact of products can be evaluated. In Germany and Canada, the government performs this function. In the United States, the federal government has no national environmental product evaluation program in place. However, two nonprofit organizations—Scientific Certification Systems (originally known as Green Cross) and Green Seal—systematically evaluate the environmental impact of products from production to disposal.

While the concept of green marketing is growing in acceptance, there are several noticeable concerns. Consumer willingness to pay a premium for green products declined during the 1990's, probably as a reaction to the unfounded earlier claims and the difficulty in actually determining the environmental impact of one good in comparison with others. Surveys conducted during the 1990's also indicated a decline in the attention that consumers paid to the environmental characteristics of the products they bought. For example, during the late

1980's McDonald's restaurants were widely criticized for using polystyrene foam packaging for its sandwiches. As public pressure increased, McDonald's signed an agreement with the Environmental Defense Fund to provide technical assistance in improving the company's environmental performance. Through these efforts, McDonald's switched to a less harmful wrapper and made other adjustments to reduce its impact. These changes, which were highly publicized, have not altered McDonald's market position relative to its major competitors, and none of the competitors have created waste reduction programs to match that of McDonald's.

Allan Jenkins

SUGGESTED READINGS: A collection of articles discussing the emergence of a new corporate perspective on the environment is provided in *Business and the Environment* (1996), edited by Richard Welford and Richard Starkey. Bob Frause and Julie Colehour, *The Environmental Marketing Imperative* (1994), is a how-to guide for businesses wishing to begin using green marketing strategies. In *The Bottom Line of Green Is Black* (1993), Tedd Saunders and Loretta McGovern describe the green tactics used by corporations and provide addresses for environmental organizations.

SEE ALSO: Coalition for Environmentally Responsible Economies; Dolphin-safe tuna; Green movement and Green parties; Recycling.

Green movement and Green parties

CATEGORY: Philosophy and ethics

The Green movement seeks to change certain fundamental values in Western society, particularly those that during the second half of the twentieth century appeared to create threats to humanity and the larger nonhuman environment, such as unrestricted technological progress and economic development. Although varied in their origins and range of interests, Green parties were started in order to urge such changes through existing democratic institutions.

The Green movement evolved from the protest movements of the 1960's. Specifically, Rachel Carson's *Silent Spring* (1962) awakened people to the health hazards of industrial pollution and encouraged them to consider humanity and the environment as interdependent. In addition, the fear that superpowers might resort to nuclear weapons during the Cold War prompted antinuclear, propeace demonstrations. The peace movement in Europe produced the first formal Green parties in the 1970's, organized on the model of two precursors: New Zealand's New Values Party, started in 1972, and Great Britain's Ecology Party, started in 1973, both of which sought to formulate a new electoral strategy with environmental issues.

The Greens first organized in the United States in 1984 as the Green Committees of Correspondence, taking the German Green Party as its model. The organization consisted of state parties and otherwise unaligned individual members. From the outset, however, Green parties were state controlled; the national organization was a loose confederation. The first individual Green candidate in the United States appeared on a ballot in 1986, and in 1990 the Alaska Green Party was the first to achieve ballot status, followed by the California party in 1992. The coalition of Green parties for the first time fielded a presidential candidate, Ralph Nader, in the 1996 election and attracted 1 percent of the votes. Despite interstate disputes over tactics during the early 1990's, twenty-five state parties formed the Association of State Green Parties (ASGP) in 1996 in order to prepare for national elections in 2000 and after.

Some degree of international affiliation exists among national parties. The ASGP, for instance, is a member of the Federation of the Green Parties of the Americas, based in Mexico City, and is associated with the thirty-member European Federation of Green Parties.

CORE PRINCIPLES

Central to the basic tenets of Green parties in all countries is the belief that the world order must be reshaped, with emphasis given to local governance. At the same time, the transformation must involve a shift in people's interest from the immediate future and self-centered satisfaction to the long-range future and sustainable production of material needs. The often-repeated slogan "think globally, act locally" reflects the spirit of the movement. To this end, the European Greens published four basic principles in the 1970's, to which Greens in the United States added six more in the 1980's.

The "four pillars" of the early European Greens were ecology, social justice, grassroots democracy, and nonviolence. By ecology (or ecological wisdom, the American term) is meant a redirection of mentality: People should consider themselves part of nature, not controllers of it, and so live in harmony with it. Practically, this entails devoting technology to achieving an energy-efficient economy and minimizing extraction of nonrenewable resources. Social justice encompasses universal equal rights, dignity, and social responsibility based upon the values of simplicity and moderation. Greens advocate community-controlled, free education and programs to prevent crime. Grassroots democracy, in the American interpretation, would distribute most state and federal power to local elected officials and mediating institutions, such as neighborhood organizations, church groups, voluntary associations, and ethnic clubs. The goal is to restore civic vitality by involving as many people as possible in decision making and avoiding reliance on lawyers, legislators, and bureaucrats. Nonviolence not only seeks to end patterns of conflict in the family and the community but also involves eliminating nuclear weapons. Worldwide, Greens opposed the 1991 Persian Gulf War and cited the environmental disaster of burning oil wells and oil spills in the Gulf as evidence that military action is counterproductive at best and likely to be ruinous.

The six key values added by American Green parties elaborate on issues inherent in the four pillars. First, respect for diversity would honor cultural, racial, sexual, and religious differences, while insisting that citizens bear individual responsibility to all beings for their actions. Second, feminism would replace traditions of dominance by a particular ethnic group, social class, or gender with ethics based upon cooperation and respect for the contemplative, intuitive capacities of all people. Community-based economics calls for employee ownership of businesses, workplace democracy, and equal distribution of wealth to ensure basic economic security for all. Similar to grassroots democracy, decentralization would give power to economically defined localities and ecologically defined regions; it entails redesigning institutions so that control over regulations and money is greatest at the community level, rather than the national level, and control over environmental policy is greatest at the regional level. Personal and global responsibility encourages wealthy communities to assist grassroots groups in developing countries—directly rather than through their governments—in order to make them self-sufficient. To produce the means for such aid, the Greens want to decrease the national defense budget, although not to the point of compromising American security. Finally, future focus, or sustainability, requires all economic, scientific, and cultural policies to be formulated with careful attention to their long-range effects and not just to their immediate benefits. For this reason, European and American Greens denounced such scientific developments as genetic engineering and nuclear power.

The European and American versions of the Green movement contain potential conflicts. For example, the European Greens call for internationalized security to prevent war; in particular they want to give control over military action to the United Nations after it is reformed so that each nation has equal voting power. They also want to replace regional trade treaties, such as the North American Free Trade Agreement (NAFTA), with treaties negotiated and monitored by the United Nations. The American goal of decentralization does not clearly accord with this vision of world order, nor is it clear how community-based economics would admit the European goal of planetwide economic solidarity.

POLITICAL POWER

In the 1980's and 1990's, doctrinal divisions, already latent among the Greens, produced contention and sometimes disaffection, especially in the European organizations. The division falls between moderates, open to compromises with other political parties in order to gain power, and radicals, for whom compromise is unacceptable. The moderates, sometimes referred to as "light green," tend to espouse the anthropocentric view: The Greens should help safeguard the human environment, although preferably not at the expense of other organisms. This faction seeks reform of existing social and economic institutions. The radicals, known as "dark green," are biocentric, holding that all creatures have equal natural rights to life and that humankind should not consider itself a favored species. They wish to end the affluent, technological, expansionist, service-providing orientation of society.

At the same time, Green parties achieved modest political success, especially in Western Europe, in placing candidates in office. Greens constituted 10 percent of some parliaments and entered into coalitions that formed ruling governments. In nearly all countries, they influenced environmental legislation. Although Green candidates did not win any national or state offices in the United States, by 1999 they had been elected to sixty-three local offices in fifteen states, mostly for such nonpartisan agencies as planning groups and school boards. Arcata in Northern California was the first municipality to have a Green majority on its city council.

The power of Green parties manifested more in influencing policy and in education about environmental and social issues than in direct legislative or administrative action because opposition to their principles is formidable in most countries. Critics in the United States accuse the Green movement of elitism, the project of well-educated, middle-class white people. The Green reconception of society appears to leave little room for individualism, which also runs counter to mainstream American sensibilities, and the proposal to give international organizations control over security, particularly the United Nations, raises fears among nationalists in all countries that cultural identity and sovereignty will be lost. Perhaps most of all, the redistribution of wealth and sustainability would dismantle the free market economies that created an economic boom during most of the 1990's. Affluence

resists basic change. Nonetheless, the Green movement succeeded in a primary objective: to end the status of technological, economic expansion as an unquestioned value and place the burden of proof on its proponents to demonstrate that specific projects will not harm humanity or nature.

Mainstream parties, such as the Democrats in the United States, complain that Green parties are spoilers, enticing away liberal votes for candidates that stand little chance of winning elections. For that reason, liberal parties adopted some of the Green environmental goals in their own platforms, weakening the attraction of the Green parties in the late 1990's.

Roger Smith

SUGGESTED READING: John J. Audley offers a case study of the Green movement's practical political influence during the negotiations for the North American Free Trade Agreement in *Green Politics and Global Trade: NAFTA and the Future of Environmental Politics* (1997). Jacqueline Vaughn Switzer examines the opposition to Green goals, especially concerning land rights and resource extraction, in *Green Backlash: The History and Politics of Environmental Opposition in the U.S.* (1997). David Pepper's *Modern Environmentalism* introduces readers to the political and philosophical development of environmentalism and the Green movement in particular. *Green Plans: Greenprint for Sustainability* (1997), by Huey D. Johnson, discusses the political background and implementation of Green initiatives in New Zealand, Canada, and the Netherlands. Benjamin Kline discusses the philosophical background of the Green movement, as well as briefly surveying its U.S. development, in *First Along the River: A Brief History of the U.S. Environmental Movement* (1997). Andrew Dobson's *Green Political Thought* (1995) is an academic, political analysis of the Greens in which he stresses the difference between Green ideology and environmentalism; the book is suited to readers interested in political philosophy.

SEE ALSO: Antinuclear movement; Environmental ethics; Environmental justice and environmental racism; Environmental policy and lobbying.

Green Plan

DATE: begun December 11, 1990
CATEGORY: Preservation and wilderness issues

The Green Plan, launched by the Canadian federal government in 1990, is a national strategy and action plan to provide Canada with a cleaner, safer, and healthier environment, as well as a sound and prosperous economy.

After many years of extensive consultations with Canadians who represented the government, business, interest groups, and the public, Canada's internationally acclaimed Green Plan emerged in December, 1990. The overall goal of the plan was to ensure that current and future generations would enjoy a safe, healthy environment and a sound economy. Although the Green Plan focuses on a wide range of environmental issues, it also incorporates the fundamentals of sustainable development into all aspects of decision making at all levels of society.

Since the Green Plan is an umbrella document, many of the details were left to work out during implementation. Many programs were initiated that affect various aspects of the lives of all Canadians, including the air they breathe, the water they drink, and the food they eat. For example, numerous full assessments of priority toxic substances were performed, and in 1992 the number of full-time Canadian environmental inspectors and investigators was increased from forty-nine to seventy.

The Green Plan established a series of sustainable development goals for Canadians, which serve as benchmarks for measuring progress and mobilizing collective, nationwide efforts in this direction. The first goal is to assure that current citizens and future generations have clean air, water, and land, which are all essential to sustaining human and environmental health. As part of this goal, ground-level ozone (smog) is being reduced below the threshold of adverse health effects, and Canada's generation of waste is being reduced by 50 percent.

The second goal is the sustainable use of renewable resources, which involves shifting forest management from sustained yield to sustainable development. Some of the key decisions involve harvesting practices (particularly of old-growth forests), reforestation, and the use of forest pesticides. Answers are being provided through the creation of a network of model forests and the creation of Tree Plan Canada.

The third goal is the protection of special spaces and species. The federal government has set aside 12 percent of the country as protected space for new parks, wildlife areas, and ecological reserves. Similarly, the fourth goal focuses on preserving and enhancing the integrity, health, biodiversity, and productivity of Canada's Arctic ecosystems. Waste cleanup and assessment has been carried out at numerous sites in the Yukon and the Northwest Territories.

The fifth goal of the Green Plan is Canada's commitment to global environmental security. For example, plans were implemented to phase out human-made chlorofluorocarbons (CFCs), methyl chloroform, and other major ozone-depleting substances by the year 2000. In addition, national emissions of carbon dioxide and other greenhouse gases are being stabilized at 1990 levels. The Green Plan also includes goals for minimizing the impact of environmental emergencies and making environmentally responsible decisions. These goals are being accomplished by implementing plans for quick, effective responses to environmental emergencies and by providing accurate, accessible information about the environment to all Canadians.

Alvin K. Benson

SEE ALSO: Conservation; Nature preservation policy; Nature reserves; Preservation; Wilderness areas.

Green Revolution

CATEGORY: Agriculture and food

The Green Revolution implemented advances in agricultural science to raise food production levels, particularly in developing countries. The revolutionary nature of this process is associated with the spreading use of high-yield varieties (HYV) of wheat, rice, and maize (corn) devel-

oped through advanced methods of genetics and plant breeding.

The Green Revolution can be traced back to a 1940 request from Mexico for technical assistance from the United States to increase wheat production. By 1944, with the financial support of the Rockefeller Foundation, a group of U.S. scientists began to research methods of adapting the new high-yield wheat that had been successfully used on American farms in the 1930's to

The Green Revolution involved the use of genetics, advanced plant breeding techniques, fertilizers, and pesticides to increase yields of crops such as wheat. (Ben Klaffke)

Mexico's varied environments. A major breakthrough is attributed to Norman Borlaug, who, by the late 1940's, was director of the research in Mexico. For his research and his work in the global dissemination of the Mexican HYV wheat, Borlaug won the 1970 Nobel Peace Prize.

From wheat, research efforts shifted to rice production. Through the work of the newly created International Rice Research Institute in the Philippines, an HYV rice was developed. This so-called miracle rice was widely adopted in developing countries during the 1960's. More recent research has sought to spread the success of the Green Revolution to other crops and to more countries.

Approximately one-half of the yield increases in food crops worldwide since the 1960's are attributable to the Green Revolution. Had there not been a Green Revolution, the amount of land used for agriculture would undoubtedly be higher today, as would the prices of wheat, rice, and maize, three species of plants that account for more than 50 percent of total human energy requirements. There is a concern, however, that the output benefits of the Green Revolution have had some negative equity and environmental effects.

In theory, a small farmer will get the same advantages from planting the HYV seeds as a large farmer. In practice, however, small farmers have had more difficulty in gaining access to the Green Revolution. To use the new seeds, farmers need adequate irrigation and the timely application of chemical fertilizers and pesticides. In many developing countries, the small farmers' limited access to credit makes the variety of complementary inputs difficult to obtain.

The Green Revolution has promoted input-intensive agriculture, which has, in turn, created several problems: Greater usage of fertilizers is associated with rising nitrate levels in water supplies, pesticides have been linked to community health problems, and long-term, intensive production has resulted in compaction, salinization, and other soil-quality problems. Because agriculture is increasingly dependent on fossil fuels, food prices will become more strongly linked to energy supplies of this type, raising concerns about the sustainability of the new agriculture. Biotechnological approaches to generating higher yields, the expected future path of the Green Revolution, will raise an additional set of equity and environmental concerns.

Bruce Brunton

SEE ALSO: Agricultural revolution; Borlaug, Norman; Genetically engineered foods; High-yield wheat; Irrigation; Pesticides and herbicides; Sustainable agriculture.

Greenbelts

CATEGORY: Land and land use

Greenbelts (also known as greenways) are tracts of open space preserved to control urban growth patterns. They are generally linear in form and follow natural landscape features.

Throughout the twentieth century, society urbanized at ever-increasing rates. As a result of the automobile, the United States and Canada in particular have been subjected to uncontrolled growth, resulting in the phenomenon referred to as "urban sprawl." As development moves outward from the central city, prime agricultural and forested lands are converted to more intensive uses, resulting in a significant loss of wildlife and plant habitat. This decrease in natural areas also leads to a subsequent degradation of air and water quality.

The concept of creating greenbelts developed as a grassroots response to address these problems. With limited public funds for open-space preservation, greenbelt proponents have focussed attention on "left-over" or abandoned lands. These parcels are often found along ridgelines and streams, areas that are too steep or too wet for development. Abandoned railroad and utility rights of way have become important potential resources as well. All of these areas have common physical characteristics. They are long, thin tracts of land that relate to the topography, threading through land more suitable for development.

Greenbelts, as linear open-space corridors, provide several important benefits. First, they enable urban areas to retain their biodiversity. This is important for maintaining plant and animal habitats, as well as establishing a source of protection for air and water quality. The natural corridors provide migration routes for species interchange. This movement of plant and wildlife along a natural pathway is particularly significant, since it may determine the ability of the species to survive in the area. Second, the retention of undeveloped, vegetated lands allows surface water to be returned naturally to the water table, minimizing surface runoff, erosion, and subsequent stream sedimentation.

Greenbelts offer many recreational opportunities as well. They usually include a system of trails that may link larger, more intensive recreational facilities or provide people with access to natural amenities from urban areas. By connecting different sorts of facilities, in essence creating a system or network of urban parks, greenbelts increase the aggregate benefit to the community. Because of the linear nature of greenbelts, they have more edge area than other kinds of parks or open spaces. This characteristic maximizes the available open space and provides potential access to a greater number of people.

The economic benefits of greenbelts are also significant. As left-over or derelict lands, suitable parcels may be purchased relatively inexpensively. Thus, minimum expenditure is often required to purchase lands for the development of a greenbelt system. The aesthetic improvement of the green edge often enhances the value of adjacent properties.

Steven B. McBride

SEE ALSO: Olmsted, Frederick Law, Sr.; Open space; Urban planning; Urbanization and urban sprawl.

Greenhouse effect

CATEGORY: Weather and climate

The greenhouse effect is a natural process of atmospheric warming in which solar energy that has been absorbed by the earth's surface is reradiated and then absorbed by particular atmospheric gases, primarily carbon dioxide and water vapor. Without this warming process, the atmosphere would be too cold to support life. Since 1880, however, the surface atmospheric temperature appears to be rising, paralleling a rise in the concentration of carbon dioxide and other gases produced by industrial activities.

Since 1880, carbon dioxide, along with several other gases—chlorofluorocarbons (CFCs), methane, hydrofluorocarbons (HFCs), perfluourocarbons (PFCs), sulfur hexafluoride, and nitrous oxide—have been increasing in concentration and have been identified as likely contributors to a rise in global surface temperature. These gases are called "greenhouse gases." The temperature increase may lead to drastic changes in climate and food production, as well as widespread coastal flooding. As a result, many scientists, organizations, and governments have called for curbs on the production of greenhouse gases. Since the predictions are not definite, however, debate continues about the costs of reducing the production of these gases without being sure of the benefits.

GLOBAL WARMING AND HUMAN INTERFERENCE

The greenhouse effect occurs because the gases in the atmosphere are able to absorb only particular wavelengths of energy. The atmosphere is largely transparent to short-wave solar radiation, so sunlight basically passes through the atmosphere to the earth's surface. Some is reflected or absorbed by clouds, some is reflected from the earth's surface, and some is absorbed by dust or the earth's surface. Only small amounts are actually absorbed by the atmosphere. Therefore, sunlight contributes very little to the direct heating of the atmosphere. On the other hand, the greenhouse gases are able to absorb long-wave, or infrared, radiation from the earth, thereby heating the earth's atmosphere.

Discussion of the greenhouse effect has been confused by terms that are imprecise and even inaccurate. For example, the atmosphere was believed to operate in a manner similar to a greenhouse whose glass would let visible solar energy in but would also be a barrier preventing the heat energy from leaving. In actuality, the reason that the air remains warmer inside a greenhouse is probably because the glass prevents the warm air from mixing with the cooler outside air. Therefore the greenhouse effect could be more accurately called the "atmospheric effect," but the term greenhouse effect continues to be used.

Even though the greenhouse effect is necessary for life on earth, the term gained harmful connotations with the discovery of apparently increasing atmospheric temperatures and growing concentrations of greenhouse gases. The concern, however, is not with the greenhouse effect itself, but rather with the intensification or enhancement of the greenhouse effect, presumably caused by increases in the level of gases in the atmosphere resulting from human activity, especially industrialization. Thus the term "global warming" is a more precise description of this presumed phenomena.

A variety of other human activities appears to have contributed to global warming. Large areas of natural vegetation and forests have been cleared for agriculture. The crops may not be as efficient in absorbing carbon dioxide as the natural vegetation they replaced. Increased numbers of livestock have led to growing levels of methane. Several gases that appear to be intensifying global warming, including CFCs and nitrous oxides, also appear to be involved with ozone depletion. Stratospheric ozone shields the earth from solar ultraviolet (short-wave) radiation; therefore, if the concentration of these ozone-depleting gases continues to increase and the ozone shield is depleted, the amount of solar radiation reaching the earth's surface should increase. Thus, more solar energy would be intercepted by the earth's surface to be reradiated as long-wave radiation, which would presumably increase the temperature of the atmosphere.

U.S. Greenhouse Gas Emissions
In Millions of Metric Tons

TYPE AND SOURCE	1988	1989	1990	1991	1992	1993	1994
Carbon dioxide (carbon content)	1,376.2	1,385.6	1,373.3	1,360.4	1,380.8	1,406.2	1,430.0
Methane gas	27.56	27.60	27.95	27.94	27.96	26.62	(NA)
Nitrous oxide	.416	.431	.438	.446	.444	.459	(NA)
Nitrogen oxide	21.05	21.08	21.02	20.83	20.84	21.22	(NA)
Nonmethane volatile organic compounds (VOCs)	22.64	21.52	22.01	21.32	20.88	21.14	(NA)
Chlorofluorocarbons (CFCs)	.278	.272	.231	.210	.187	.166	.133
Hydrochlorofluorocarbons (HCFCs)	.074	.076	.084	.091	.102	.112	.135

Source: U.S. Department of Commerce, *Statistical Abstract of the United States, 1996*, 1996.

However, whether there is a direct cause-and-effect relationship between increases in carbon dioxide and the other gases and surface temperature may be impossible to determine because the atmosphere's temperature has fluctuated widely over millions of years. Over the past 800,000 years, the earth has had several long periods of cold temperatures—during which thick ice sheets covered large portions of the earth—interspersed with shorter warm periods. Since the most recent retreat of the glaciers around 10,000 years ago, the earth has been relatively warm.

PROBLEMS OF PREDICTION

How much the temperature of the earth might rise is not clear. So far, the temperature increase of around 1 degree Fahrenheit is within the range of normal (historic) trends. The possibility of global warming became a serious issue during the late twentieth century because the decades of the 1980's and the 1990's included some of the hottest years recorded for more than one century. On the other hand, warming has not been consistent since 1880, and for many years cooling occurred. The cooling might have resulted from the increase of another product of fossil fuel combustion, sulfur dioxide aerosols, which reflect sunlight, thus lessening the amount of solar energy entering the atmosphere. Similarly, in the early 1990's temperatures declined, perhaps because of ash and sulfur dioxide produced by large volcanic explosions. In the late 1990's temperatures appeared to be rising again, thus indicating that products of volcanic explosions may have masked the process of global warming. The Environmental Protection Agency (EPA) states that the earth's average temperature will probably continue to increase because the greenhouse gases stay in the atmosphere longer than the aerosols.

Proper analysis of global warming is dependent on the collection of accurate temperature records from many locations around the world and over many years. Since human error is always possible, "official" temperature data may not be accurate. This possibility of inaccuracy compromises examination of past trends and predictions for the future. However, the use of satellites to monitor temperatures has probably increased the reliability of the data.

Predictions for the future are hampered in various ways, including lack of knowledge about all the components affecting atmospheric temperature. Therefore, computer programs cannot be sufficiently precise to make accurate predictions. A prime example is the relationship between ocean temperature and the atmosphere. As the temperature of the atmosphere increases, the oceans would absorb much of that heat. Therefore, the atmosphere might not warm as quickly as predicted. However, the carbon dioxide absorption capacity of oceans declines with temperature. Therefore, the oceans would be unable to absorb as much carbon dioxide as before, but exactly how much is unknown. Increased ocean temperatures might also lead to more plant growth, including phytoplankton. These plants would probably absorb carbon dioxide through photosynthesis. A warmer atmosphere could hold more water vapor, resulting in the potential for more clouds and more precipitation. Whether that precipitation would fall as snow or rain and where it would fall could also affect air temperatures. Air temperature could lower as more clouds might reflect more sunlight, or more clouds might absorb more infrared radiation.

To complicate matters, any change in temperature would probably not be uniform over the globe. Since land heats up more quickly than water, the Northern Hemisphere, with its much larger land masses, would probably have greater temperature increases than the Southern Hemisphere. Similarly, ocean currents might change in both direction and temperature. These changes would affect air temperatures as well. In reflection of these complications, computer models of temperature change range widely in their estimates. Predicted increases range from 1.5 to 11 degrees Celsius (3 to 20 degrees Fahrenheit) over the early decades of the twenty-first century.

MITIGATION ATTEMPTS

International conferences have been held and international organizations have been established to research and minimize potential detriments of global warming. In 1988 the United Nations Environment Programme and the World Meteorological Organization established the International Panel on Climate Change (IPCC). The IPCC has conducted much research on climate change and is now considered an official advisory body on the climate change issue. In June, 1992, the United Nations Conference on Environment and Development, or Earth Summit, was held in Brazil. Participants devised the Framework Convention on Climate Change, considered the landmark international treaty. It required signatories to reduce and monitor their greenhouse gas emissions.

A more advanced agreement, the Kyoto Accords, was developed in December, 1997, by the United Nations Framework Convention on Climate Change. It set binding emission levels for all six greenhouse gases over a five-year period for the developed world. Developing countries do not have any emission targets. It also allows afforestation to be used to offset emissions targets. The Kyoto agreement includes the economic incentive of trading emissions targets. Some countries, because they have met their targets, would have excess permits, which they might be willing to sell to other countries that have not met their targets.

Margaret F. Boorstein

SUGGESTED READINGS: Dean Edwin Abrahamson's *The Challenge of Global Warming* (1990) discusses the causes and effects of global climate change. Thomas Karl, Neville Nicholls, and Jonathan Gregory, "The Coming Climate," *Scientific American* (May, 1997), examines the greenhouse effect in light of meteorological records and computer models. R. A. Houghton and G. M. Woodwell, "Global Climatic Change," *Scientific American* (April, 1989), discuss the greenhouse effect and possible controls. *Global Warming: The Complete Briefing* (1997), edited by J. T. Houghton, examines global warming and its possible environmental and political consequences. John Firor, *The Changing Atmosphere*

(1990), connects atmospheric problems to economic and political issues in emphasizing the consequences of human actions and attempts at correction. Stephen Schneider, *Global Warming: Are We Entering the Greenhouse Century?* (1989), describes climate mechanisms, predicts what might happen in the future, and offers possible solutions and means of cooperation by business, government, and individuals. The Web pages of the Environmental Protection Agency (http://www.epa.gov) and the National Oceanic and Atmospheric Administration (http://www.noaa.gov) offer explanations and statistics of the greenhouse effect and global warming.

SEE ALSO: Air pollution; Chlorofluorocarbons; Climate change and global warming; Ozone layer and ozone depletion; Rain forests and rain forest destruction.

Greenpeace

DATE: established 1971
CATEGORY: Animals and
    endangered species

In 1971 the Don't Make a Wave Committee based in Vancouver, Canada, renamed its organization Greenpeace. The organization's mission statement identifies it as "an independent, campaigning organization which uses non-violent, creative confrontation to expose global environmental problems and to force the solutions which are essential to a green and peaceful future."

In September, 1971, twelve Canadian volunteers sailed a small boat, named the *Greenpeace*, into the United States atomic test zone on the island of Amchitka, off the coast of Alaska. They were trying to stop the United States from conducting a nuclear test, which they believed could destroy the wildlife haven, cause an earthquake, or create a tidal wave. Since that first

voyage, Greenpeace protesters have often taken their boats into dangerous zones, placing themselves at risk to save a whale from a harpoon or block toxic or radioactive wastes from being released into the ocean. Within twenty years of its founding, Greenpeace had offices in thirty countries, owned six ships, and had more than four million members worldwide. It is mainly funded through small, individual donations. Its work is based on a belief in the importance of the role of individuals and of the visibility of high-profile actions. In 1993, in their publication *Actions Speak Louder*, Greenpeace stated that its mission was to save the rain forests, avert global warming, oppose toxic pollution, and stop "nuclear folly."

Greenpeace activists hang a banner on an observation tower in Niagara Falls, New York, in 1998. The organization advocates the use of direct, nonviolent action to protest the destruction of the environment. (AP/Wide World Photos)

Greenpeace's ships, operated by Greenpeace Marine Services, travel the world to bring public and media attention to local and regional environmental problems. The ships have been positioned between whaling fleets and the whales they are trying to catch and have been used to stop the killing of seals. In 1980 the Greenpeace ship *Rainbow Warrior* was bombed and sunk in Auckland Harbor, New Zealand, by French intelligence agents because it was shadowing French nuclear vessels to protest nuclear testing in French Polynesia.

The main decision-making body of Greenpeace is the Council, which consists of a representative from every Greenpeace office around the world. The Council meets once per year to set policy and approve the annual budget. Over the years the organization has undergone changes and disputes within its leadership. In 1996 the articles of association were changed to allow for more efficient allocation of resources. As a result of the complexity of the issues with which it must deal, Greenpeace has been conducting more scientific research, using its ships as mobile laboratories. It is expanding its offices and work in Asia and Latin America.

Greenpeace has been successful in many of its campaigns against nuclear testing, preserving populations of whales and seals, and protecting the fragile environment of Antarctica. Volunteers in the organization intend to continue their work, emphasizing local decision making and activities.

Colleen M. Driscoll

SEE ALSO: Antinuclear movement.

Groundwater and groundwater pollution

CATEGORY: Water and water pollution

Groundwater is the water found below the surface of the earth. Many public and private water supplies rely on wells that tap important groundwater reserves. Pollution of groundwater leads to changes in water quality that can affect groundwater use for a given purpose.

Humans require vast amounts of fresh water for use in homes, livestock operations, agriculture, and industrial processes. Groundwater is an important source of fresh water. The pollution of groundwater by human activity can contaminate water-supply wells, making them unacceptable for providing water for drinking and other purposes. This can lead to a need for new water supplies that may not be readily available or easily accessible. In some instances polluted groundwater interacts with surface water, thus contaminating the surface water environment as well.

Groundwater constitutes a small but significant portion of the world's overall water supply. Much of the earth's surface is covered by water, but an estimated 97.2 percent of it exists as salt water. Since fresh surface water may account for as little as 0.009 percent of the earth's water, groundwater is a significant source of readily available fresh water. Groundwater occurs in the saturated zone of the earth, which is the area below the surface where pores between particles—void spaces in the soil or rock—are filled with water. In some places groundwater may be encountered near the surface, but in other areas, such arid regions, it can be quite deep. Groundwater flows from areas of high hydrostatic head to areas of low hydrostatic head. Shallow groundwater often mimics topography, flowing downhill toward streams and lakes.

The soil and rock through which groundwater flows consists of particles of varying size, which help determine the classification of the soil or rock and how well water will move through the material. Sand-sized particles are seen in unconsolidated sandy soils or sandstones. Smaller particles may form silty or clayey soils or their bedrock equivalents of siltstones and shales. In the saturated zone, groundwater saturates the pores and voids between the particles. The size of the pores and the degree to which they are interconnected affects hydraulic conductivity—a measure of the ability of water to move through the rock or soil.

Transmissivity is the measure of the ability of an aquifer to transmit water and is a measure of the hydraulic conductivity multiplied by the saturated thickness of the aquifer. Therefore, a thick aquifer with relatively poor hydraulic conductivity might be able to transmit as much water as a thinner aquifer composed of materials with greater hydraulic conductivity. Groundwater is recharged by rainwater percolating through the soil, snowmelt, and rivers and streams.

Humans produce a wide array of pollutants and combinations of pollutants. The degree and extent to which individual pollutants can affect groundwater quality is dependent on a large number of variables, which can include the amount of contaminant introduced into the environment, the time frame in which it was introduced, its toxicity, its mobility, whether it will readily degrade in the environment, and the chemical and physical characteristics of the soil or rock through which it will pass.

Even something as common as nitrogen can lead to pollution in groundwater. Nitrogen can be mobile in the environment in the form of dissolved nitrates and nitrites. Sources for pollution include septic tanks, leaks from sewage treatment plants and lagoons, and animal wastes. Nitrogen is also an important component of many fertilizers used in agriculture and may become dissolved by rainwater and percolate down into groundwater. In high enough concentrations, nitrates can make water unacceptable for human consumption. At even higher concentrations, the water can become unacceptable for livestock and other animals.

Gasoline spills or leaks from underground storage tanks are a relatively common source of groundwater pollution. Some of the dissolved phase components of gasoline are quite mobile in the environment; however, many are also susceptible to biological degradation. Gasoline and other substances less dense than water can float on the surface of the groundwater, but seasonal fluctuations in the water table can smear the contaminant in the soil, potentially making it more difficult to remove. Other contaminants, such as chlorinated solvents, can be denser than water and have the capacity to sink into aquifers. Although metals as a group are generally not considered very mobile and tend to be adsorbed onto soils, some are quite mobile, and contamination by heavy metals can be a relatively common form of groundwater contamination.

Although less common, radiological contamination of groundwater can be a concern. Groundwater often moves slowly, but radioactive half-lives can be quite long.

Ray Roberts

SUGGESTED READINGS: A significant work on the subject of groundwater is R. Allan Freeze and John A. Cherry's *Groundwater* (1979). Herman Bouwer's *Groundwater Hydrology* (1978) contains some useful discussions on groundwater flow and contamination. An informative presentation on groundwater can be found in Ralph C. Heath's *Basic Groundwater Hydrology* (1991), U.S. Geological Survey Water-Supply Paper 2220.

SEE ALSO: Acid mine drainage; Aquifer and aquifer restoration; Sewage treatment and disposal; Soil contamination; Water pollution; Wells.

Group of Ten

CATEGORY: Ecology and ecosystems

The Group of Ten was a short-lived but influential coalition of representatives from the leading U.S. environmental groups.

On January 21, 1981, leaders of nine U.S. environmental organizations met at a Washington, D.C., restaurant to discuss commomnn goals. In attendance were J. Michael McCloskey, the chief executive officer of the Sierra Club; John Hamilton Adams of the Natural Resources Defense Council; Janet Brown of the Environmental Defense Fund; Thomas Kimball of the National Wildlife Federation; Jack Lorenz of the Izaak Walton League; Russell Wilbur Peterson, the former governor of Delaware, of the National Audubon Society; William Turnage of the Wilderness Society; Louise C. Dunlap of the Environmental Policy Center; and Rafe Pomerance of Friends of the Earth. Although he was not in attendance at the original meeting, Paul Pritchard of the National Parks and Conservation Association was later invited to join the group, which thus became known as the Group of Ten.

The activists who formed the group hoped to

send newly inaugurated President Ronald Reagan and his incoming administration a signal that American environmental organizations would fight to preserve the environmental legislation of the 1960's and 1970's. In addition, the attendees hoped to protect and enhance environmental quality by promoting public awareness and corporate social responsibility.

The group's leaders thus conducted meetings with the heads of major corporations including Du Pont, Exxon Chemical, Union Carbide, Dow, American Cyanamid, and Monsanto; a number of these companies were sensitive to unfavorable publicity that they had been receiving for the environmental consequences of their businesses. In the meetings, Group of Ten representatives urged the corporate leaders to assume greater social responsibility for the protection of the environment and for the welfare of future generations.

In 1985, the Group of Ten published *An Environmental Agenda for the Future*, a detailed plan for the worldwide environment's preservation and renewal. With the waning of the Reagan administration's assault on domestic environmental legislation in the late 1980's, the Group of Ten disbanded as a formal organization, but its legacy of united action in defense of the environment continued to prove influential.

Glenn Canyon

SUGGESTED READINGS: John H. Adams, et al, *An Environmental Agenda for the Future* (1985) is a detailed history of the Group of Ten and its environmental objectives. Robert Gottlieb's *Forcing the Spring: The Transformation of the American Environmental Movement* (1993) contains a discussion of the Group of Ten from its inception to its dissolution at the end of the 1980's. Jonathan Lash, Katherine Gillman, and David Sheridan's *A Season of Spoils: The Reagan Administration's Attack on the Environment* (1984) is a harsh critique of the environmental policies of the federal government during the early years of the Reagan administration, the era that gave rise to the Group of Ten.

SEE ALSO: Environmental policy and lobbying; Green movement and Green parties; Sierra Club; Wilderness Society.

Gulf War oil burning

DATE: August, 1990-November, 1991
CATEGORY: Atmosphere and air pollution

During their retreat from occupied Kuwait, Iraqi armed forces damaged an oil pipeline and set fire to more than five hundred oil wells, creating the worst oil-field disaster in history.

The Persian Gulf is a shallow, northwest-trending body of water that covers an area of about 260,000 square kilometers (100,400 square miles). The Gulf is actually a large bay about 800 kilometers (500 miles) long, 201 kilometers (125 miles) wide, and 91 meters (300 feet) deep at the deepest point. It is bordered by Iran, Iraq, Kuwait, Saudi Arabia, Bahrain, Qatar, Oman, and the United Arab Emirates. The Shatt-al-Arab, a river formed by the merging of the Tigris and Euphrates Rivers, has created a combined river floodplain and delta region at the head of the Persian Gulf that covers more than 3,200 square kilometers (1,235 square miles).

The Persian Gulf is teeming with wildlife. The coastal mangrove swamps and coral reefs provide habitat for birds and fish. Hundreds of species of fish, including mackerel, snapper, and mullet, live in the region and feed on the abundant algae. Wading birds such as shanks and sand plovers feed along the coastal mudflats. Also, valuable crustaceans such as prawn and shrimp are farmed along the Persian Gulf's shores.

The region holds more than one-half of the world's proven reserves of oil and natural gas. The Persian Gulf also provides the world's major shipping lanes for oil tanker traffic. According to Jane Walker in *Man-made Disasters: Oil Spills* (1993), approximately 40 million liters (10 million gallons) of oil spills or leaks into Persian Gulf waters each year. The results of these oil releases are usually absorbed by the environment without significant ecological damage. However, in August of 1990 the area was hit by an environmental disaster in the form of the Persian Gulf War.

On August 2, 1990, Iraqi forces invaded the small, oil-rich neighboring nation of Kuwait. On

An American tank passes a burning oil field in Kuwait in March of 1991. Iraqi forces retreating from Kuwait set oil wells on fire and broke an oil pipeline, causing immense amounts of environmental damage to the Persian Gulf region. (Reuters/ Clauda Salhani/Archive Photos)

November 29, 1990, the United Nations voted to permit the use of force to expel Iraqi forces from Kuwait. On January 16, 1991, air attacks on Iraqi targets began, followed on February 23 with a ground attack against Iraq by a coalition of forces from twenty-eight countries, including the United States, Saudi Arabia, Egypt, and Great Britain.

During the ground war, which lasted only one hundred hours, Iraqi soldiers departing from Kuwait damaged an oil pipeline and set fire to more than five hundred Kuwaiti oil wells. These damaged wells poured hundreds of thousands of liters of oil into the surrounding countryside, forming large lakes up to 1.6 kilometers (1 mile) long and nearly 1 meter (3 feet) deep. More than 800 million liters (211 million gallons) of oil issued from the damaged wells each day. The spilled oil threatened to pollute the water supply of Kuwait City and other inhabited areas, and the contaminated soil and vegetation threatened to harm wildlife. The soil is adversely affected by the oily mist, which forms a thin film over the topsoil and reduces the amount of oxygen and water that penetrates the soil profile. This seal reduces the activity of a number of microbes and earthworms that help to keep the soil fertile.

Because currents in the upper Persian Gulf generally move counterclockwise around the bay, a thick oil slick spread northwest to southeast along the west coast from disrupted oil terminals, pipelines, and individual wells in the vicinity of Kuwait City to the lower part of Saudi Arabia; a thin oil film spread from the coastal region to the middle of the upper Persian Gulf. The colossal oil slick threatened the entire ecosystem of the Persian Gulf. The slick resulted in the deaths of twenty thousand birds, including some flamingos, by soaking their feathers and causing them to drown or die from exposure. The Persian Gulf's population of sea cows (dugongs), dolphins, and green turtles were at risk, as were the endangered hawksbill turtles that lay eggs on the local islands.

A number of the wells that were ignited released poisonous gasses into the air that endangered people (including military personnel) and wildlife. The smoke released by the fires contained noxious gases such as deadly carbon monoxide and sulfuric acid. The sky was darkened by the clouds of toxic black smoke, causing temperatures to drop 10 degrees Fahrenheit lower than normal. The Burgan Oil Refinery complex south of Kuwait City was devastated. Most damage occurred within an 800-kilometer (500-mile) range that included most of Iran, Iraq, and Kuwait. Soot and acid rain clouds extended nearly 1,920 kilometers (1,200 miles) away from the sabotaged oil fields. In March, 1991, *The New York Times* reported that the pollution "rained on Turkey and reached the western shore of the Black Sea, touching Bulgaria, Rumania, and the southern Soviet Union, becoming more prevalent over Afghanistan and Pakistan." *The New York Times* also indicated that the toxic clouds resulted in a significant increase in respiratory diseases among the elderly and the very young. In *Oil Spills* (1993), Jane Walker reported that many cattle and sheep "died in Kuwait either from breathing oil droplets in the air, or from eating oil-covered grass."

The cleanup operation was undertaken in two phases: extinguishing the burning wells and controlling the oil slick. The oil field fires were eliminated in less than one year by several professional fire-fighting groups. Among the techniques used to combat the fires were cooling the well equipment with water, removing the well debris, cutting off oxygen to the fire by blowing out the flame with an explosive (usually dynamite), capping the well head with a stinger (plug), and attaching a new valve assembly to shut off flow.

The main technique used to confine and recover oil from Persian Gulf waters was skimming the greasy layer off the water's surface. Oil-skimming ships recovered between 20,000 and 30,000 barrels (about 3 to 5 million liters) of oil per day. The Saudi Arabian Company (Saudi Aramco) placed about 40 kilometers (25 miles) of floating booms along the periphery of the spill and sent more than twenty oil-recovery craft to the area. Although many scientists believe that most marine life requires only three years to recover from the effects of exposure to crude oil, marine life exposed to refined oil, especially in enclosed areas such as the Persian Gulf, may require ten years or longer to recover.

Donald F. Reaser

SUGGESTED READINGS: Jane Walker, *Man-made Disasters: Oil Spills* (1993), reviews the overall effects of the Persian Gulf disaster. K. A. Browning, et al., "Environmental Effects from Burning Oil Wells in Kuwait," *Nature* 351 (May 30, 1991), is a moderately technical summary that concludes that the oil fires had measurable local effects but were unlikely to have significant global effects. T. M. Hawley, *Against the Fires of Hell: The Environmental Disaster of the Gulf War* (1992), provides an assessment of the environmental damage caused by the war. John Horgan, "Burning Questions: Scientists Launch Studies of Kuwait's Oil Fires," *Scientific American* 265 (July, 1991), is a nontechnical summary of some of the results of the Kuwait oil fires. A sidebar discusses allegations that some environmental data had been censored or withheld.

SEE ALSO: Nuclear winter; Oil spills.

H

Hanford Nuclear Reservation

CATEGORY: Nuclear power and radiation

Hanford stores a greater volume of liquid, high-level radioactive waste than any other site in the United States. Procedures at the reservation allowed radioactive waste to be discharged directly into the soil, slightly radioactive cooling water to be discharged into the Columbia River, and radioactive iodine to be released into the atmosphere.

The Hanford Nuclear Reservation is a roughly circular area 40 kilometers (25 miles) in diameter lying on the Columbia River in south-central Washington State. As a key part of the Manhattan Project to build the atomic bomb, several nuclear reactors were constructed at Hanford beginning in 1943. Plutonium produced in these reactors was extracted from spent nuclear fuel and then shipped off-site for fabrication into nuclear weapons. Although the river water used for cooling spent only one or two seconds in the reactor core, the intense neutron bombardment made some trace elements radioactive. Radioactivity was also picked up from uranium fuel elements whose aluminum cladding had corroded. After leaving the reactor, cooling water was held in basins to allow short-lived radioactive elements to decay, after which it was returned to the river.

The Hanford Nuclear Reservation, whose disposal policies allowed radioactive water to be dumped into the Columbia River. (AP/Wide World Photos)

Radioactive elements detected downstream from Hanford include sodium 24, phosphorus 32, and zinc 65. The latter two were concentrated in the flesh of fish and waterfowl. People who ate fish and drank water from the Columbia River were estimated to experience a maximum increased dose from this radioactivity of about 10 percent above natural background radiation, a value believed to be safe.

Chemical procedures used to extract plutonium from spent fuel released significant amounts of radioactive iodine 131 into the air, particularly in 1945. This was believed to be safe because iodine 131 has a short half-life (eight days), and its concentration is greatly diluted as it disperses across the countryside. It has since been learned that cows and goats concentrate iodine in their milk if they are fed grass contaminated with iodine and that it is concentrated further in the thyroids of people drinking this milk. Infants and small children who drank a great deal of milk from grass-fed cows or goats near Hanford may have received harmful doses in 1945, but no such individuals have yet been identified.

Hanford stores 256,000 cubic meters (9 million cubic feet) of highly radioactive liquid waste in 177 concrete-encased steel tanks. The chemical mixtures in some tanks may become explosive under certain conditions, so great vigilance is required. Approximately one-half of the waste is stored in tanks with only a single steel liner; over the years, some of these tanks have leaked an estimated 4 million liters (1 million gallons) of highly radioactive waste into the soil. In addition, 800 billion liters (200 billion gallons) of low-level radioactive waste were discharged directly into the soil with the belief that it would not migrate far. In fact, some elements have migrated farther than anticipated, but only low levels of radioactivity have migrated off-site. A planned extensive clean-up of the site calls for stabilization or removal of contaminated soil and restrictions on groundwater usage. Liquid waste is to be solidified, vitrified (incorporated into glass), and taken to a permanent storage site.

Charles W. Rogers

SEE ALSO: Groundwater and groundwater pollution; Nuclear and radioactive waste; Radioactive pollution and fallout.

Hardin, Garrett

BORN: April 12, 1915; Dallas, Texas
CATEGORY: Ecology and ecosystems

Garrett Hardin is a major proponent of the need to control human population growth. He first received public attention in 1968 through the publication of his article "The Tragedy of the Commons" in the journal Science.

The title of Garrett Hardin's article "The Tragedy of the Commons" was taken from the concept of the English commons. Everyone in a typical English community could graze their animals on pasture that was held in common by all the inhabitants. The nature of the commons made it possible for its users to overexploit the pasture because any extra animals that one person raised were available for that user's benefit alone, while the grazing resources would be lost to all users of the commons. When the total population is low, any misuse of a commons can be compensated for by moving to a new area. When the total population has increased to the point where no new land is available, the resource held in common will be depleted. The results can be seen in acid rain and ozone depletion (in which the air is treated as a commons), overgrazing (in which the land is treated as a commons), and the whaling and commercial fishing industries (in which the resources of the sea are treated as a commons).

Hardin notes that each problem could be treated as though it were separate and distinct from every other, but he believes that this approach simply hides the underlying cause of these problems. The pivotal commons is the freedom to breed, particularly when the costs of shaving children are spread over the entire population and are not restricted to the parents. According to Hardin, the single root cause that lies behind problems such as excessive resource consumption, war, starvation, ethnic cleansing, poverty, noise pollution, foreign aid, and simple traffic jams is human overpopulation.

In *The Limits to Altruism* (1977), Hardin states that every nation has the responsibility to reduce its population to the level it can support using its own resources. Special privileges enjoyed by a few people today should be tolerated if they contribute to national self-reliance in the future. If all nations do not become self-reliant, the resulting population crash will be so devastating that civilization may never recover even if small enclaves of humans survive.

In *Living Within Limits* (1993), Hardin compares the United States to a lifeboat in a sea of human misery and concludes that the United States will only survive a coming era of resource scarcity by banning immigration and stabilizing its population. Hardin denies the workability of the global village and offers no apology to kindhearted people for his lack of compassion. In the absence of human moderation, nature will once more assert dominance in controlling human population size through disease, starvation, and war.

Gary E. Dolph

SEE ALSO: Population-control movement; Population growth.

Hawk's Nest Tunnel

DATE: March 31, 1930-November, 1931
CATEGORY: Human health and the environment

The building of the Hawk's Nest Tunnel, one of the deadliest engineering projects in the history of the United States, caused the deaths of hundreds of workers from acute silicosis.

In 1917 the Electro-Metallurgical Company of Fayette County, West Virginia, merged with three other companies and became the Union Carbide and Carbon Corporation. Interested in developing the water resources of West Virginia, Union Carbide, through the New Kanawha Power Company, filed its plan to build dams, tunnels, and power stations with the federal government. The Federal Power Commission took no action, the climate of the times being that government hesitated to interfere with the wishes of industry. Union Carbide rushed the building of the Hawk's Nest Tunnel out of fear that the government's attitude would change.

Built near Gauley Bridge, West Virginia, in 1930-1931, the purpose of the 5-kilometer-long (3-mile-long) tunnel was to divert a portion of the New River underground to produce hydroelectric power.

The rushed construction of the tunnel resulted in the deaths of hundreds of workers. Although accidents, such rock falls, accounted for some deaths, the majority resulted from silicosis, a lung disease that occurs when silica particles are inhaled. Silica particles small enough to inhale are released when mining processes pulverize granite and sandstone. The entire length of the Hawk's Nest Tunnel was dug through granite and sandstone, and much silica dust was produced. Although mining techniques that reduce the dust were known at the time, they were not used because they would have slowed productivity. Although Union Carbide admitted that 109 men died as a result of working on the tunnel, some estimates place the death toll at closer to 700 workers if the deaths from silicosis that occurred over the next five years are included.

The Hawk's Nest Tunnel disaster served to bring the plight of mine workers to national attention. Dozens of personal lawsuits were filed against Union Carbide. The lawsuits claimed that the mining company knew the dangers of silicon dust and that they knew of alternate, but slower, mining methods that would have protected the workers. Although Union Carbide won most of these lawsuits, a congressional subcommittee that convened in 1936 to study the issue did raise awareness about the dangers of silica.

The tunnel disaster also highlighted the environmental issue of water rights. In 1934 the Federal Power Commission sued Union Carbide and claimed full jurisdiction over the New River and all of the artifacts that affected its flow. The attorney general of the state of Virginia, siding with industry, argued the case of Union Carbide before the United States Supreme Court. The federal government decided that the case was critical to its future jurisdiction over water rights and continued to battle Union Carbide for the next thirty years. In 1964 the federal government finally won, forcing Union Carbide to install stream gauges and fishways at the Hawk's

Nest Dam. Throughout the legal battle, the federal government was only concerned with the cost of water. The human cost of disease and damage to the waterways and the surrounding environment was not an issue.

Louise Magoon
SEE ALSO: Environmental illnesses; Silicosis.

Hazardous and toxic substance regulation

CATEGORY: Human health and the environment

The regulation of hazardous and toxic substances has become necessary in order to protect the environment from the by-products of industrialization. Policymakers must find a balance between safeguarding against undesirable health or environmental effects and reaping the benefits of the activities that produce such hazards.

The survival of the human species is inseparable from the preservation of the environment. This concern may be divided into three categories: the health of the general population, occupational hazards, and the availability of food and water supplies that are pure and free from pollutants. Within these categories, industrial development has created new problems and environmental pressures, even as technological advances improve the ability to detect, study, and analyze environmental changes.

Since ancient times humans have been concerned with issues surrounding public health. The most common concerns before the Industrial Revolution of the nineteenth century were related to food and water supplies. Tasks involving occupational hazards were left for slaves and the lower classes and did not, therefore, receive much recognition. After the Industrial Revolution, contact with occupational hazards and toxic substances become much more commonplace.

By the early twentieth century, conditions had worsened for the average working person, and those not faced with occupational hazards were beginning to feel the effects of an industrialized society that gave rise to hazardous by-products

and toxic waste. Consciousness of environmental problems began to increase, but the rate of industrial development outstripped societal awareness of the hazards it was creating.

Although the United States Congress made attempts to regulate the hazardous and toxic wastes being generated from mining, agriculture, and industry, progress was slow because there was little public interest in the situation. Awareness of the negative impact of industrialization substantially increased during the 1960's as environmental activists such as Barry Commoner and Rachel Carson succeeded in publicizing issues relating to the deterioration of the environment. Public interest reached a peak with the celebration of the first Earth Day in 1970. Congress began to respond to public pressure by strengthening the laws that had previously been passed to regulate air and water pollution.

Efforts to decrease hazardous and toxic waste levels in the environment were relatively ineffective until August 2, 1978, when New York State officials ordered the emergency evacuation of 240 families living within two blocks of the Love Canal chemical waste dump in Niagara Falls, New York. Headlines throughout the nation declared Love Canal the largest human-made environmental disaster in decades, and Americans began paying attention to hazardous and toxic waste issues. The general public held policymakers accountable for the disaster and the prevention of similar occurrences in the future.

DEVELOPMENT OF LEGISLATION

The U.S. regulatory style is considered to be one of open conflict, with interest groups, various media, legislators, and the courts playing an important role in the development of laws. Policymakers must also rely on the consensus of scientists in understanding the risks and magnitude of the hazardous and toxic waste problem. This reliability complicates and considerably slows the regulation process. Policymakers must also make tough decisions on the health benefits versus economic costs of controlling hazardous and toxic wastes.

This difficult task of creating legislation that works to protect the environment and the health of citizens in a hazardous society can be broken

A project engineer for the U.S. Army Corps of Engineers inspects a site formerly classified as one of the most polluted hazardous waste areas in the United States. Legislation required that the toxic soil be removed, decontaminated, and replaced. (AP/Wide World Photos)

down into a four-step process: creating a law, putting the law to work, creating a regulation, and enforcing the law. The collaboration of Congress, government agencies, and societal awareness and responsibility is needed to create effective legislation.

To create new legislation, a member of Congress must propose a bill, which, if approved by both the House of Representatives and the Senate and signed by the president, becomes a new law. The act is codified by the House of Representatives and published in the United States Code. To put the law to work, regulations for the law must be created by government agencies authorized by Congress. Regulations are rules about the law, specifying what is legal and what is not. To create a regulation, the authorized government agency, usually the Environmental Protection Agency (EPA) in the case of hazardous waste, determines the need to form a regulation. The regulation is proposed on the federal register, and members of the public are allowed to provide input in the form of comments and sug-

gested modifications. Revisions to the regulation may be made accordingly. Once a completed regulation is finished, it is published in the Code of Federal Regulations (CFR). Twice per year each agency publishes a comprehensive report that describes all the regulations it is working on or has recently finished. Laws and regulations are enforced by the government agency that put them into effect.

IMPORTANT LEGISLATION

The responsibility for regulating and enforcing hazardous materials laws in the United States is primarily split among three or four federal regulatory agencies. There is some overlap between the fields that regulate hazardous and toxic waste materials. The government agencies that are involved in regulating hazardous and toxic substances are the EPA, the Occupational Safety and Health Act (OSHA), and the Department of Transportation (DOT).

Several environmental laws affect hazardous and toxic waste policy in the United States.

OSHA, passed in 1970, provides standards of allowable exposure to toxic chemicals in the workplace. This law establishes the standards for more than twenty toxic or hazardous substances and more than four hundred toxic air contaminants. OSHA also establishes the labeling standards for equipment, standards for personal protection, and monitoring requirements for the health of workers. The Federal Insecticide, Fungicide, and Rodenticide Act (FIFRA) of 1972 established a regulatory program for the EPA to control the manufacture of potentially harmful pesticides. This law was created to prevent the adverse environmental effects posed by new pesticides and to ensure safety standards for those using pesticides.

The Safe Drinking Water Act (SDWA) of 1974 was established to protect groundwater and drinking water sources from contamination by hazardous chemicals. The act sets two levels of standards to limit the amount of contamination that might be found in drinking water: primary standards with a maximum contaminant level (MCL) to protect human health and secondary standards that relate to color, taste, smell, or other physical characteristics. The Hazardous Materials Transportation Act (HMTA) of 1975 provides a high level of environmental protection during hazardous waste transportation managed by the DOT. By requiring special packing and routing, the act ensures the careful shipment of hazardous substances.

The 1976 Resource Conservation and Recovery Act (RCRA) and its amendments deal with the ongoing management of solid wastes throughout the United States. These acts treat hazardous waste with a "cradle-to-grave" management system designed to protect groundwater supplies by focusing on the treatment, storage, and disposal of such substances. The RCRA focuses on five main areas for hazardous waste management: identification and classification of hazardous waste; requirements for generators of hazardous waste to identify themselves so that hazardous waste activities and standards of operation for generators may be identified; adoption of standards for the transportation of hazardous wastes; standardization of treatment, storage, and disposal facilities; and provisions for enforcement of the standards with a program of legal penalties for noncompliance. The RCRA also defines four characteristic properties for waste materials: ignitability, corrosivity, reactivity, and toxicity.

The Toxic Substances Control Act (TSCA) of 1976 requires that all chemicals produced in or imported into the United States be tested, regulated, and screened for toxic effects prior to commercial manufacture. This law bans the manufacture of polychlorinated biphenyls (PCBs) and regulates asbestos. The EPA works with other federal agencies under this law to fill in the gaps of the other acts that attempt to manage hazardous materials. Additional laws under which the EPA acts include the Clean Air Act (CAA) and amendments; the Clean Water Act (CWA) and amendments; the Comprehensive Environmental Response, Compensation, and Liability Act (CERCLA) of 1980, and the Superfund Amendments and Reauthorization Act (SARA) of 1986.

Marcie L. Wingfield and Massimo D. Bezoari

SUGGESTED READINGS: *Hazardous Waste in America* (1982), by Samuel S. Epstein, Lester O. Brown, and Carl Pope, provides a comprehensive look at every aspect of hazardous waste and how it affects people's lives. *Principles of Hazardous Materials Management* (1988), by Roger D. Griffin, provides adequate information for those seeking general knowledge of hazardous materials management. For an easy-to-read, alternative approach to waste regulation, consult *Using Economic Incentives to Regulate Toxic Substances* (1992), by Molly K. Macauley, Michael D. Bowes, and Karen L. Palmer. An interesting comparison between the U.S. and Canadian approaches to toxic substance regulation is contained in *Risk, Science, and Politics: Regulating Toxic Substances in Canada and the United States* (1994), by Kathryn Harrison and George Hoberg. *Hazardous Waste Regulations* (1981), by Alex Maslow, is an interpretive guide for examining difficult regulation material. For an overview of environmental laws, see James A. Stimson, *Guide to Environmental Laws: From Premanufacture to Disposal* (1993).

SEE ALSO: Hazardous waste; Nuclear and radioactive waste; Right-to-know legislation; Solid waste management policy; Superfund.

Hazardous waste

CATEGORY: Waste and waste management

Hazardous wastes are products of industrial society that pose dangers to human health and the environment. In the United States they are legally defined as materials that have ignitable, corrosive, reactive, or toxic properties.

In the early 1990's approximately 97 percent of all hazardous waste in the United States was produced by 2 percent of the waste generators. Remediation and cleanup of these wastes involve substantial economic cost. Since the 1970's the United States and other Western democracies have tried to regulate hazardous waste disposal. Hazardous wastes are also a serious problem in the former Soviet Union and other Eastern European nations.

Improper disposal of hazardous waste can lead to the release of chemicals into the air, surface water, groundwater, and soil. High-risk wastes are those known to contain significant concentrations of constituents that are highly toxic, persistent, highly mobile, or bioaccumulative. Examples include dioxin-based wastes, polychlorinated biphenyls (PCBs), and cyanide wastes. Intermediate-risk wastes may include metal hydroxide sludges, while low-risk wastes are generally high-volume, low-hazard materials. Radioactive waste is a special category of hazardous waste, often presenting extremely high risks, as do biomedical and mining wastes.

Hazardous waste presents varying degrees of health and environmental hazards. When combined, two relatively low-risk materials may pose a high risk. Factors that affect the health risk of hazardous waste include dosage received; age, gender, and body weight of those exposed; and weather conditions. The health effects posed by hazardous waste include cancer, genetic defects, reproductive abnormalities, and central nervous system disorders.

Environmental degradation resulting from hazardous waste can render various natural resources, such as cropland or forests, useless and can harm animal life. For example, chemicals can leach out of improperly stored waste and into groundwater. Hazardous wastes may gener-

Disposal of Hazardous Waste Products

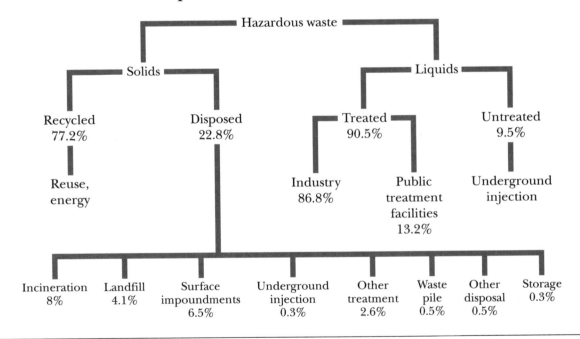

A sign at the Brookhaven National Laboratory in Upton, New York, warns people about the presence of hazardous waste. (AP/Wide World Photos)

ate long-lasting air and water pollution or soil contamination. Because the amount of waste produced in any period is based on the amount of natural resources used, the generation of both hazardous and nonhazardous waste poses a threat to the sustainability of the economy.

Because there were no standards for what constituted a hazardous waste in the past, such materials were often buried or stored in unattended drums or other containers. This situation created a threat to the environment and human health when the original containers began to leak or the material leached into the water supply.

The technology for dealing with hazardous solid and liquid waste continues to evolve. Several solutions have had positive impacts on the environment and consumption of natural resources. One solution is to reduce the volume of the waste material by generating less of it. The second approach is to recycle hazardous material as much as possible. A third means of deal-

ing with hazardous waste is to treat it to render it less harmful and often reduce its volume. The least-preferred solution is to store the waste in landfills. The Environmental Protection Agency (EPA) has established standards for responsibility and tracking of hazardous wastes, based on the principle that waste generators are responsible for their waste "from cradle to grave." This principle has involved extensive record-keeping by waste generators and disposal sites.

The U.S. Congress's 1984 Resource Conservation and Recovery Act (RCRA) revisions involved a thorough overhaul of hazardous waste legislation. Previously exempt sources that generated between 100 and 1,000 kilograms of hazardous waste per month were brought under the RCRA provisions. Congress further tried to force the EPA to adopt a bias against the landfilling of hazardous waste with a "no land disposal unless proven safe" provision. Congress also added underground storage tanks for gasoline, petroleum, pesticides, and solvents to the list of sources to be regulated and remediated. In addition to the RCRA, Superfund (from the Comprehensive Environmental Response, Compensation, and Liability Act of 1980) provides for the cleanup of all categories of abandoned hazardous waste sites except for radioactive waste. Several other statutes (and ensuing EPA regulations) deal with these aspects of the hazardous waste problem.

The costs for the cleanup and remediation of hazardous waste are substantial and are likely to continue to grow. This situation is particularly true in Eastern Europe and the former Soviet Union, where the magnitude of past dumping of hazardous materials is slowly becoming apparent. Meanwhile, less industrialized nations are ignoring the hazardous waste issue and are therefore setting themselves up for future difficulties.

The waste minimization philosophy expressed in the RCRA is a sound long-range strategy for dealing with hazardous waste. However, some materials will continue to be deposited in landfills. Incineration offers some solutions to the problem of volume of material but poses issues of air quality and highly toxic ash. As some firms have found, minimizing their waste stream affords them economic benefits while conserving natural resources. Household waste, which is not regulated by the RCRA, often includes small quantities of hazardous materials such as pesticides. Most of this waste was still being landfilled in the late 1990's. The cleanup of existing sites will continue to be a troubling problem, while the cleanup and disposal of radioactive waste will be a major issue for the future.

John M. Theilmann

SUGGESTED READINGS: Useful discussions of various aspects of hazardous waste are included in Michael D. LaGrega, Philip L. Buckingham, and Jeffrey C. Evans, *Hazardous Waste Management* (1994) and Aaron Wildavsky, *But Is It True?* (1995). Other helpful works are Joe Grisham, *Health Aspects of the Disposal of Waste Chemicals* (1986) and Michael B. Gerrard, *Whose Backyard, Whose Risk* (1994).

SEE ALSO: Hazardous and toxic substance regulation; Landfills; Nuclear and radioactive waste; Superfund.

Heat islands

CATEGORY: Weather and climate

Heat islands are warm-air domes found over nearly all cities caused by efficient absorption of solar radiation and the release of artificial heat. Heat islands increase temperatures, concentrate pollution, and may affect precipitation in urban environments.

Five factors are responsible for heat island formation: urban fabric (the rocklike material of a city's buildings and streets, which conducts heat three times faster than nonurban sandy soil); city structure (a complex web of multiple reflections and energy exchanges); artificial heat production, which occurs mainly in winter when artificial heating is highest; urban water balance, marked by rapid drainage and reduced humidity; and urban air pollution, which retains heat within the dome.

The impacts of heat islands on urban environments are many. The retention of heat leads to temperatures that are several degrees higher in town compared to the surrounding nonurban outskirts. Temperature differences tend to be greatest in winter and at night, but heat islands also exist during summer. Higher temperatures lead to longer freeze-free seasons in cities, with first freezes in the fall occurring about one month later and last freezes in the spring occurring about one month earlier on average than the surrounding rural areas.

Pollution concentration in urban areas is greater because of the local circulation produced within a city and the polluting industries situated there. The local circulation of a heat island results from warm air rising near the central business district (CBD), spreading out at the top of the dome, descending at the outer edges, and flowing back into the city center. Pollution tends to attenuate incoming short-wave radiation and reduce visibility. Pollution levels tend to be highest in winter when the sun angle is low and on workdays. Heat islands are associated with pollution disasters such as those that occurred in London, England, in 1952; Donora, Pennsylvania, in 1948; and the Meuse Valley of

Belgium in 1930. Each event occurred during the cold season.

The heat island's impact on precipitation remains in question. Some investigations indicate greater precipitation amounts over cities largely because of air rising in the city center and the greater abundance of condensation nuclei over the city. Others argue that the number of condensation nuclei is so vast that cloud drops are too tiny to grow to raindrop size, thus reducing the amount of rain. However, a consensus that the number of rainy days in cities is slightly more than in the country is accepted. Most urban climatologists have reported higher frequencies of hail and thunderstorms in cities and that fog is more prevalent. The heat island effect generally produces less snow and more rain during a storm. Humidity, both relative and absolute, tends to be slightly lower in the urban environment.

Ralph D. Cross

SEE ALSO: Air pollution; Black Wednesday; Donora, Pennsylvania, temperature inversion; London smog disaster; Smog.

Heavy metals and heavy metal poisoning

CATEGORY: Pollutants and toxins

Dense metallic chemical elements that are released into the environment by mining, industry, and disposal of goods can poison living organisms by interfering with the body's metabolic functions, often by substituting for benign metallic elements that have similar atomic sizes and ionic charges. Such poisoning can cause impaired nervous system functioning, birth defects, and death.

The chemical elements lead (Pb), mercury (Hg), cadmium (Cd), and thallium (Tl) are located together at the central bottom portion of the periodic table. Although mercury, thallium, and cadmium are fairly rare, lead is an abundant element on earth. All are dense, soft metals (mercury is a liquid at room temperature) that

have a high affinity for chemically bonding with the element sulfur (S). They occur on or near the earth's surface as sulfur-containing minerals (sulfides) that are insoluble in water. This lack of water solubility kept the heavy metals isolated from life forms as they were evolving prior to the onset of the technological era, at which point humans began to mine and purify these useful elements. Once the heavy metals and their ions became more abundant in the environment, mine workers, industrial workers, and even the general public were at risk for toxic exposure to elements their bodies were not well equipped to process.

The ions of heavy metals resemble beneficial lighter metals such as zinc, calcium, magnesium, and iron in terms of their diameters and charges. This enables heavy metals to substitute for the beneficial elements and thus reside in the body over time. These lighter metals are much more abundant in the environment, and so life evolved with them, even employing them in critical roles.

The strong ability of heavy metals to bond to sulfur is significant because body biochemicals known as metalloproteins are composed of lighter metals such as zinc bonded to sulfur. (Sulfur is a normal constituent element of proteins.) In cases of heavy metal intoxication, defective metalloproteins are biosynthesized, employing an incorrect toxic metal that is strongly bound to sulfur. Although the bonding ability and even the size and charge of the heavy metal are appropriate for fitting into the structure of the biochemical, it fails to function as required, leaving the organism deficient in some vital way.

The degree of poisoning depends upon the level and duration of exposure to the toxic element. Organisms have limited detoxification defenses, including proteins called metallothioneins, which scavenge metal atoms by virtue of having a large number of sulfur atoms in their structures. Medical intervention is possible in certain cases; for example, chelating agents are drugs designed to scavenge metal atoms and make them easier to eliminate from the body in the urine.

Lead has been mined since antiquity and occurs in one form as the common mineral galena (PbS). The use of lead in paints and as an antiknock agent in gasoline has caused it to be spread throughout the environment, although its use in these capacities has been largely phased out in the United States. It was also used extensively in plumbing, and older structures still have drinking water conveyed via lead pipes. Lead is a soft, easily worked metal that was used for food containers in Roman times. Acidic wine and other food is thought to have leached lead from the vessels to cause lead poisoning. Historians believe the resulting neurological damage (Saturnism) contributed to the decline of the Roman Empire. Lead is currently used on a vast scale in lead batteries. Lead in soil is absorbed by crops and works its way up the food chain. Lead dust from crumbling paint has been implicated in learning impairment and violent behavior in children. The nature of this neurotoxicity is not well understood. Other effects of chronic lead exposure include kidney damage and anemia. Anemia results from lead's interference with the production of hemoglobin, the iron-containing component of blood. Lead is able to displace calcium in bone, where it can remain in the body for years.

Cadmium, a fairly rare metal, occurs in zinc-containing ores and is found in high concentration near zinc smelters. It is used industrially in batteries and metal coatings. Airborne cadmium gets into soil, where, like lead, it gets into crops, including tobacco. Smokers consume about twice the cadmium each day that nonsmokers receive from the environment. Results of cadmium poisoning include lung and kidney damage, and painful damage to the joints known in Japan (where cadmium pollution has been severe) as *itai-itai* (ouch-ouch) disease. A diet high in calcium helps limit cadmium poisoning.

Mercury is used in industry in many ways. In the past, improperly disposed mercury caused poisoning, for example, at Minamata Bay in Japan in the 1950's. Mercury use is being phased out where other compounds can safely be used, such as in pesticides. Many mercury compounds are volatile and easily absorbed by the body. Several can cross the blood-brain barrier; therefore, key symptoms of mercury poisoning involve vision and hearing disturbance, loss of coordina-

tion, and tremors. Other results involve miscarriage and birth defects. Leaching of mercury from dental amalgam fillings is believed not to be significant.

The effects of thallium poisoning are similar to the other heavy metals. It can behave chemically similar in some ways to the vital element potassium. Unlike potassium, however, once it has been absorbed from the environment, it binds to sulfur, from which it will be slowly released, effectively exposing the individual over long periods. It was once used as a rat poison, but its toxic nature has precluded such domestic use. It is a by-product of zinc and lead production.

Wendy Halpin Hallows

SUGGESTED READINGS: For an overview of the biological activity of chemical elements, see *Bioinorganic Chemistry: Inorganic Elements in the Chemistry of Life* (1991), by Wolfgang Kaim and Brigitte Schwederski, and *The Biological Chemistry of the Elements: The Inorganic Chemistry of Life* (1991), by J. J. R. Frausto da Silva and R. J. P. Williams. For an account of the distribution and uses of the elements, see *The Elements on Earth: Inorganic Chemistry in the Environment* (1995), by P. A. Cox.

SEE ALSO: Lead poisoning; Mercury and mercury poisoning; Oak Ridge, Tennessee, mercury releases; Minimata Bay mercury poisoning.

Hetch Hetchy Dam

DATE: completed 1925
CATEGORY: Preservation and wilderness issues

The controversy over the building of the Hetch Hetchy Dam in Yosemite National Park in the early twentieth century was the first environmental issue to be argued on the national stage in the United States. Although environmentalists were not successful in stopping the construction, they developed strategies and gathered support that became useful in later battles.

In 1890, eighteen years after Congress named Yellowstone the first national park in the United

The Hetch Hetchy Dam was built during the 1920's to supply water to San Francisco, 150 miles away. The dam's reservoir submerged a valley whose beauty rivaled that of nearby Yosemite Valley. (Douglas Long)

States, Yosemite was named the second. Occupying some 3,100 square kilometers (1,200 square miles) in the Sierra Nevada in eastern California, the new park featured giant redwoods and sequoias, and two great scenic mountain valleys less than twenty miles apart: Yosemite Valley (which technically remained under state control for several more years) and Hetch Hetchy Valley. Both valleys offered breathtaking wilderness: flowering meadows surrounded by sheer cliffs of colorful granite punctuated by dramatic waterfalls. Writers such as John Muir, John Burroughs, and Mary Austin tramped through both valleys and the surrounding glacier-scoured mountains, bringing back descriptions of awe-inspiring beauty. Many thought that of the two, the oddly named Hetch Hetchy was the more beautiful. The 5.6 kilometers (3.5 miles) of the flat valley floor were traversed by a clear, clean river, and its granite walls were straight and steep. Because Yosemite's state control was less stringent

than Hetch Hetchy's federal control, concessions and tourist businesses sprang up around Yosemite Valley, making it the more popular attraction.

San Francisco, 150 miles to the west, was one of the fastest-growing cities in the United States. As its population increased, so did its need for fresh water and electricity. At the time there was no public water supply, and customers were at the mercy of private companies who were not always responsive and responsible. By the beginning of the twentieth century, the needs were desperate. In 1901, under pressure from California legislators, the U.S. Congress passed the Right of Way Act, giving local governments the right to use national park lands for water projects such as dams and reservoirs if they were in the public interest. San Francisco officials wasted no time in declaring their intention to build a dam at the narrow end of Hetch Hetchy Valley, flood the valley, and create a reservoir to supply city water. The very things that contributed to the valley's natural beauty—the flat valley floor, the steep cliffs, and the purity of the river—also made it an ideal spot for a reservoir.

When the city first applied for a right of way in 1903, the request was denied by the secretary of the interior, who felt a dam was not in the public interest. After San Francisco's devastating earthquake and fire of 1905, the city's efforts intensified. Gifford Pinchot of the U.S. Forest Service was sympathetic to San Francisco's plan, and in 1906 he urged the city to reapply for a right of way. The new secretary of the interior, it appeared, was inclined to grant permission this time.

Immediately Muir and the Sierra Club sprang into action to prevent the project. Muir wrote a personal letter to his old hiking companion, President Theodore Roosevelt, asking that other locations be developed instead. Roosevelt agreed that other rivers and dams should be exploited first but did not completely rule out the eventual flooding of Hetch Hetchy. Sierra Club members and others wrote letters to the editors of major newspapers on both coasts and garnered enough support for preserving Hetch Hetchy to stall the project in Congress.

For the next six years the issue was debated in public, the Congress, newspapers and newsletters, and public addresses. On the one hand, San Francisco's need for water was real; however, some thought that the need should be met without destroying irreplaceable wilderness. Some argued that a reservoir in the valley would actually be more beautiful and attract more tourists than the wild valley. For others, the issue was a matter of private versus public control of the city's water supply. Still others expressed the debate in terms of Pinchot's preservationism and the Sierra Club's conservationism.

Several congressional hearings on the right of way were held between 1908 and 1913. A brochure written by Muir in 1911 titled *Let Everyone Help to Save the Famous Hetch-Hetchy Valley and Stop the Commercial Destruction Which Threatens Our National Parks* included tips on lobbying Congress—a strategy that would become common for environmentalists over the rest of the twentieth century. Although the conservationists at-

tracted an impressive amount of support across the nation, they were ultimately defeated. In 1913 Congress passed the Raker Act, giving permission for the construction of a dam at Hetch Hetchy Valley. The dam was finished in 1925, but the flooded valley never did attract many tourists, even in the late 1990's when Yosemite Valley was so crowded that access by car was restricted. The water from the reservoir did help meet the needs of San Francisco, but only after control over its distribution was turned over to a private utility.

Cynthia A. Bily

SUGGESTED READINGS: The most thorough analysis of the Hetch Hetchy incident for general readers is *John Muir and the Sierra Club: The Battle for Yosemite* (1965), by Holway R. Jones. A shorter version of the story may be found in Frank Graham, Jr., *Man's Dominion: The Story of Conservation in America* (1971), which clearly shows how the incident was influenced by contemporary politics and how it influenced later conservation efforts. John Muir's *Yosemite* (1912) contains beautiful descriptive passages about Hetch Hetchy before the dam was built and an appeal for the area's preservation that was as impassioned as it was futile. Although it may be difficult to find, Ray W. Taylor's *Hetch Hetchy: The Story of San Francisco's Struggle to Provide a Water Supply for Her Future Needs* (1926) is a rare proconstruction account of the controversy written by an engineer who worked on the project.

SEE ALSO: Muir, John; Pinchot, Gifford; Roosevelt, Theodore; Sierra Club; Yosemite.

High-yield wheat

CATEGORY: Agriculture and food

High-yield wheat has helped increase food production and change the structure of agriculture worldwide.

Wheat (*Triticum sativum*) is the world's most important grain crop because it serves as a major food source for most of the world's population.

Large portions of agricultural land are devoted to the production of wheat worldwide. Wheat constitutes a large part of the domestic economy of the United States, contributes a large part to the nation's exports, and serves as the national bread crop. Wheat is the national food staple for forty-three countries and about 35 percent of the people of the world, and it provides 20 percent of the total food calories for world's population.

No one knows for certain when wheat was first cultivated, but by six thousand years ago, humans had discovered that seeds from wheat plants could be collected, planted in land that could be controlled, and later gathered for food. As human populations continued to grow, it was necessary to select and produce higher yielding wheat. The Green Revolution of the twentieth century helped to make this possible. Agricultural scientists developed new, higher yielding varieties of numerous crops, particularly the seed grains such as wheat that supply most of the calories necessary for maintenance of the world's population.

Wheat, like other major crops, originated from a low-yield native plant, but it has been converted into one of the highest yielding crops in the world. There are two major ways to improve yield in seed grains such as wheat. One way is to produce more seed per seed head, and the second way is to produce larger seed heads. Both of these approaches have been utilized to produce high-yield wheat.

Numerous agricultural practices are required to produce higher yields, but one of the most important is the selection and breeding of genetically superior cultivars. When a grower observes a plant with a potentially desirable gene mutation that produces a change that improves a yield characteristic, the grower collects its seed and grows additional plants, which produces higher yields. This selection process remains one of the major means of improving yield in agricultural crops. Advances in the understanding of genetics have made it possible to breed some of the desirable characteristics that have resulted from mutation into plants that lacked the characteristic. In addition, the advent of recombinant deoxyribonucleic acid (DNA) technology makes it possible to transfer genetic characteristics from one plant to any other plant.

While tremendous increases in the world's food supply have resulted from high-yield crops such as wheat, such crops have also had an impact on the environment. The new crop varieties have led to an increased reliance on monoculture, the practice of growing only one crop over a vast number of acres. The production of high-yield wheat in modern agricultural units is highly mechanized and thereby uses large amounts of energy, devotes large amounts of land to the production of only one crop, and is highly reliant on agricultural chemicals such as fertilizers and pesticides.

D. R. Gossett

SEE ALSO: Agricultural revolution; Biotechnology and genetic engineering; Borlaug, Norman; Green revolution; Pesticides and fertilizers; Sustainable agriculture.

Homestead Act

DATE: Passed 1862
CATEGORY: Land and land use

> *The Homestead Act granted large sections of the American West to settlers and farmers. It opened much U.S. territory to development but also led to rampant speculation and to land and resource abuse.*

The Homestead Act of 1862 granted 160 acres of land to potential farmers free of charge. To obtain full ownership, settlers were required to farm the land for five years. By 1867, at least 2.5 million acres had been allotted under the act.

When Congress created the Homestead Act, it failed to classify the available land in terms of farming, mining, cattle raising, or timber regions. All plots were open to interested settlers, regardless of their intentions, and the government employed too few people to monitor the uses of the land. The Homestead Act, in theory, was designed to deal only with farmland, which left the other lands unprotected. The General

Reform Act of 1891 and the Desert Land Act of 1877 repaired the omissions of the Homestead Act and accounted for the diversity of the western United States.

The Homestead Act's commutation clause, which stated that after six months the land could be purchased for $1.25 per acre, helped lead to speculation in natural resources. Between 1881 and 1904, settlers gathered 23 percent of the Homestead lands by commutation; companies and speculators, in particular, used the clause to buy large plots. In Aberdeen in the Dakota Territory, for example, speculators claimed 75 percent of the land. With thousands of settlers moving west, the demand for firewood and building materials greatly increased. Because of speculation to meet the settlers' needs, forests in Minnesota, Wisconsin, and Michigan were almost destroyed. Congress hired special agents to investigate timber misuse and illegal land claims, but these officials were responsible for large regions and had little effect.

By the 1880's, timber misuse had become a problem. Congress was forced to enact the 1873 Timber Culture Act, granting an additional 160 acres to interested settlers. These farmers were required to plant at least forty acres of timber for ten years, but 90 percent of the land was used to produce timber for sale, not to replace the misused forest regions. In Kansas, Nebraska, and the Dakota Territory, however, the Timber Culture Act replaced a small number of trees. The Timber Culture Act not only responded to the loss in forest lands because of the Homestead Act but also was designed to reduce winds, increase rainfall, and attract new settlers.

Although speculators used the Homestead Act to their advantage, its opening of the frontier also prompted settlers to make use of lands that were previously vacant. During the Civil War, for example, Homestead farmers helped to produce enough food for 1.5 million soldiers. Allied with improvements in farming methods such as irrigation, the settlement of areas opened by the Homestead Act made the West a new source of food production for the nation.

Keith E. Rolfe

SEE ALSO: Dust Bowl; Forest and range policy; Land use policy; Range management.

Hunting

CATEGORY: Animals and endangered species

Hunting originated as a means of subsistence, but it seldom has that status today. Hunting persists as a sport and a tool for wildlife or environmental regulation. The assessment of hunting from the standpoint of values is politically and philosophically controversial.

When hunting was necessary for survival, there were no real environmental issues, merely expediencies. Native Americans burned land to make it more attractive to buffalo and deer, but this was done because such conditions favored human subsistence. When European colonizers traveled to North America between the sixteenth and eighteenth centuries, American Indians were able to collaborate in the fur trade and rise above a subsistence economy, but at a price that they could not foresee. Commercially desirable species were soon trapped out of settled regions and on the frontiers, forcing Native Americans who relied on such species to move to new areas.

British colonies began to regulate hunting by enforcing laws that limited the take of desirable species of mammals, birds, and fish and also by offering bounties on predators that threatened livestock. Laws seldom prevented, and sometimes were not intended to prevent, market hunters from using guns, nets, and traps to kill wildlife and then sell their take in public markets. Initially, settlers had little time to contemplate hunting as a sport, even if they enjoyed doing it, because it was an important supplement to farming.

In Europe, from which the settlers came, hunting was often the exclusive right of royalty and aristocracy, who hunted exclusively for sport. Such a formal class system did not take hold in the Americas, and class restriction of hunting never developed. However, property owners often felt that they had special rights to the wildlife on their lands and waters. Therefore, colonies, and later states, had to legislate the hunting, fishing, and trapping rights of property owners and others. Although some species were depleted in the East, they seemed abundant both in the West and in northern Canada, and therefore most people did not worry about depletions.

Eventually, naturalists, such as Pehr Kalm in the 1740's and Alexander Wilson in the early nineteenth century, realized that Americans were too complacent about their wildlife heritage and began writing against reckless killing of animals. By the mid-nineteenth century nature writers were awakening concerns for wildlife, and after the Civil War, when sports hunters rode the transcontinental railroad west to kill buffalo, the carnage was reported in newspapers and magazines. However, there were also commercial hunters who used the railroads to take buffalo hides and tongues to eastern markets, and settlers on the plains believed that farms and ranches were incompatible with freely roaming buffalo and American Indians. Thus, al-

though several bills protecting buffalo were introduced into Congress during the 1870's, none became law. By the 1880's the buffalo was nearly extinct. Congress reacted by establishing the National Zoological Park in 1889 to help breed endangered species.

By 1900 the ethics of hunting and preservation of game species were serious public concerns. However, the ways these matters have been understood and evaluated have changed over time. Initially, abundance was thought to depend upon passing and enforcing hunting restrictions. This approach was later supplemented by the establishment of federal wildlife refuges and national parks, which preserved habitats.

The ethical issue that dominated the late nineteenth and early twentieth centuries was whether market hunting of mammals and birds should continue. The White House Conference

A hunter in Pennsylvania loads a dead deer into the back of his pickup truck. Hunting became an important tool for the management of wildlife after natural predators were hunted into near extinction by humans. (AP/Wide World Photos)

on Conservation (1907) emphasized the need for state wildlife or conservation departments to study and manage hunting and fishing in each state. In that new climate, market hunting was outlawed, except for species trapped for their fur. Yet nonhunting nature lovers asked whether sport hunting was any more defensible on moral grounds than market hunting. Nevertheless, opposition to sport hunting was muted by the fact that sport hunters were active in habitat preservation and supporting enforcement of wildlife laws. Poachers were enemies of both groups. Aldo Leopold, an avid hunter and pioneer wildlife manager, also helped bridge the gap between these groups by developing a land ethic that became influential after his death in 1948. Wildlife managers also discovered that in environments transformed by civilization, hunting can become one of the tools for wildlife management.

After World War II the numbers of nonhunting nature hobbyists greatly increased, whereas the numbers of hunters declined as a percentage of the total population. Hunters constitute only about 10 percent of the population, but nature hobbyists are several times as many. Along with these changes, there has been an increased interest in environmental ethics. At the practical level, deer hunters argue that they are needed to control deer populations now that wolves and American Indians no longer do so, because deer will degrade their own environment if their numbers are not limited to the environment's carrying capacity. However, some environmental ethicists, at the philosophical level, respond that no one imagines that hunters go to all the trouble they do in order to do the deer a favor. Many environmentalists have suggested that deer populations could be better controlled if their natural predators, which were removed by hunters in the first place, were reintroduced into their native habitats. Such programs were implemented in Yellowstone National Park and the southwestern United States during the 1990's.

Frank N. Egerton

SUGGESTED READINGS: The ethics of hunting at different times in history is surveyed in Matt Cartmill, *A View to a Death in the Morning* (1993). Aldo Leopold's contributions to both wildlife management and environmental ethics are discussed in Curt Meine, *Aldo Leopold: His Life and Work* (1988). The contribution of hunters to wildlife conservation is defended by James B. Trefethen in *An American Crusade for Wildlife* (1975).

SEE ALSO: Animal rights movements; Leopold, Aldo; Predator management; Wolf reintroduction.

Hydroelectricity

CATEGORY: Energy

Hydroelectricity is electricity produced by the energy of falling water. In comparison with other sources of electricity—such as coal, oil, gas, and nuclear power—hydroelectric power generation has been regarded as a safe, clean, reliable, and inexpensive technology.

Hydroelectrical operations are characterized by the abilities to quickly start power generation and rapidly adjust power output. Since the essential natural source of the hydroelectricity is water, hydroelectric power plants are built near rivers or waterfalls. In order to store enough water, water flows are either channeled into reservoirs or intercepted by dams to form reservoirs. Once a river is altered or interrupted, however, its natural flow seasons and wildlife cycles will be altered.

In the United States, water resources are grouped into twenty regions, such as the Great Lakes Region, the Upper Mississippi Region, the Lower Mississippi Region, and the Missouri Region. The hydroelectric power generation industry is well regulated, and the plant operation licenses are issued by the Federal Energy Regulatory Commission.

Among existing hydroelectric projects, the Niagara Power project has been regarded as one of the most successful. Water from the Niagara River and Niagara Falls is channeled into a reservoir, with the maximum flow rate of 375,000 gal-

Hydroelectricity is clean and reliable, but harnessing water power usually involves altering watercourses in environmentally harmful ways. (AP/Wide World Photos)

lons per second. However, the actual flow rate is carefully adjusted so that the natural beauty of the Niagara Falls can be preserved. The water storage volume is related to the area occupied by the reservoir and the height of the dam. When water enters the power plant intake, it drops thirty stories through a pipeline. The potential energy of the water is then transformed into the kinetic energy. Rapid water flow turns a turbine engine, which spins an electrical generator. Leaving the power plant, the water flows downstream and can be utilized again to generate electricity.

With several power generators in the Niagara Power plant, the maximum electrical power generation capacity is 2.4 gigawatts. The Niagara water resource has been a valuable treasure to the economy in New York state. There is another power plant along the Niagara River in Canada. In the United States and Canada, hydroelectric-

ity is the most commonly utilized renewable energy source in comparison to other renewable sources such as solar energy, wind energy, and biomass. Between the two countries, the combined hydroelectricity consumption is approximately one-quarter of the total world consumption. Even though hydroelectricity is a mature technology, the balance between electricity generation and environmental preservation is rather delicate.

Many hydroelectric reservoirs can also be utilized for water management purposes such as flood control and irrigation. On the other hand, reservoirs are artificial lakes that change the ecosystems in their surrounding areas. If a dam is built on a river, the natural flow will be interrupted. Fish can no longer swim back and forth along the original waterway. Ducks and other animals are often forced to move to other habitats.

Each reservoir is characterized by its maximum water level, which is always lower than that of the dam. Houses below or near the waterline will be considered unhabitable. Sometimes thousands of people must be relocated before a reservoir can be formed. Such relocation or migration is a difficult task. Furthermore, trees and plants below the waterline will be destroyed, while soils and rocks will be eroded. Such erosion contributes to sand sedimentation in the reservoir. Another source of sedimentation is the sand brought into the reservoir from upstream. Sedimentation reduces the effective storage volume of the reservoir and may damage the turbine engines if it is not filtered. Besides sediment problems, water erosion may also cause landslides. Since water has weight, it will also apply pressure to the basin of a reservoir. Such pressure will affect the geological formation of the basin. Many scientists believe that reservoirs built near faultlines apply pressure that may induce earthquakes.

Among the hydroelectric projects underway around the world, the largest one is the Three Gorges Dam along the Yangtze River in China. The massive dam is about 185 meters (610 feet) high and 2 kilometers (1.2 miles) long. The maximum electrical power generation capacity will be approximately 18 gigawatts, and the maximum water storage capacity will be approximately 39 billion cubic meters. The cost of the project is estimated to be US$20 billion, with a targeted completion date of 2010. Since the reservoir site is located in one of the most populated provinces in China, it is estimated that more than one million people will be relocated.

Xingwu Wang

SUGGESTED READINGS: A summary of hydroelectricity in the United States is provided in *Hydroelectric Power Resources of the United States: Developed and Undeveloped* (1992), published by the Federal Energy Regulatory Commission. A good discussion of environmental issues related to dams is Patrick McCully's *Silenced Rivers: The Ecology and Politics of Large Dams* (1996). An earlier discussion may be found in John D. Echevrria et al., *Rivers at Risk: The Concerned Citizen's Guide to Hydropower* (1989).

SEE ALSO: Alternative energy sources; Dams and reservoirs; Flood control; Irrigation; Power plants; Tidal energy.

I

Indoor air quality

CATEGORY: Atmosphere and air pollution

Indoor air has been found to contain carbon monoxide, smoke, and other combustion products, as well as microorganisms, cooking and body odors, solvent vapors, and other pollutants. These pollutants are believed to cause thousands of deaths each year—mainly from lung cancer caused by radon and carbon monoxide poisoning—as well as a considerable amount of illness and discomfort.

Since most people spend more time indoors than outdoors, air quality in homes, offices, stores, and other buildings has a greater effect on human health than the quality of outdoor air. The term "sick building syndrome" is used when a majority of a building's occupants experience health and comfort problems caused by a variety of pollutants that are difficult to identify.

Outdoor air is one source of indoor air pollution because the ventilation of a building brings in air from the outside. Fortunately, some pollutants are trapped as they enter a building; particulate matter, for example, may impinge on walls and stick to them. Far more important are activities that occur inside the building. The major contributor to indoor air pollution is indoor combustion by appliances and smokers. An unvented or improperly vented furnace or water heater may allow combustion products such as carbon monoxide to enter the living space; this can also occur because of the cracking of the heat exchanger in a furnace or some similar malfunction. Carbon monoxide, which can also enter a building from a garage in which a motor vehicle engine is running, causes hundreds of accidental deaths annually in the United States and produces a large amount of often-unrecognized illness. Tobacco smoke is also known to be harmful not only to smokers but also to nonsmokers exposed to a significant amount of secondhand smoke; tobacco smoke is also a major source of annoyance and discomfort to most people.

Carbon dioxide, normally regarded as nontoxic, causes nausea and headaches at elevated levels and should not exceed 5,000 parts per million (ppm). Even at 1,000 ppm, however, a room will seem stuffy. Unvented gas or kerosene space heaters should never be used indoors because they necessarily lead to high carbon dioxide levels and often produce high levels of carbon monoxide, nitrogen oxide, and sulfur oxide (the latter is a particularly severe problems with kerosene heaters since kerosene contains sulfur).

Building materials can also contribute to indoor air pollution. Asbestos, a known cause of lung cancer, may be present in the air if insulation or other materials containing asbestos have broken down and can be especially dangerous during renovation or removal. Formaldehyde, a major component of urea-formaldehyde foam insulation, particle board, and some packaging materials, can produce acute eye, nose, and throat irritation at levels below 1 ppm. However, formaldehyde is mainly a concern during the first few years after building construction or renovation, after which it eventually disappears.

A variety of other volatile organic compounds may contribute to poor indoor air quality. These include acetone and other ketones, alcohols, aromatic hydrocarbons (such as benzene and toluene), and halogenated hydrocarbons (such as methylene chloride) found in adhesives, household cleaners, enamels, glues, paints, solvents, and varnishes. Indoor hobbies or renovations involving such compounds should only take place in a well-ventilated space.

Microorganisms such as bacteria, fungi, molds, and viruses can be a danger in buildings that are not kept clean. They appear to be most troublesome in buildings with low relative humidity, but some thrive in moist areas. The organism responsible for Legionnaires' disease has been thought to grow in cooling and ventilation systems.

The ventilation rate of a building plays an important role in indoor air quality. The concentration of air pollutants can rise at a rapid rate in a poorly ventilated building because the pollutants generated inside the building are not being removed quickly enough. This problem has been aggravated since the energy crisis of the 1970's by the practice of making buildings airtight in order to reduce energy costs associated with the heating or cooling of outdoor air that has entered a building. It is possible to ventilate a building well and yet have low energy costs if an air-to-air heat exchanger is used; this device allows the incoming and outgoing air streams to pass near each other across a thin conducting barrier, so that a large fraction of the heat from the outgoing air in winter is transferred to the incoming air. On the other hand, a building in which there is little indoor generation of air pollutants can be airtight and still have superior air quality; increasing the ventilation rate in such a building may actually decrease indoor air quality by bringing in outdoor pollution.

Radon levels are particularly and subtly dependent on the way a building is ventilated. Radon is normally present in underground air in concentrations sufficient to cause concern if even a small percentage of the air in the building comes from underground; the ventilation of the lower level of a building may actually increase this percentage.

Laurent Hodges

SUGGESTED READINGS: Donald W. Moffat's *Handbook of Indoor Air Quality Management* (1997) is an excellent resource with particular emphasis on what should be done in commercial buildings to ensure good air quality. *Indoor Air Quality and Control* (1993), by Anthony L. Hines, Tushar K. Ghosh, Sudarshan K. Loyalka, and Richard C. Warder, Jr., is especially strong on methods of measuring pollutant concentrations and strategies for controlling specific pollutants that are found to be troublesome. The *Indoor Air Quality Design Guidebook* (1990), edited by Milton Meckler, is a good source of information about sources,

concentrations, and health effects of indoor air pollutants and methods to avoid or control them.

SEE ALSO: Air pollution; Particulate matter; Radon; Secondhand smoke; Sick building syndrome.

Integrated pest management

CATEGORY: Agriculture and food

Integrated pest management (IPM) is the practice of integrating insect, animal, or plant management tactics, such as chemical control, cultural control, biological control, and plant resistance, to maintain pest populations below damaging levels in the most economical and environmentally compatible manner.

In the past, pest management strategies in agriculture primarily focused on eliminating all of a particular pest organism from a given field or area. These strategies depended on the use of chemical pesticides to kill all of the pest organisms. Prior to the twentieth century, farmers used naturally occurring compounds such as kerosine or pyrethrum for this purpose. During the latter half of the twentieth century, synthetic pesticides began playing a prominent role in controlling crop pests. After 1939 the use of pesticides such as dichloro-diphenyl-trichloroethane (DDT) was so successful in terms of controlling pest populations that farmers began to substitute a heavy dependence on pesticides for sound pest management strategies. The more pesticides the farmers used, the more dependent they became. Soon, pests in high-value crops became resistant to one pesticide after another. In addition, outbreaks of secondary pests occurred because either they developed resistance to the pesticides or the pesticides killed their natural enemies. This supplied the impetus for chemical companies to develop new pesticides, to which the pests also eventually developed resistance.

Certain pests have developed resistance to all federally registered materials designed to control them. In addition, many pesticides are toxic to humans, wildlife, and other nontarget organisms and therefore contribute to environmental

Integrated Pest Management in the United States, 1986

CROP	TOTAL ACRES PLANTED	ACRES PLANTED USING IPM	PERCENTAGE OF TOTAL ACRES USING IPM
Alfalfa	26,748,000	1,273,000	4.7
Apples	461,000	299,000	65.0
Citrus	1,057,000	700,000	70.0
Corn	76,674,000	15,000,000	19.5
Cotton	10,044,000	4,846,000	48.2
Peanuts	1,572,000	690,000	43.8
Potatoes	1,215,000	196,000	16.1
Rice	2,401,000	935,000	38.9
Sorghum	15,321,000	3,966,000	25.8
Soybeans	61,480,000	8,897,000	14.4
Tomatoes	378,000	312,000	82.5
Wheat	72,033,000	10,687,000	14.8
Totals	269,384,000	47,801,000	17.7

Source: U.S. Department of Agriculture.

pollution. Also, it is now very expensive for chemical companies to put a new pesticide on the market. For these reasons, many producers are now looking at alternative strategies such as IPM for managing pests. The driving force behind the development of IPM programs is concern for the contamination of groundwater and other nontarget sites, adverse effects on nontarget organisms, and development of pesticide resistance. Pesticides will probably continue to play a vital role in pest management, even in IPM, but it is believed that the role will be greatly diminished over time.

An agricultural ecosystem consists of the crop environment and its surrounding habitat. The interactions among soil, water, weather, plants, and animals in this ecosystem are rarely constant enough to provide the ecological stability of nonagricultural ecosystems. Nevertheless, it is possible to utilize IPM to manage most pests in an economically efficient and environmentally friendly manner. IMP programs have been successfully implemented in the cropping of cotton and potatoes, and they are being developed for other crops.

There are generally three stages of development associated with IPM programs, and the speed at which a program progresses through these stages is dependent on the existing knowledge of the agricultural ecosystem and the level of sophistication desired. The first phase is referred to as the pesticide management phase. The implementation of this phase requires that the farmer know the relationship between pest densities and the resulting damage to crops so that the pesticide is not applied unnecessarily. In other words, farmers do not have to kill all of the pests all of the time. They must use pesticides only when the economic damage caused by a number of pest organisms present on a given crop exceeds the cost of using a pesticide. This practice alone can reduce the number of chemical applications by as much as one-half.

The second phase is called the cultural management phase. Implementation of this phase requires knowledge of the pest's biology and its relationship to the cropping system. Cultural management includes such practices as delaying planting times, utilizing crop rotation, altering harvest dates, and planting resistance cultivars. It is necessary to understand pest responses to other species as well as abiotic factors, such as temperature and humidity, in the environment. If farmers know the factors that control population growth of a particular pest, they may be able to reduce the impact of that pest on a crop. For example, if a particular pest requires short days to complete development, farmers might be able to harvest the crop before the pest has a chance to develop.

The third phase is the biological control phase, which involves the use of biological organisms rather than chemicals to control pests. This is the most difficult phase to implement because farmers must understand not only the pest's biology but also the biology of the pest's natural enemies and the degree of effectiveness with which these agents control the pest. In general, it is not possible to completely rely on biological control methods. A major requirement in using biological agents is to have sufficient numbers of the control agent present at the same time that the pest population is at its peak. It is sometimes possible to change the planting dates so that the populations of the pests and the biological control agents are synchronized. Also, there is often more than one pest species present at the same time within the same crop, and it is extremely difficult to simultaneously control two pests with biological agents.

D. R. Gossett

SUGGESTED READINGS: *Entomology* (1998), by William S. Romoser and John G. Stoffolano, Jr., contains an excellent overview of the general aspects of IPM. An insightful discussion of the strategies utilized in IPM can be found in *Entomology and Pest Management* (1996), by Larry P. Pedigo. A good discussion of the implementation of IPM can be found in *Integrated Pest Management Systems and Cotton Production* (1989), ed-

ited by J. L. Crawford, C. M. Bonner, and F. G. Zalom. *Integrated Pest Management for Potatoes* (1986) is part of a series of books compiled by the IPM Manual Group that contain useful information on IPM in a variety of crops, including potatoes, citrus, cotton, rice, and tomatoes.

SEE ALSO: Agricultural chemicals; Agricultural revolution; Biopesticides; Organic gardening and farming; Pesticides and herbicides; Sustainable agriculture.

Intergenerational justice

CATEGORY: Philosophy and ethics

Intergenerational justice is the sense of obligation or fair play that one generation of humanity holds toward the generations that follow or precede it.

In addition to societal issues such as how the young should treat the elderly or whether one generation should pay for the education of the next, intergenerational justice includes numerous questions regarding the environment. Is it fair or just for the current generation to exploit natural resources to the point where those resources may become exhausted? Is it fair or just for today's society to fill landfills with garbage or the atmosphere with pollutants that tomorrow's citizens will have to clean up? Although the answers to such questions regarding an implicit social contract reaching across generations would seem self-evident, not everyone agrees that each generation has a moral obligation to leave the world a better place than it found it.

Some economists and policy analysts have argued in favor of what might appear to be short-sighted selfishness on the part of the current generation of humanity. They point to past ecologically unsound practices, such as overreliance on fossil fuels, and claim that technological progress can be attributed to responding to problems created by the selfish behavior of past generations. Using this line of reasoning, they claim that it is unnecessary for current generations to preserve natural resources, curb population growth, or reduce industrial pollution. Frequently coupled to this argument is the statement that past generations showed no restraint or consideration for intergenerational justice and thus current generations should be equally free to engage in selfish behavior. This latter argument is sometimes referred to as mutual unconcern between generations.

The flaw in pursuing a policy of mutual unconcern is that it is based on an assumption of continual technological and scientific progress. While it may be historically true that technological advances allowed past generations to substitute new resources for depleted ones, such as the substitution of coal for fuel when deforestation rendered charcoal scarce in Great Britain during the Industrial Revolution, humans cannot presume that science will always provide technical solutions to environmental problems. The historical record is rife with examples of technical solutions that, in the long run, generated more problems than they solved.

Further, engaging in unsound or damaging practices while arguing that the next generation will find a way to clean up the resulting mess fails on moral grounds. People should recognize that current actions do have significant impacts on the future. The fact that current generations may live to see the consequences of their actions should not release them from the moral obligations implicit in the social contract. The idea of distributive justice within a generation suggests, for example, that it is immoral for the wealthy to exploit the poor; that same concept of distributive justice suggests that rather than pursuing a policy of mutual unconcern, intergenerational justice mandates mutual concern, particularly regarding the environment.

Nancy Farm Männikkö

SEE ALSO: Environmental ethics; Sustainable development.

Intergovernmental Panel on Climate Change

DATE: established 1988
CATEGORY: Weather and climate

Intergovernmental Panel on Climate Change consists of a group of scientific experts from many nations who are charged with evaluating humanity's impact on global climate.

During the 1980's scientists, governments, policy experts, and others began to seriously consider the possibility that human activities might affect the earth's climate, which could have a broad impact on vital natural and human-managed systems.

In 1988 the World Meteorological Organization (WMO) and the United Nations Environment Programme (UNEP) established the Intergovernmental Panel on Climate Change (IPCC) to assess scientific and technical information and policy alternatives related to climate change and its possible impacts. Scientists from many nations were invited to participate, with the objective of evaluating the best available scientific research relevant to national and international policy. The IPCC seeks balanced geographic and national representation in its governing bureau and working groups, subject to the need for specific expertise.

The IPCC originally established three working groups to carry out its mandate. The working group missions were restructured in 1993 and again in 1997. The charge of working group 1 remained unchanged throughout, focusing on the science of the earth's climate system. The 1997 reorganization changed the charge of working group 2 to include assessment of scientific, environmental, economic, and social impacts—both negative and positive—on ecological and socioeconomic systems and human health, emphasizing regional and sectoral analyses. Working group 3 was given the task of assessing all aspects of mitigation strategies and response alternatives.

In general, the IPCC's analyses have provided strong support for the theories and projected broad impacts of climate warming. The IPCC published its First Assessment Report in 1990, which was followed by a supplementary report in 1992. The IPCC reports provided a useful summary of current scientific information in advance of the negotiation of the 1992 Framework Convention on Climate Change. The IPCC Second Assessment Report was published in 1995. One sentence from the 530-page working group 1 report has been widely quoted: "The balance of evidence suggests that there is a discernible human influence on global climate." The IPCC has also published several supporting documents and technical reports addressing specific scientific and policy issues. In the late 1990's the IPCC announced that it planned to release a Third Assessment Report in 2001.

The IPCC's rules of procedure require an extensive scientific peer-review process, as well as governmental review of IPCC documents prior to publication. A concerted effort is also made to achieve consensus before publication. According to IPCC, 350 to 400 experts coordinate in authoring major reports, and 2,500 to 3,000 others contribute via the peer-review process. In all, scientists from roughly 130 countries actively participate.

Although IPCC findings in particular and global warming theory in general are widely accepted in the scientific community, a small number of skeptics have received wide press attention. They point to what they regard as flaws in IPCC analysis and contradictory scientific evidence. Their arguments have been addressed and largely dismissed by mainstream climate scientists. However, because of the long time frames involved in climate change and because the systems and variables influencing climate are extremely complex and difficult to model, uncertainty remains regarding the timing, magnitude, and regional effects of climate change.

Phillip A. Greenberg

SEE ALSO: Climate change and global warming; Greenhouse effect; United Nations Environment Programme.

International Biological Program

DATE: 1964-1974
CATEGORY: Ecology and ecosystems

The International Biological Program (IBP) was an international coalition of scientists who studied ecosystems and evaluated the impact of natu-

ral and human-made changes to larger ecological biogeographic zones called biomes.

The research stage of the U.S. IBP lasted from July 1, 1967, to June 30, 1974. There were two foci: environmental management and human adaptability. The environmental half was much better funded, largely by the National Science Foundation (NSF). W. Frank Blair, an American animal ecologist and administrator for the U.S. IBP from 1968 to 1974, helped set up the multidisciplinary teams that studied five biomes: grasslands (the largest and most successful biome project), eastern deciduous forest, coniferous forest, desert, and tundra. Joseph S. Weiner, a British biological anthropologist, organized the planning for the human adaptability section. U.S. investigators studied five populations that were experiencing dramatic change: South American Indian, circumpolar, high altitude, migrant, and nutritionally stressed peoples.

Modeling complete ecosystems was a primary goal of the biome studies, but modeling subsystems was more practical. Many researchers gained new skills. The biome studies also provided more balanced coverage than nonbiome research in the four areas traditionally examined by ecologists—abiotic, biotic producers (green plants), biotic consumers (animals), and biotic decomposers (bacteria, fungi and some animals)—in that more attention was paid to nutrient cycling and decomposition. Multidisciplinary cooperation resulted in standardized methodologies and techniques, and the human studies provided baseline information on child development, nutrition, and inherited variation that has been used by international agencies for planning. However, the interaction between investigators from the environmental management and human adaptability sides of the IBP was minimal, and they essentially separated study of natural systems from human impacts such as agriculture. The human adaptationists were also criticized for neglecting biocultural interactions.

The U.S. IBP officially ceased its studies in 1974, but some monies were transferred to a new NSF program called Ecosystem Studies. In addition, the human adaptability program served as the framework for and was superseded by the Man and the Biosphere Programme (MAB) established during the General Conference of the United Nations Educational, Scientific, and Cultural Organization (UNESCO) in 1970. The MAB differs from the IBP in its intergovernmental structure and because its primary objective is to use interdisciplinary teams to determine how ecosystems function under different levels of human impact. The MAB's activities include training, informational exchange, and support for research in developing areas. An important element is the incorporation of community input. The MAB also helps fund the setup and conservation of more than three hundred biosphere reserves in seventy-five countries.

The U.S. Department of State established the U.S. counterpart to the MAB in 1974. Six program directorates were established: a biosphere reserve program and five research foci on ecosystems (high latitude, human-dominated, marine and coastal, temperate, and tropical). Although the IBP no longer exists, its vision is being carried forward.

Joan C. Stevenson

SEE ALSO: Biodiversity; Biosphere concept; Biosphere reserves; Ecology; Ecosystems; Global Biodiversity Assessment.

International Union for the Conservation of Nature/ World Wildlife Fund

DATE: IUCN founded October 5, 1948; WWF founded 1961
CATEGORY: Preservation and wilderness issues

The International Union for the Conservation of Nature and Natural Resources (IUCN) works with the World Wildlife Fund (WWF) and many other conservation organizations throughout the world. Their goals include protecting all species and ecosystems, as well as encouraging sustainable use of natural resources.

The World Wildlife Fund (WWF) originally served as the fund-raising arm of the International Union for the Conservation of Nature and Natural Resources (IUCN). Today the WWF not only raises money for the protection and education programs but also puts the programs into motion. The IUCN, commonly known as the World Conservation Union, focuses on research and compiles the *Red List* of endangered and rare species.

The World Conservation Union was established on October 5, 1948, as the International Union for the Protection of Nature, shortly to be renamed the IUCN. Originally headquartered in France, the IUCN was the first organization that pledged to protect endangered species of wildlife on a worldwide scale. It brought together the best scientific minds and focused on gathering data, identifying species and wildlife areas in need of protection, and determining ways to sustain the earth's resources for present and future generations. During the 1950's the group focused on research projects. As a scientific organization, it did not have funds to devote to other projects, nor did it have the capability to raise funds.

By 1960 funds were needed for more than research projects. Many prominent scientists noticed and wrote about the decline in the number of animals in West Africa. Business executive Victor Stolan wrote to Sir Julian Huxley, a notable zoologist and United Nations adviser on wildlife conservation, stating that "vigorous and immediate" international fund-raising was required to alter wildlife decline in Africa. Huxley contacted Sir Peter Scott, a notable ornithologist and naturalist who served as vice president of IUCN. Ensuing meetings led to the birth of the WWF. The IUCN would share its scientific knowledge, and representatives of the WWF would raise money to finance projects and make contact with top-level government representatives to promote wildlife-friendly legislation. A simple but striking black-and-white panda logo was created so that the WWF would be easily identified. It was based on sketches of Chi-Chi, a popular giant panda from China at the London Zoo. Headquarters for both IUCN and WWF were moved to Switzerland in 1961, the official birth year of the WWF.

The first order of business was to get the word out about the state of the earth and its animals.

Participants at a press conference in Argentina in 1998 sit behind a banner displaying the World Wildlife Fund's distinctive giant panda logo. (AP/Wide World Photos)

A lengthy article appeared in Britain's *Daily Mirror* describing the animals and species doomed to disappear because of human activity. The 1961 article resulted in the donation of a total of $100,000 to the WWF. Using these funds, the IUCN and the WWF set up national organizations in four countries and initiated their first projects. These included creating a footpath in a reserved forest of Madagascar, transporting eight endangered white rhinos from South Africa to what is now Zimbabwe for breeding, and saving the rare Arabian oryx from extinction.

Their first major campaign, Operation Tiger, focused on saving tigers and their habitats in Asia. Once the WWF succeeded in getting tiger hunting banned in India, they helped tag and track remaining tigers to help increase populations. They also helped improve or create havens for tigers in Indonesia, Thailand, and India. Other animals helped through WWF campaigns include whales, rhinos, and pandas.

Another important campaign focused on ending the smuggling of endangered animals and plants. Exotic birds and plants often die at the hands of people who do not know how to care for them. In many parts of the world people believe that medicine made from exotic animal parts, such as rhino horns or monkey brains, will cure illnesses. Other animals, such as elephants, are killed for their ivory. Such illegal killing, or poaching, has reduced many species to near extinction. To end this type of trade, the WWF set up the Trade Records Analysis of Flora and Fauna in Commerce (TRAFFIC) in 1976. TRAFFIC investigators found that smuggling and poaching occurred throughout the world, even in countries that had signed treaties opposing such activities. The WWF's attention to the smuggling problem encouraged local governments to work with TRAFFIC to stop illegal commerce in exotic animals and plants.

By 1980 the IUCN and the WWF had formally broadened their outlook to include other aspects of life on Earth. They banded together with like-minded organizations around the globe, such as Greenpeace, to strengthen efforts in scientific research, lobby for laws to protect wildlife, stop the illegal trade of endangered species, and educate the public on conservation and environmental issues. They continue to work with government groups and other research organizations to protect specific areas of the earth, such as when the IUCN teamed with the National Cancer Institute to identify helpful rain forest plants. The WWF also created the Guardians of the Rainforests campaign to help reduce the destruction of tropical forest ecosystems.

Lisa A. Wroble

SUGGESTED READINGS: Information about the World Wildlife Fund, as well as other conservation and protection groups, is available in *The Nature Directory: A Guide to Environmental Organizations* (1991), by Susan D. Graham-Lanier, and the *Encyclopedia of the Environment*, edited by Ruth A. Eblen and William R. Eblen for the René Dubos Center for Human Environments. More detailed information about the history and programs of the WWF, as well as IUCN's role in conservation worldwide, is available in *World Wildlife Fund* (1994), by Peter Denton. Though less information is devoted to the IUCN in books, their official Web site (http://www.iucn.org/info_and_news/about_iucn/index.html) is helpful. The IUCN-U.S. homepage (http://www.iucnus.org/) also helps put the union's objectives in perspective. World Wildlife Fund also maintains a U.S. Web site (http://www.wwfus.org/who/index.html).

SEE ALSO: Endangered species and animal protection policy; Environmental policy and lobbying; Wildlife refuges.

International Whaling Ban

DATE: implemented in 1986
CATEGORY: Animals and endangered species

The International Whaling Ban was a prohibition on commercial whaling approved by the International Whaling Commission (IWC) in 1982 and implemented in 1986. The ban was a response to the dramatic decline in the world's whale population that resulted from the excesses of the whaling industry.

By the 1920's the whaling industry had decimated certain species of whale. Fearing a collapse of the whaling industry, which was an essential economic activity in nations such as Norway, some countries implemented laws banning the catch of certain species or of females with calves. However, international cooperation proved to be difficult, if not impossible. The International Whaling Commission (IWC), which was founded on December 2, 1946, in Washing-ton, D.C., could do little to slow the continuing destruction of the world's whale population.

During the 1960's growing interest in protecting the environment led many people to turn their attention to the problems facing whales, especially the impact of the whaling industry. Public interest led participants at the United Nations Environmental Conference, held in Stockholm, Sweden, in 1972, to call for a ten-year moratorium on commercial whaling. That same year the U.S. Congress passed the Marine Mammal Protection Act, which ended U.S. involvement in commercial whaling and allowed the United States to press for a global ban at IWC meetings.

Despite public pressure and the advocacy of the United States, the IWC resisted efforts to adopt a moratorium. However, by the early 1980's it became abundantly clear that the whale population had diminished to the point where commercial whaling was fast becoming an industry unable to make profits. The IWC responded to the realities of the situation on July 23, 1982, when it approved a ban on commercial whaling to commence during the 1986 whaling season.

Although environmentalists greeted the passage of the International Whaling Ban with enthusiasm, several whaling nations, including Japan, Norway, and the Soviet Union, all of which had voted against the ban, sought to create loopholes that would allow them to continue their whaling activities. By 1987 declining profits led the Soviet Union to support the ban, but Japan requested exemptions that allowed whaling near its coastline. When the United States threatened economic sanctions, the Japanese agreed to end commercial whaling. Because the IWC did permit "research" whaling for scientific purposes, Japan was able to continue whaling on a limited basis. While the ostensible reason for whaling was scientific study, the Japanese used the whale meat as a food resource.

The original ban called for a review based on studies of whale populations. Although some species did increase in number during the period of the ban, in 1993 most IWC nations agreed to continue the prohibition. Japan and Norway vigorously protested, and Norway, which had abandoned commercial whaling for economic reasons in 1967, announced its intention to begin hunting minke whales. The Norwegians claimed that minkes had populations large enough to permit whaling to take place without any negative environmental impact. Environmentalists condemned the move, but the IWC had no mechanism for enforcing the voluntary ban.

The continuation of the ban threatened the existence of the IWC. Iceland quit in 1992, and other members expressed growing dissatisfaction with the prohibition when evidence existed that many species had undergone significant increases in population. During the 1997 IWC meeting, members debated implementation of the Revised Management Procedure (RMP), which would end research whaling but would allow Norway and Japan to conduct coastal whaling for domestic consumption with strict quotas. While some observers praised the RMP because it closed the research whaling loophole and kept Japan and Norway on the IWC, environmental groups condemned it as the first step toward a return to full-scale commercial whaling. The RMP remained a matter of debate during the 1998 meeting.

Another issue that raised objections from environmentalists was that of aboriginal whaling. The 1986 ban permitted indigenous people to continue whaling for subsistence and ceremonial purposes. For example, the Inuit were allotted an annual quota of bowhead whales. Opponents argued that because they employed modern whaling techniques, their activities no longer met the standards of aboriginal whaling. The issue came to the fore again in 1998, when the Makah in Washington State received permission to hunt five California gray whales, which had been removed from the endangered species list in 1993. Although the Makah had not hunted whales since 1928, a nineteenth century treaty with the U.S. government guaranteed their whaling rights. Environmentalists contended that the U.S. government's support for Makah whaling was intended as a signal to Japan that the United States would support efforts to resume commercial whaling.

Although the ban had a symbolic significance, its importance as an environmental measure was limited. It was enacted only after commercial whalers experienced declining profits, a result of decades of hunts that had depleted whale populations. As such, it merely acknowledged the reality that whaling had been devastating for whales. The ban's loopholes regarding scientific and aboriginal whaling allowed hunts to continue, and whalers killed an estimated eighteen thousand whales during the first twelve years of the ban. Lacking any enforcement mechanisms, the IWC depended on the voluntary cooperation of its member states, many of which adamantly opposed the continuation of the ban during the 1990's. One reason the ban remained in place during this period was the United States' use of economic sanctions against nations that violated the ban; however, the United States failed to impose sanctions on Norway when it resumed minke whaling in 1993. While the United States officially supported a continuation of the ban, it nonetheless participated in efforts to implement the RMP.

Thomas Clarkin

SUGGESTED READINGS: J. N. Tonnessen and A. O. Johnsen provide a detailed account of efforts to regulate the whaling industry prior to the ban in *The History of Modern Whaling* (1982). Richard Ellis's *Men and Whales* (1991) offers a highly critical analysis of the IWC. Peter Stoett's *Atoms, Whales, and Rivers* (1995) places the ban in the larger context of international regulation. David Day's *The Whale War* (1989) focuses on the efforts of antiwhaling activists.

SEE ALSO: International Whaling Commission; Whaling.

International Whaling Commission

DATE: founded December 2, 1946
CATEGORY: Animals and endangered species

The International Whaling Commission was established to regulate whaling and ensure the conservation of whales.

During the 1930's concern over declining whale stocks and the near extinction of some species prompted efforts to regulate the whaling industries. These efforts at international cooperation met with little success, but as World War II drew to a close, proponents of regulation pressed for a regulatory framework. In November, 1945, several whaling nations gathered in Washington, D.C., for the International Whaling Conference. Although some delegates argued for an agency under the direction of the United Nations, the conference led to the creation of an autonomous organization called the International Whaling Commission (IWC) in 1946. Great Britain, Norway, the Soviet Union, and the United States were among the fifteen charter nations. Japan joined in 1950, and by 1982 membership had climbed to thirty-seven nations.

The stated mission of the IWC was to regulate international whaling to promote the increase of whale stocks, thereby ensuring the continued existence of the lucrative whaling industry. The commission established a whaling season and created sanctuary zones in which whaling was not permitted. Defined in terms of blue whale units (BWUs), the whaling season ended when that year's quota had been met. However, the IWC was virtually powerless, and it allowed member nations a multitude of opportunities to bypass the regulations it implemented. Furthermore, there was no mechanism for punishing nations that violated IWC regulations. As a result, the IWC did little to slow the destruction of whale populations during the first several decades of its existence.

Despite its ineffectiveness, the IWC became an arena of conflict and disagreement among the member states. Because the quota system created a free-for-all environment in which each nation attempted to catch as many whales as possible before the season ended, many nations argued that they did not have an opportunity to catch their fair share of the total number of BWUs available. However, efforts to assign quotas to individual nations proved impossible, as each nation jockeyed for a high percentage of the total at the expense of other nations. Debates grew so acrimonious that the IWC closed its meetings to the public and reporters for a time during the 1960's.

As a regulatory body, the IWC failed to live up to its promise. Quotas were based on estimates of whale populations that were too high. In addition, the commission proved both unwilling and unable to rein in the whaling industries of member nations. In 1982 the IWC voted for the International Whaling Ban, which was implemented in 1986. This act reflected the declining economic vitality of the whaling industry more than a concern for the environment or the fate of the whales. As such, the ban came far too late to ensure the survival of many species. After the implementation of the ban, the IWC became a forum for nations attempting to circumvent or weaken restrictions on whaling rather than an organization committed to the preservation of whales.

Thomas Clarkin

SEE ALSO: International Whaling Ban; Whaling.

Introduced and exotic species

CATEGORY: Animals and endangered species

An exotic species is one that has been displaced from its native or original geographic range to another area, generally by human activity. An introduced species is one that has been transplanted for some intended purpose. Such displacement often disrupts the environment into which the species has been displaced or introduced.

All introduced species may be considered exotic, but not all exotic species are introduced. Natural range extensions, although they may have been assisted indirectly by human activities, do not qualify for either category, and a species transported by a natural event (such as a storm or flood) may be considered exotic for a time while its fate is being determined by natural selection. Transplants may be accidental or purposeful. Accidental transport is usually suspected to occur by ship or other means of commercial conveyance. Since most ecological communities have each evolved containing a unique assemblage of species with its own unique system of interactions and interdependencies, seldom can an-

Frequency of Clam Species Collection in the Saline and Ouachita Rivers, 1976-1990

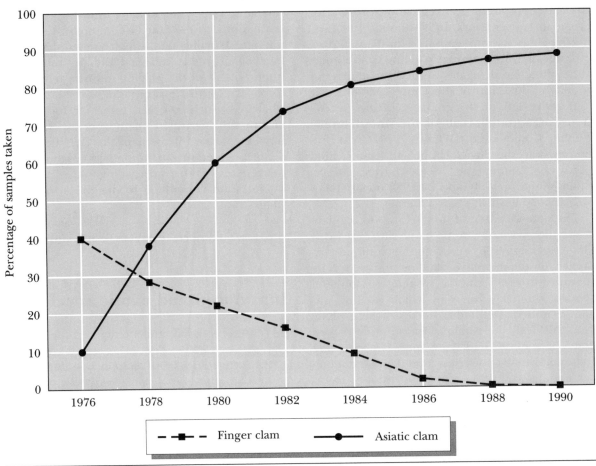

The frequency with which researchers collected these clam species in two Arkansas rivers shows that the introduced Asiatic species had virtually eliminated the native species by 1990.

other species be added without causing one or more significant disruptions in some community pattern (competition, food web, or predator-prey relationships). While some introduced species have been beneficial, many have had one or more detrimental effects.

ASIATIC CLAM

The Asiatic clam (*Corbicula fluminea*) was apparently accidentally deposited from a ship's bilge near the upper West Coast of the United States in the mid-1930's. From there it rapidly spread into virtually all river systems of the West Coast. By the late 1960's it had spread over all major portions of the lower Mississippi River drainage. By 1983 the Asiatic clam had infested nearly all rivers in the western and southern states and most of the eastern half of the United States.

The Asiatic clam has exhibited two significant detrimental effects. Its high reproductive capacity, its somewhat mysterious ability to rapidly disperse (even upstream), and its rapid particle-filtering rate give the Asiatic clam the ability to compete with native species of fingernail clams (*Sphaerium* spp.). In the Saline and Ouachita Rivers of south Arkansas between 1976 and 1988, the Asiatic clam virtually eliminated the fingernail clam.

The second effect is based on its affinity for hiding in dark places, particularly water intake structures. If eggs or small juveniles are drawn into water systems having some minimal flow, they may attach, filter particulates, mature, reproduce, and cause occlusion of the pipes. During the 1970's and 1980's, this caused considerable consternation among industrial plant managers and caused significant sums of money to be invested in the control and subsequent prevention of Asiatic clam reproduction.

Zebra Mussel

Larvae of the zebra mussel (*Dreissena polymorpha*) were apparently flushed from a ship's ballast in St. Clair Lake (between Michigan and Ontario, Canada) in the mid-1980's. Subsequent studies have revealed one of the most phenomenal dispersal rates of all time. By 1991 the zebra mussel had been found in the St. Lawrence River and the Hudson River, and by 1992 in the Mississippi River system. By early 1994 it had spread to virtually all tributaries of the Mississippi system navigable by commercial ships or barges.

Whereas the Asiatic clam is a more passive filterer and prefers smaller streams with sandy or gravelly substrate into which it can burrow, the zebra mussel is a more positive, pumping filterer that prefers large rivers and lakes with hard, rocky substrate to which it can attach. The zebra mussel's dispersal pattern coincides with previously unoccupied freshwater niches in North America. Consequently, the zebra mussel disperses more rapidly than and successfully competes with native species of fresh-water mussels, which are more passive filterers. Since zebra mussels must attach to hard surfaces, they often attach to each other, producing two or more layers of mussels. Adult densities up to 100,000 per square meter and veliger densities up to 350,000 per cubic meter have been reported.

Several concerns have arisen regarding the zebra mussel. Their positive filtering mechanism and high densities may remove a significant portion of the plankton and particulates from a water column, thus robbing other species of the resource. Their need to settle and attach to hard surfaces has made the mussels native to much of the eastern two-thirds of the United States particularly vulnerable to being covered by zebras because the partially exposed shells of mussels are often the only hard substrate in the lower reaches of most of the rivers of the region. When a native mussel is settled upon by zebra mussels, it is partially smothered, so it pulls itself farther out of the substrate only to be covered more extensively by the zebras. Finally the zebra mussels may completely cover the larger clam and smother it. This is a source of grave concern in all aspects of the shelling industry and involves several nations. The zebra mussels' need to settle on hard surfaces also increases the probability that they will soon (if not already) be transported on sporting boat hulls from the large rivers to recreational lakes and reservoirs. Resource managers are concerned that zebra mussels will significantly alter the biological community and reduce its values for angling and possibly other types of water recreation.

Exotic Fishes

More than fifty species of foreign origin have been introduced into North American waters. Most of the introductions have been done intentionally to provide additional food or sporting opportunities. Among the most notable are the common carp (*Cyprinus carpio*) and the grass carp (*Ctenopharyngodon idella*). The common carp possessed such an omnivorous talent and high reproductive rate that it quickly became well established in all types of aquatic habitats, preferring larger rivers and lakes. In its European homeland it is prized for food, but in North America people are more interested in other, more desirable native species as a food source. On the negative side, the common carp's tendency to disturb sediments and destroy vegetation while feeding causes problems with high turbidity and community disruption. For a time, resources managers feared they would overpopulate and out-compete some native species.

Great caution accompanied the grass carp's introduction in the late 1960's. There were still concerns about the common carp's potential, and many feared that the grass carp would adapt and reproduce as well if not better. This concern was grave enough to stimulate the legislatures of several states in the Mississippi River system to ban its import. The Arkansas Game and Fish Commission sponsored research to learn how to control its propagation and dispersal; as a result, sterile grass carp are now produced and widely marketed as vegetation control agents. As it turns out, most of the concerns about overpopulation were unfounded.

Perhaps the best-known example of exotic fish introduction occurred when the Welland Canal was built to bypass Niagara Falls for shipping into Lake Erie and points beyond. The sea lamprey (*Petromyzon marinus*) discovered the canal and invaded Lake Erie and displaced its lake trout (*Salvelinus namaycush*) and whitefish (*Coregonus* spp.) populations. Canada and the United States have invested millions of dollars to discover a control for the lamprey. Currently, sea lampreys are controlled by a selective larval toxin and long-term community adaptation. The role of the lake trout in commercial and sport fisheries has been replaced by the coho salmon (*Oncorhynchus kisutch*), another exotic species that is nonnative to the Great Lakes.

Aquaculture has matured as an industry partly because marine fisheries resources have significantly declined. Many different exotic species have been brought from other countries or transplanted from one basin to another within North America in the ongoing effort to provide variety and find more efficient subjects on which to practice aquaculture, to provide broader culinary and sporting opportunities, or to satisfy the aquarium enthusiasts. The rainbow trout (*Oncorhynchus mykiss*) has been moved from its native range in western North America to many points east, particularly to dam tailwaters for sporting purposes. Brook trout (*Salvelinus fontinalis*) have been transplanted from their native streams in the northeastern United States to numerous Rocky Mountain lakes. Brown trout (*Salmo trutta*) were carried from Europe to North America and successfully added to the native salmonid fauna.

Goldfish (*Carassius auratus*), guppies (*Poecilia reticulata*), variable platy (*Xiphophurus variatus*), oscars (*Astronotus ocellatus*), and many other species have been imported to the United States for the aquarium trade. The primary problem that

has resulted is the ignorant and random release of exotic species simply because the aquarist may become tired of keeping it. The bighead carp (*Hypophthalmichthys nobilis*), ruffe (*Gymnocephalus cernuus*), and several cichlids (*Cichlasoma* spp. and *Tilapia* spp.) were introduced for possible food fish. The importation of an exotic species is sometimes done illegally, outside the supervision of any regulating agency. The potential for biological community disruption has been serious enough to stimulate the American Fisheries Society to establish an Exotic Fishes Section and many state game and fish agencies (or natural resource departments) to set up special committees or enact regulations attempting to control these importations.

OTHER EXOTIC SPECIES

Most ecologists are aware of the detrimental effects of the chestnut blight and Dutch elm disease in the United States. It is still unknown whether the targeted species will eventually adapt to resist the disease or succumb to it. Other problems include the overpopulation of rabbits (*Sylvilagus* sp.), in the absence of adequate predators, following their introduction into Australia. After mongooses (*Herpestes* sp.) were introduced into some of the Caribbean islands to control snakes, they apparently ran out of snakes and started eating nonintended prey items.

An often-overlooked introduced species that has had a profound impact on the history and development of the Americas is the domestic horse (*Equus caballus*). Although originally native to North America, horses migrated to Asia (presumably across the Bering land bridge during the final Pleistocene glaciation), became extinct in America, and were reintroduced by the Spanish during early explorations. During the horse's absence from America, biotic communities would have certainly evolved horse-absent features so that a reintroduction could well have caused ecological perturbations. On the other hand, their actual role in a natural community as a primarily domesticated species is debatable.

An important part of biogeographic studies includes a determination of the native range of a species, which is then used to help explain dispersal routes and mechanisms and possibly predict future range extensions. Biogeographic studies are also correlated with continental drift patterns and distributions of the fossil record. When transplant records are never made, or are lost or inaccessible, it may be difficult for biogeographers to determine the status of certain species.

John Rickett

SUGGESTED READINGS: Among the sources that discuss the introduction of fish into North America are C. E. Bond, *Biology of Fishes* (1996), and C. R. Robins et al., "Common and Scientific Names of Fishes from the United States and Canada," *American Fisheries Society Special Publication* 20 (1991). The spread of Asiatic clams is covered in C. L. Counts III, "The Zoogeography and History of the Invasion of the United States by *Corbicula fluminea* (Bivalvia: Corbiculidae)," *American Malacological Bulletin* 2 (1986), and B. G. Isom, "Historical Review of Asian Clam (*Corbicula*) Invasion and Biofouling of Waters and Industries in the Americas," *American Malacological Bulletin* 2 (1986). Journal articles that explore problems that have accompanied the displacement of zebra mussels include S. Filipek, "Mississippi River Zebras," *Arkansas Zebra Mussel Newsletter* (1994), and L. E. Johnson and J. T. Carlton, "Post-establishment Spread in Large-scale Invasions: Dispersal Mechanisms of the Zebra Mussel, *Dreissena polymorpha*," *Ecology* 77 (1996).

SEE ALSO: Balance of nature; Ecosystems; Food chains.

Irrigation

CATEGORY: Water and water pollution

The demand of feeding and clothing the rapidly expanding world population requires the production of increasing amounts of food and fiber. One important strategy has been the use of irrigation techniques to supply additional water to arid and semiarid regions where few, if any, crops could otherwise be grown.

Approximately 350 million acres of land worldwide are irrigated. In the United States more than 10 percent of the crops, encompassing approximately 50 million acres, receive water through irrigation techniques; 80 percent of these are west of the Mississippi River. In certain other countries, including India, Israel, North Korea, and South Korea, more than one-half of food production requires irrigation. From 1950 to 1980, the acreage of irrigated cropland doubled worldwide; increases since then have been more modest.

An often-cited example of irrigation success is that of the Imperial Valley of Southern California. The valley, more than 12,900 square kilometers (5,000 square miles) in size, was originally considered to be a desert wasteland. The low annual rainfall resulted in a typical desert with cacti, lizards, and other arid-adapted plants and animals. In 1940, however, engineers completed the construction of the All-American Canal, which carries water 130 kilometers (80 miles) from the Colorado River to the valley. The project converted the Imperial Valley into a fertile, highly productive area where farmers grow fruits and vegetables all year.

Successful agriculture in Israel also requires irrigation. As a result of continuing settlement of the area throughout the twentieth century, large amounts of food must be produced. To fulfill this need, a system of canals and pipelines carries water from the northern portion of the Jordan Valley, where the rainfall is heaviest, to the arid south.

Irrigation is an expensive operation that requires advanced technology and large investments of capital. In many cases, irrigation systems convey water from sources hundreds of miles distant. Such vast engineering feats are largely financed by taxpayers. Typically, water from a river is diverted into a main canal and from there into lateral canals that supply each farm. From the lateral canals, various systems are used to supply water to the crop plants in the field.

Flood irrigation supplies water to fields at the surface level. Using the sheet method, land is prepared so that water flows in a shallow sheet from the higher part of the field to the lower

A sprinkler system irrigates potatoes in Idaho. (Ben Klaffke)

part. This method is especially suitable for hay and pasture crops. Row crops are better supplied by furrow irrigation, in which water is diverted into furrows that run between the rows. Both types of flood irrigation cause erosion and loss of nutrients. However, erosion can be reduced in the latter type by contouring the furrows.

Sprinkler irrigation, though costly to install and operate, is often used in areas where fields are steeply sloped. Sprinklers may be supplied by stationary underground pipes or a center pivot system in which water is sprinkled by a raised horizontal pipe that pivots slowly around a pivot point. Besides its expense, another disadvantage of sprinkler irrigation is loss of water by evaporation. In drip irrigation, water is delivered by perforated pipes at or near the soil surface. Since water is delivered directly to the plants, much less water is wasted by evaporation as compared to other methods.

Much of the water utilized in irrigation never reaches the plants. It is estimated that most practices deliver only about 25 percent of the water to the root systems of crop plants. The remain-

ing water is lost to evaporation, supplies weeds, seeps into the ground, or runs off into nearby waterways.

As fresh water evaporates from irrigated fields over time, a residue of salt is left behind. The process, called salinization, results in a gradual decline in productivity and can eventually render fields unsuitable for further agricultural use. Correcting saline soils is not a simple process. In principle, large amounts of water can be used to leach salt away from the soil. In practice, however, the amount of water required is seldom available, and if it is used, it may waterlog the soil. Also, the leached salt usually pollutes groundwater or streams. One way to deal with salinization is to use genetically selected crops adapted to salinized soils.

As the number of acres of farmland requiring irrigation increases, so does the demand for water. When water is taken from surface streams and rivers, the normal flow is often severely reduced, changing the ecology downstream and reducing its biodiversity. Also, less water becomes available for other farmers downstream, a situation that often leads to disputes over water rights. In other cases water is pumped from deep wells or aquifers. Drilling wells and pumping water from such sources can be expensive and may lead to additional problems, such as the sinking of land over aquifers. Such land subsidence is a major problem in several parts of the southern and western United States. Subsidence in urban areas can cause huge amounts of damage as water and sewer pipes, highways, and buildings are affected. In coastal areas, depletion of aquifers can cause the intrusion of salt water into wells, rendering them unusable. The federal government spends millions of dollars to repair damage to irrigation facilities each year.

Like many modifications to natural ecosystems, the use of water for irrigation achieves some remarkable but temporary advantages that are complicated by long-term environmental problems. Assessments of total financial costs and environmental impacts are continuously weighed against gains in production.

Thomas E. Hemmerly

SUGGESTED READINGS: For a broad overview of environmental issues relating to water, see the *National Geographic* special edition *Water, the Power and Turmoil of North America's Fresh Water* (November, 1993). Textbooks in environmental science that cover irrigation problems include *Environmental Science* (1998), by Daniel B. Botkin and Edward A. Keller, and *Environmental Science* (1999), by G. Tyler Miller.

SEE ALSO: Agricultural revolution; Erosion and erosion control; Soil, salinization of; Sustainable agriculture; Water use.

Italian dioxin release

DATE: 1976
CATEGORY: Human health and the environment

On July 10, 1976, a chemical-reaction chamber exploded at a factory in Seveso, Italy, about 21 kilometers (13 miles) north of Milan. The resulting cloud of dioxin adversely affected the health of five thousand people.

The Seveso, Italy, chemical factory, owned by Industrie Chimiche Meda Societa Anonima (ICMESA), produced the phenoxy-type herbicide trichlorophenol 2,4,5-T. Developed in 1945, this herbicide was used to control broadleaf weeds and brush along highways and railroads, in rangeland and forests, and in wheat, rice, corn, and sugarcane fields. Agent Orange, a defoliant that received widespread use during the Vietnam War, is made from a combination of trichlorophenol 2,4,5-T and 4,4-D. The U.S. Environmental Protection Agency (EPA) restricted use of 2,4,5-T to rice fields and rangeland in 1979 and banned the herbicide entirely in 1985. One of the chief complaints was that 2,4,5-T contains dioxin, minute quantities of which cause cancer in animals. The toxic cloud that settled over Seveso contained an estimated 1.8 kilograms (4 pounds) of dioxin.

The disaster occurred when the chemical mixture in a reactor became seriously overheated and exploded, releasing a white plume that shot into the atmosphere to a height of about 45 meters (150 feet). The plume turned into a cloud that drifted in a southerly direction before cooling and settling over a 700-acre area. Five thousand people were exposed. Those living closest to the ICMESA factory experienced nausea and skin irritation on exposed parts of their bodies. People who were outside during the explosion developed burnlike sores on their faces, arms, and legs. Widespread cases of chloracne, an acnelike symptom occurring during industrial exposure to dioxin, also occurred. All residents developed headaches, dizziness, and diarrhea. Birds were especially vulnerable to the chemicals and died quickly, often before they could fly out of the area. Approximately 3,300 small animals such as rabbits, mice, chickens, and cats sickened and died soon after the accident.

Five days after the explosion, authorities instructed the population not to eat fruit or vegetables from the most affected areas. Nineteen children were hospitalized and later evacuated. Two weeks after the disaster occurred, Milan's health administration determined that dioxin contamination from the cloud was serious enough to warrant evacuation of the population.

This began on July 26 with the relocation of more than seven hundred people living in Zone A, a 267-acre area directly south of the ICMESA plant. People in the surrounding 665-acre Zone B were allowed to remain in their homes, although the area was sealed off from nonresidents. Italian army troops strung barbed wire around Zone A and patrolled the perimeter. To prevent toxins from entering the food chain, emergency slaughtering of farm animals was decreed, and by 1978 some 77,000 animals had been killed. Scientific evidence that dioxin causes cancer in humans is mixed, but an ongoing study of Seveso residents indicates an elevated risk of several cancers.

Peter Neushul

SEE ALSO: Chloracne; Dioxin; Pesticides and herbicides; Times Beach, Missouri, evacuation.

Ivory and the Ivory Trade

CATEGORY: Animals and endangered species

The international trade in ivory decimated the world's once-abundant elephant populations, but a 1990 international ban on the trade has done much to restore the species.

Ivory is made up of enamel, a resilient material found in mammalian teeth. The principal source of ivory is elephant tusks, although the teeth of the hippopotamus and walrus and of some types of whales and boars are also considered to be ivory. Tusks, also called "raw" ivory, grow on the male Indian, or Asian, elephant *Elaphus maximus* and on both sexes in the African elephant *Loxodonta africana*. Diverse human societies have used ivory for centuries to make crafts and medicines and in rituals. The impact on elephant populations was originally minimal. Trade in ivory in the eighteenth and nineteenth centuries was limited by the logistics of obtaining ivory and because of low demand. The twentieth century witnessed an explosion in ivory trade driven by the enormous profits involved in exporting ivory to western and eastern countries from India and, especially, Africa. Such profits soon precipitated a large illegal trade in ivory, and poaching became rampant.

In both Asia and Africa, the elephant population was decimated by ivory poaching. The African elephant used to roam most of the continent but has now been reduced to isolated populations in protected national parks and wildlife reserves. The African population was estimated at 1.3 million in 1979; it had dropped to 750,000 in 1988 and 600,000 by 1992. At such rates, scientists projected, the species would be extinct by 2025.

THE IVORY MARKET

Massive hunting of African elephants started in 1970 in order to satisfy the great demand for raw ivory in consumer countries, where it was converted into expensive crafts and medicines. Hong Kong, Japan, Taiwan, and China were the major markets. The price of raw ivory rose from three dollars a pound in 1975 to $125 a pound by 1987. The slaughter of elephants in producer countries was accelerated by poverty and human-elephant conflict. A single elephant could fetch up to $3,600, more than a typical person in a many impoverished countries could make in five years. Moreover, most of the countries involved did not have the personnel and equipment needed to protect elephants, and corrupt government officials exacerbated the problem by participating in the ivory trade.

The effects of the widespread hunting on the species were significant. Since poachers went for the elephants with the largest tusks, the average tusk size decreased from 9.8 kilograms to 4.7 kilograms. Elephants are slow to reproduce and take years to mature; the relentless killing reduced the average elephant's life span from sixty years to only thirty years. The herds' tight matriarchal social systems were also disrupted. Selective removal of large bulls created a dearth of breeding males and lowered genetic diversity.

BANNING THE IVORY TRADE

Scientists and nonscientists alike predicted the demise of the largest terrestrial animal unless immediate measures were implemented to save it. In the 1970's therefore, a global effort was initiated to halt the decline in elephant populations. In 1978, the Convention on International Trade in Endangered Species (CITES) called for protection of elephants and their habitats, and elephants were placed on the agreement's list of threatened species. In June, 1989, President George Bush authorized a moratorium on all ivory imports into the United States; and in October, elephants were moved to the CITES list of endangered species. Congress passed the African Elephant Conservation Act the same year. Other countries joined the moratorium, and a total ban on the ivory trade by 103 CITES signatory nations was declared in 1990. Botswana, Malawi, South Africa, Zambia, Zimbabwe, and Hong Kong voted against the ban, but Kenya, one of the countries where poaching was rampant, burned two thousand tusks valued at three million dollars in support of the measure.

In the wake of the ban's adoption, the price of ivory plummeted to one dollar a pound. As profits declined, so did poaching, and elephant populations in southern African countries increased by as much as 25 percent, sometimes

exceeding the local carrying capacities. Habitat destruction, especially in woodland areas, occurred because elephants strip and knock down trees in search of succulent bark and leaves. Increased elephant-human interaction also led to stresses, as rogue elephants can tear down fences, raid farms, and occasionally kill people. As a result, measures to cull herds were adopted in Zimbabwe and South Africa. Meat from culled elephants is distributed to local communities.

The ban has not been embraced by all countries. Japan and several ivory-producing southern African countries proposed reversing the ban and returning the elephant to the less restrictive threatened list during the 1992 CITES meeting in Kyoto, Japan. For example, representatives from Zimbabwe argued that the country's elephant population was healthy and that the ban was depriving the nation of vital resources required to protect wildlife. Proponents of the ban argued that any trade in ivory would spur illicit traffic in ivory, however, and the ban on the ivory trade was maintained.

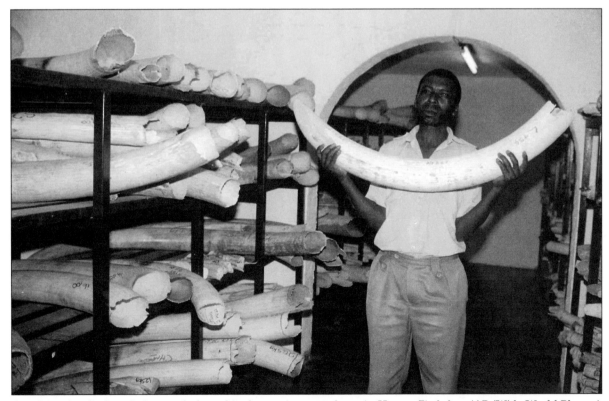

A park warden displays elephant tusks housed in the state ivory warehouse in Harare, Zimbabwe. (AP/Wide World Photos)

OTHER MEASURES

In 1997, several southern African countries were allowed to sell their legal stockpiles of ivory to Japan. The ivory was required to be coded and marked by country of origin, and the codes were entered into a database that would enable the ivory to be tracked by the Ivory Monitoring Unit. However, genetic and isotope markers used on "worked ivory" are not foolproof, raising the possibility of a renewed illegal trade in ivory. In the former ivory-producing nations of Africa, the emphasis is now on tourism as a means of generating revenue, and limited hunting, in which tourists pay up to ten thousand dollars to shoot an elephant, is allowed in Zimbabwe, Namibia, and South Africa.

Joseph M. Wahome

SUGGESTED READINGS: A detailed analysis of projected trends in African elephant population is given by Graeme Caughley, Holly Dublin, and Ian Parker Ian in "Projected Decline of the African Elephant," *Biological Conservation* 54 (1990). In *Economics for the Wild: Wildlife, Diversity, and Development* (1992), T. M. Swanson and E. B. Barbier provide an informative account of the conflict between conservation and politics. A critical look at conservation of the African elephant is provided by Bonner Raymond in *At the Hand of Man: Peril and Hope for Africa's Wildlife* (1993).

SEE ALSO: Convention on International Trade in Endangered Species; Ecotourism; Endangered species; Hunting; Poaching.

K

Kaibab Plateau deer disaster

DATE: 1906-1939
CATEGORY: Animals and endangered species

Between 1906 and 1939, a significant population eruption and crash was reported for mule deer on the Kaibab Plateau in Arizona. The disaster has frequently been cited as an illustration of the adverse impact humans can have, particularly through predator control, on the natural balance between an ungulate (hoofed animal) population and its environment.

The Kaibab Plateau, part of the Kaibab National Forest, lies on the North Rim of the Grand Canyon in Arizona. In 1906 President Theodore Roosevelt set aside the Grand Canyon National Game Preserve, including the Kaibab Plateau, for the protection of game animals. As part of this protection, government trappers removed thousands of predators from the region between 1906 and 1931, including mountain lions, coyotes, bobcats, and the last remaining gray wolves. Livestock grazing by sheep, cattle, and horses was also greatly reduced between 1889 and 1931.

Following this reduction of predators and competing livestock, the mule deer (*Odocoileus hemionus*) population of the plateau was described as having increased from 4,000 in 1906 to 100,000 in 1924. After depleting its food supply and experiencing two consecutive harsh winters (1924-1925 and 1925-1926), approximately 60 percent of the herd purportedly died, and the population reportedly declined to fewer than 10,000 in 1939.

Researcher Graeme Caughley later evaluated the available information and challenged both the events described and their interpretation. Caughley noted that most population estimates were based on poor data. For example, estimates of the peak population in 1924 by different biologists and visitors to the area varied from 30,000 to 100,000 deer, and none of these estimates were grounded in sound scientific studies. The purported 60 percent decline (from 100,000 to 40,000 deer) was actually an extrapolation of one visitor's estimate of a 60 percent decline from 50,000 to 20,000 deer between 1924 and 1926. The only deer population estimates with any continuity over this period were made by the forest supervisor, who reported a general increase from 4,000 deer in 1906 to approximately 30,000 deer in 1923; notably, he also reported a stable population of approximately 30,000 deer over the time span (1924 to 1929) of the purported population crash. Thus, much of the data on the Kaibab deer herd between 1906 and 1939 is judged to be unreliable and inconsistent.

Overall, there is reasonable field evidence for a significant population increase, some starvation, and a population decline among the deer of the Kaibab Plateau between 1906 and 1932. In a general fashion, the increase in deer numbers coincided with a predator control program and a reduction of livestock grazing in the region. Caughley believed that the Kaibab deer eruption was similar to that observed in other ungulates and was most likely linked to changes in food or habitat and was terminated by depletion of food supplies. In summary, however, the magnitude, timing, and causes of the Kaibab deer population eruption and crash are unclear. There are few scientific conclusions about human impacts on ungulate populations that can confidently be inferred from these events.

Richard G. Botzler

SEE ALSO: Balance of nature; Grand Canyon; Hunting; Predator management.

Kesterson National Wildlife Refuge poisoning

DATE: 1983
CATEGORY: Animals and endangered species

Transfer of subsurface drainage water from the San Joaquin Valley of California into the Kesterson National Wildlife Refuge led to widespread death and deformity of migratory birds. The toxicity was linked to dissolved selenium in the drainage water.

Selenium is a naturally occurring trace element that is essential for health in concentrations ranging from 0.05 to 0.3 parts per million (ppm) but becomes increasingly toxic at concentrations exceeding these low dietary levels. Selenium is found in igneous rocks, sedimentary rocks, and fossil fuels. Some soils contain naturally high selenium concentrations, and selenium occurs in the drainage from these soils because its ionic forms are readily soluble.

The Kesterson Reservoir in Merced County, California, was a 500-hectare series of holding ponds that collected the surface and subsurface drainage from irrigated agriculture in the western San Joaquin Valley. Because development had eliminated more than 90 percent of California's wetlands and because of the scarcity and high price of water in California, the U.S. Bureau of Reclamation and the U.S. Fish and Wildlife Service mutually agreed to incorporate the reservoir and its waters as part of the national wildlife refuge system in the mid-1970's.

By 1982 virtually all the water entering the Kesterson Reservoir was from subsurface drainage, and in 1983 scientists found that aquatic and migratory birds nesting in the refuge had grossly deformed embryos and high embryo mortality rates. By 1984 dead adult birds were being discovered in unusually high numbers. Although agricultural chemicals were suspected, tissue analysis of these birds revealed that they had selenium concentrations one hundred times in excess of normal concentrations because of selenium bioaccumulation in the food chain.

In 1986 discharge into the Kesterson Reservoir was halted, and the reservoir was allowed to drain naturally. In 1988, to protect migratory waterfowl from nesting in periodically flooded zones of the reservoir, about 60 percent of the lowest lying parts of the reservoir were filled with enough selenium-free backfill to ensure that the average seasonal groundwater level remained 15 centimeters (6 inches) below the soil surface. These steps were taken because wildlife biologists determined that the potential for selenium contamination of wildlife was much more limited in dryland environments than wetland environments.

The Kesterson Reservoir continues to be monitored because studies indicate that selenium concentrations in plant and wildlife from this site will remain elevated for decades. Bioremediation schemes were also investigated, and it was observed that microorganisms, particularly fungi, were able to volatilize up to 50 percent of the selenium in their vicinity within one year.

Mark Coyne

SEE ALSO: Agricultural chemicals; Biomagnification; Bioremediation; Food chains; Groundwater and groundwater pollution; Runoff: agricultural; Wetlands; Wildlife refuges.

Killer bees

CATEGORY: Animals and endangered species

Killer bees are a genetic strain of honeybees introduced to Brazil in 1957 in an attempt to hybridize the familiar European honeybee with an African subspecies. Since their introduction to Brazil, killer bees have spread across much of South America, Central America, Mexico, and the southern tier of United States.

Although killer bees (called Africanized bees by scientists) are no more venomous than ordinary honeybees, they exhibit lower average honey production, greater dispersal, and more aggressive behavior than their European counterparts. A person who disturbs a hive of Africanized bees may be pursued for more than 90 meters (300 feet) by thousands of bees. Each bee stings only once, but an aggravated swarm can inflict a lethal number of stings. While the popular media have exaggerated their danger, killer bees can be as deadly as their name implies. They have caused hundreds of human casualties since 1957. Most of these victims were trapped or otherwise unable to run away from the bees. In South and Central America, wild bee colonies under bridges or near farm machinery are common sources of attacks. In the United States, where most honeybees are managed in hives, the Africanized bee problem has been less severe.

An individual Africanized honeybee appears so similar to the European type that they are practically indistinguishable. Most of the differences are behavioral: In addition to their more aggressive nature, Africanized bees reproduce more often, depleting the hive's food reserves as they send out swarms to establish new colonies. Through frequent swarming, Africanized bees have expanded their range in the Americas by 400 kilometers (250 miles) per year. Because they store less honey and do not cluster for warmth in winter, Africanized bees do not survive in cold climates. They have slowed their advance at the southern third of the United States, with the limit of their range fluctuating from year to year depending on weather patterns.

The Africanized bee problem is a result of human introduction of a subspecies of honeybees from Africa to South America. Although the original goal was to improve honey harvests in the tropics, the unintended consequences of this failed experiment stand as a warning against releasing nonnative organisms into the wild. Because Africanized traits are dominant, hybridization and subsequent back-crosses produce bees with essentially African characteristics. As a result, the killer bee has not become more docile as its range has expanded. The difficulty in managing Africanized bees has threatened the multimillion dollar beekeeping industry in Mexico and the United States. Even more serious is the potential loss of pollination. Because managed bee colonies have replaced native pollinating insects in so many agricultural and natural communities, threats to beekeepers represent a global ecological concern.

Eradication of the Africanized bee is unlikely, but steps can be taken to limit its impact. Beekeepers have responded by changing management practices and maintaining tame varieties for breeding stock. Biologists continue to investigate honeybee genetics, and the public is learning to exercise caution around wild bee colonies.

Robert W. Kingsolver

SEE ALSO: Introduced and exotic species.

Kings Canyon and Sequoia National Parks

CATEGORY: Preservation and wilderness issues

Sequoia National Park and Grant Grove were established in 1890 in California's southern Sierra Mountains. In 1940 Grant Grove was incorporated into the new Kings Canyon National Park. Both parks have faced numerous environmental challenges since their creation.

The impetus for the creation of both Kings Canyon and Sequoia came from such preservationists as John Muir and the Sierra Club, as well as local inhabitants from the San Joaquin Valley. When the National Park Service was created in 1916, all national parks were given the obligation to provide public recreation as well as preserve the parks' natural environments. However, the two aims proved to be difficult to reconcile, particularly in Sequoia and Kings Canyon.

By the end of the 1920's it was apparent to some that environmental damage was already occurring in the Giant Forest area of Sequoia. Even before 1890 people were camping among the redwoods, and it was inevitable that the concessionaire would construct cabins and other buildings in the shadows of the great trees. Their fragile root systems were negatively affected by the buildings, the sewer system, and human traffic. It was not until the 1990's, however, that the buildings were actually torn down, to be relocated in areas that did not impact the redwoods.

Fire prevention was one of the prime concerns of national park officials. Here, too, time and knowledge radically altered that function. In the 1960's park officials realized that periodic fires were necessary to maintain the health of forests because they removed dead material and allowed new seedlings to sprout and flourish. A living forest required fire, and controlled burns became common in the park.

While most visitors got no further than the Giant Forest and General Grant areas, both similar in their redwood ambience, both parks were mainly wilderness areas of streams, meadows, and some of nation's most rugged mountains. Before the parks were established, the area had been overgrazed by sheep and cattle. Although the problem ended when the parks were founded, overuse by campers and backpackers continued to be a challenge. Eventually, access to the backcountry was limited through a permit system, the overused meadows were temporarily closed, and everyone was required to pack out everything that they carried in to minimize damage to the fragile wilderness environment.

Kings Canyon also faced environmental challenges. Its dramatic central valley along the Kings River was subject to numerous requests that dams be constructed, even within the park, and that the valley floor be developed for additional visitors. Here, too, preservation finally took precedence over recreation and other development, and the facilities were intentionally limited.

By the end of the twentieth century, preservation of the environment and its ecology had taken precedence over mere recreation in both parks. In general, the public has understood and supported the change in priorities. However, parks are also being threatened by forces outside the Park Service's control, particularly air pollution wafting up from the floor of the valley below and, further afield, possible dangers from the greenhouse effect.

Eugene Larson

SEE ALSO: Muir, John; National parks; Wilderness Act.

Kyoto Accords

DATE: December 11, 1997
CATEGORY: Weather and climate

The Kyoto Accords committed industrialized nations to place legally binding limits on their emissions of six greenhouse gases.

In 1988, the World Meteorological Organization and the United Nations established the Intergovernmental Panel on Climate Change (IPCC) to assess both scientific information and the environmental impacts of climate change. In 1990, the IPCC, consisting of more than two thousand leading scientists, released its first report. It concluded that human-made greenhouse gases would exacerbate the greenhouse effect, resulting in additional warming of the earth's surface by the twenty-first century unless measures were enacted to limit the emissions of these gases.

At the 1992 Earth Summit in Brazil, the United Nations Framework Convention on Climate Change (UNFCCC) was adopted. This treaty, signed by more than 150 nations, required each nation to limit its greenhouse gas emissions, with the industrialized nations taking the first step by voluntarily reducing their emissions to 1990 levels by the year 2000. The treaty was ratified by the U.S. Senate in October, 1992, and was eventually ratified by more than 160 nations.

In 1995, the IPCC released a report stating for the first time that the balance of evidence suggests a "discernible human influence on global climate." The report noted that even if emission levels remained constant, the atmospheric concentrations of carbon dioxide would approach twice the preindustrial concentration by the end of the twenty-first century, changing the earth's climate in significant ways. Between 1992 and 1995, global greenhouse emissions continued to rise, and it was agreed that the voluntary approach had not been successful. At a meeting in Berlin, Germany, in 1995, negotiators agreed that the industrialized nations, which emit the majority of greenhouse gases, would have to take the lead in adopting stronger measures. In 1996, the United States announced that future emission targets should be legally binding and challenged other industrialized nations to agree. More than one hundred nations agreed to develop legally binding targets.

In a March, 1997, meeting in Bonn, Germany, the European Union (EU) took the lead by proposing that industrialized nations reduce emissions by 15 percent from 1990 levels by the year 2010. The U.S. government proposed a system of international trading of emissions rights, in which nations could buy and sell the rights to emit greenhouse gases. The United States also proposed a "joint implementation" program, which would allow nations to earn emissions credits by implementing noncarbon-based "clean energy" projects. At the same meeting, the IPCC chairperson reported that reductions undertaken solely by industrialized nations would not be sufficient to limit global warming to environmentally sustainable levels.

As the December, 1997, Conference to the Parties to the UNFCCC in Kyoto, Japan, drew near, it appeared that reaching a consensus would be difficult, as proposed policies ranged from a 20 percent reduction (compared to 1990 levels) by 2005 to the U.S. proposal of merely stabilizing emissions at 1990 levels. Most developing nations stated that they would not commit to emissions controls until after the developed nations had acted.

As talks opened, the United States made a significant change in its position by announcing that it would support a system of flexible targets for different nations that would take into consideration the situation of each country. The United States maintained its position of zero reduction compared to 1990 levels until late in the conference, when Vice President Al Gore, Jr., instructed the U.S. delegation to be more flexible in negotiations. Gore's involvement helped to break the logjam of deliberations, and the final days of the conference were marked by around-the-clock negotiations until an agreement was finalized during the last hours.

The Kyoto Accords called for a 5.2 percent reduction in emissions carbon dioxide, meth-

Signatories to the Kyoto Accords of 1997 agreed to reduce emissions of six greenhouse gases from all sources, including industry. (Jim West)

ane, and nitrous oxide from 1990 levels by the period from 2008 to 2012, with the United States reducing emissions by 7 percent, Japan by 6 percent, and the EU by 8 percent. Three other greenhouse gases, all chlorofluorocarbon (CFC) substitutes, were to be cut by comparable levels, using 1995 rather than 1990 as the baseline. While the accords marked the beginning of a legal framework for reducing carbon emissions over the long run, the emission cuts fell far short of the levels that most climate scientists predicted would be needed to prevent significant climate change.

The accords were criticized by many environmental leaders as having major loopholes such as "flexibility measures" that could allow nations to circumvent reductions in emissions. Rules for how industrialized nations could trade or sell emission rights were not formalized. Russia, because its emissions plummeted during the collapse of its economy, was left with emission credits. These credits could be sold to countries such as the United States, with the result that there could be little or no reduction in emissions by major polluters. Also not addressed was how joint implementation would affect emissions reduction targets. Finally, no compliance mechanisms were included in the accords.

The most contentious issue resulting from the Kyoto Conference concerned the exclusion of developing countries from the accords. At the conference, a united group of 130 developing nations, led by China and India, voiced strong objections to the process of emissions trading, claiming that such trades would allow rich polluters to buy their way out of reductions. Voluntary commitments from developing nations to reduce emissions were demanded by the United States at the conference, but the proposal was rejected in the final negotiations. On a per-capita basis, emissions in developing nations were well below those of the industrialized nations. However, emissions were growing so rapidly in developing nations that they were projected to exceed those of the industrialized nations by 2025, with China projected to be the largest carbon dioxide emitter by 2015. Many industrialized nations, such as the United States, refused to ratify the accords unless the developing nations were included.

The Kyoto Accords represented the first international, legally binding attempt to prevent human activity from causing significant adverse changes to the earth's climate. For the accords to enter into force, however, they had be ratified by governments representing at least fifty-five nations, including industrialized nations representing 55 percent of all 1990 carbon dioxide emissions. Successful implementation of the accords thus awaited resolution of issues of nationalism, ideology, and economics as first steps in the international effort to reduce human influence on the climate.

Craig S. Gilman

SEE ALSO: Air pollution policy; Climate change and global warming; Intergovernmental Panel on Climate Change.

L

Lake Baikal

CATEGORY: Preservation and wilderness issues

Lake Baikal, located in Siberian Russia, is a unique freshwater lake whose pure waters and distinctive plant and animal life are threatened by industrial water pollution.

Lake Baikal is the oldest and deepest lake on earth. Almost 650 kilometers (400 miles) long, its surface area is approximately the same as that of Lake Superior, but its water volume makes up one-fifth of all fresh water on the earth's surface. The lake itself is more than 1.6 kilometers (1 mile) deep. Below the surface lies a 6.4-kilometer-deep (4-mile-deep) floor of sediment, which has drifted down from the lake during the last twenty to thirty million years. All this lies on a rift where several tectonic plates meet along a little-studied fault. The activity of these plates has apparently widened the rift and deepened the lake over many millennia.

Lake Baikal essentially has a closed ecosystem. Several hundred rivers feed into it, but the watershed consists entirely of the mountain area surrounding the lake, unconnected to other river systems. Baikal has only one outlet, the Angara River, which flows out from its southeast corner, past the old frontier city of Irkutsk, and ultimately to the Arctic Ocean. Because of the lake's isolation, many species found nowhere else live there. It is a fascinating site for biological study.

Among the lake's intriguing fauna is the silver-furred nerpa, the smallest known seal. Its closest relative, the Arctic seal, lives some 3,200 kilometers (2,000 miles) away. How the nerpa reached Baikal and adapted to fresh water is one of the lake's many mysteries. Nerpas eat an oily fish also found only in Baikal, the golomyanka. This almost transparent fish gives birth to live young, then promptly sinks and dies. Algae, plankton, and similar microscopic creatures form the bottom of the food chain. These serve as prey for a tiny crustacean called epishura, which strains Baikal's waters to a pristine clarity.

The land around Lake Baikal—consisting of taiga, or northern woodlands—shelters a variety of Siberian wildlife. Mountains ring the lake, creating spectacular scenery. Olkhon Island, located near the western shore in the midsection of the crescent-shaped lake, has a dry, almost snowless climate and contains grasslands and sand dunes. The smaller islands are seal nesting grounds. The lake surface freezes to a depth of more than 1 meter (3 feet) during the long Siberian winter.

Despite its remote location, the region was not immune to technological forces. The Trans-Siberian Railroad's builders left clear-cut areas and debris in the southern reaches, a problem repeated with the building of the Baikal-Amur Mainline paralleling the northwest shore decades later. By the early 1900's, sable in the surrounding forests had been hunted almost to extinction. Sturgeon and sturgeon eggs, valued as luxury food items, were dangerously depleted by the 1950's.

Buryat Mongols, indigenous inhabitants of the region, lived unobtrusively on the land, much like Native Americans. Russian settlers, arriving either by choice or by involuntary exile, had to make a living in the isolated region. Although fisheries could sustain operations without depleting populations of marine life, they did not always do so. Meanwhile, timbering and farming methods took a major toll in erosion.

The greatest environmental damage came with the establishment of large-scale industry in the region. During Joseph Stalin's regime as the Soviet leader (1929-1953), the Soviet Union emphasized industrial production over every other goal. His immediate successors retained this policy. A huge cellulose production plant opened at Baikalsk on the southern shore of the lake in 1966 and began spewing toxic chemicals into the lake water and murky smoke into the air. A pulp plant at Selingiske, located on one of Lake Baikal's major tributaries, and factory wastes and sewage from Ulan-Ude farther upriver created another major pollution area around the Selenga River delta. Hydroelectric dams built near Irkutsk in the 1950's brought more heavy industry and noxious by-products.

The harm wrought by these factories is unquantifiable but quite visible. In many places along the southern shore of the lake, the formerly pure water is now unfit to drink. Many square miles of lake area are now dead, with little or no aquatic life. Elsewhere, the populations of many native species have drastically shrunk. In some places newly introduced, hardier species are replacing them, with unknown consequences.

Distrust of the Baikalsk project spawned the first environmental protests in the Soviet Union. Although these did not prevent the plant's opening, a precedent and framework was created for the future. By the time of Mikhail Gorbachev's chairmanship (1985-1991), the lake's pollution was obvious, and the political scene was less repressive. Gorbachev pledged to convert the Baikalsk plant to nonpolluting activities, but political turmoil and the Soviet Union's breakup intervened, after which Lake Baikal became Russia's responsibility. The lake was embraced as a national treasure deserving protection, and plans were drawn up to lessen or reverse damage to the lake and region. Some antipollution laws were passed. Russia began to draw on outside resources, both scientific and business-oriented, for this effort. Tourism was touted as a relatively clean industry for the future.

The biggest problems of the 1990's were those of inaction caused by Russian political and economic problems. Laws and plans were adopted, but the money to implement them has been scarce. Coordination between local political entities is difficult. Declining industrial activity caused by economic slowdown may have slowed pollution more effectively than any active measures. Whether human-caused damage can be reversed in and around the lake is still an open question.

Emily Alward

SUGGESTED READINGS: *Baikal: Sacred Sea of Siberia* (1992), by Peter Matthiessen, is an insightful travel journal along with an impassioned plea to help save the lake. "Russia's Lake Baikal: The World's Great Lake," *National Geographic* (June, 1992), by Don Belt, combines information with local anecdotes and the magazine's usual excellent photographs. "No-Polluting Zone," *Scientific American* (December, 1994), by W. Wayt Gibbs, explains a plan adopted for sustainable development in the Baikal watershed. A Baikal Watch Web site can be found at http://www.igc.apc.org/ei/baikal/currents. The site has accessible information from Russian sources, although it is not always up-to-date.

SEE ALSO: Pulp and paper mills; Water pollution; Watersheds and watershed management.

Lake Erie

CATEGORY: Water and water pollution

Sustained efforts to rehabilitate the severely polluted Lake Erie began in 1965 and brought about a general recovery of the lake.

Lake Erie, the shallowest and southernmost of the Great Lakes, was far along the process of eutrophication, or natural aging, prior to settlement by Europeans. Early settlers accelerated this process by draining coastal wetlands and stripping away vegetation, which increased the amount of sediment carried to the lake. With the advent of widespread agriculture, artificial fertilizers also began to wash into Lake Erie, which contributed to overenrichment of the lake's waters. Sewage and fertilizers, such as phosphorous and nitrogen, caused the rapid growth of surface algae scums, which affected the taste and odor of drinking water supplies, clogged water intakes, and forced beach closures. More important, decaying algae consumed the water's oxygen, leading to the suffocation of bottom-dwelling organisms. Eventually, desirable fish stocking the lake were stressed by lack of adequate food, and populations declined.

Industries along the lake's main tributaries contributed to the problem by injecting industrial wastes, oil, floating solids, and heavy metals into the water supply. Heavy metals such as lead and mercury, as well as organic chemicals such as polychlorinated biphenyls (PCBs) and dioxin, magnify their concentrations as they pass up the food chain. These bioaccumulation and biomagnification effects may increase levels of toxic materials by one million times in fish such as salmon and trout. Fish consumption advisories were required for some species because of contamination levels. By 1965 Lake Erie had become so polluted that public indignation over its condition led to action by government officials.

Four states in the United States (Michigan, New York, Pennsylvania, and Ohio) and one Canadian province (Ontario) share in managing Lake Erie. The International Joint Commission was established by Canada and the United States in 1909 to arbitrate disputes over shared boundary waters. The commission established the Great Lakes Water Quality Agreement, which provided for reduction in nitrogen and phosphorus discharge in Lake Erie by improving the municipal sewage-treatment systems. The retention of sanitary and storm-sewer overflow for later treatment has greatly reduced health problems, and beach closures have become less frequent with improved treatment of sewage.

The International Reference Group on Great Lakes Pollution from Land Use, under the International Joint Commission Authority, prepared reports that provided the groundwork for the ecosystem approach to reducing pollution in Lake Erie promulgated in the 1978 Water Quality Agreement. The agreement states that the programs must center on the physical, chemical, and biological relationships among air, land, and water resources. The ecosystem approach used by the 1978 agreement means that standards and monitoring methods must take into account the air, land, and water movement of pollutants and their risks to humans and other organisms.

The Great Lakes Fishery Commission, created in 1955 by Canada and the United States, is concerned with restoring and stocking lake fish. White fish are showing signs of recovery. Lake trout and coho salmon are stocked in the lake. Walleye and yellow perch are managed for recreational and commercial fishing. Tests have also revealed that the level of PCBs in some Lake Erie fish are diminishing.

In 1985 the Great Lakes Charter was inaugurated to resist the transfer of the Great Lakes water to other areas. It authorizes the development of an information database for surface water and groundwater resources. In 1986 the Great Lakes Toxic Substances Control Agreement (GLTSCA) was formed to coordinate the actions of the Great Lakes states to reduce toxic substances in the basin. The Great Lakes Water Quality Agreement of 1987 called for forming specific ecosystem objectives and indicators. This approach enhances the evolution of full ecosystem management strategies, which incorporate mathematical modeling.

Ronald J. Raven

SEE ALSO: Biomagnification; Cultural eutrophication; Dioxin; Polychlorinated biphenyls; Water pollution.

Land-use policy

CATEGORY: Land and land use

Land-use policy refers to the way in which societies organize, plan, and manage social and physical activities on the landscape, and the issues raised in pursuit of these activities.

Human societies are organized to survive in particular environments. The adaptive nature of culture allows the society to respond to changes in the environment and cause even more change. Once sedentary existence in cities began, formal land-use policies arose, and decisions about how to manage land became critical to survival. Even during the earliest periods of the Mayan, Egyptian, and other city-state civilization, land-use control was implemented for religious, agricultural, hunting, and residential purposes. Religious proscriptions dictated the appropriate appearances of structures. Plagues, warfare, and resource distribution demonstrated the need to isolate structures as well as groups of people. Crowding and cultural conflicts made it necessary to create processes to make government decisions about who would get to use particular resources. Land-use policies are even more necessary in the contemporary world of competing interests, growing populations, and diminishing natural resources. Increased scientific knowledge about environmental impacts has fueled the need to ensure that appropriate land-use decisions are made in the interests of survival through long-term resource management.

ROOTS OF LAND-USE POLICY

The earliest land-use policies took the form of religious prohibitions and mandates as the priest class interpreted the needs of the gods for sacred space. Kings exercised a divine mandate in interpreting just what could be allowed in sacred areas, while market forces dictated the use of less important secular space. Later, after kings began losing their divine authority, they could still regulate the use of secular space in the name of promoting social order. Social class structure continued to support the allegiance to class-based authority.

After the Renaissance, European societies believed that while God no longer directly intervened in day-to-day activities, his presence was felt in the need to maintain social order through hierarchies. It was felt that this social order must

The Land-Use Planning Process

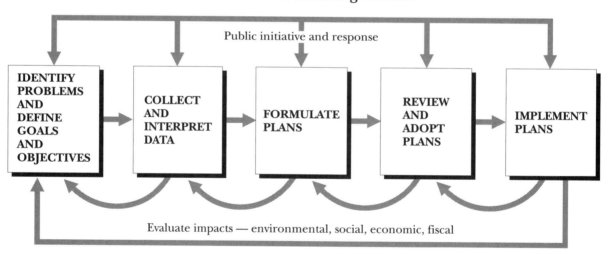

Public initiative and response

IDENTIFY PROBLEMS AND DEFINE GOALS AND OBJECTIVES → COLLECT AND INTERPRET DATA → FORMULATE PLANS → REVIEW AND ADOPT PLANS → IMPLEMENT PLANS

Evaluate impacts — environmental, social, economic, fiscal

COLLECT DATA

Earth science and other
  information
Background studies
  Existing land use
  Transportation
  Economic
  Political
  Social
Land capability studies

FORMULATE PLANS

Land use
Watershed
Natural resources
Hazard mitigation
Open space
Waste management
Public facilities

IMPLEMENT PLANS

Zoning and subdivision regulations
Erosion and sedimentation control
  ordinances
Building and housing codes
Environmental impact statements
Capital improvement programs
Health and information codes

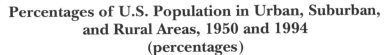

Percentages of U.S. Population in Urban, Suburban, and Rural Areas, 1950 and 1994
(percentages)

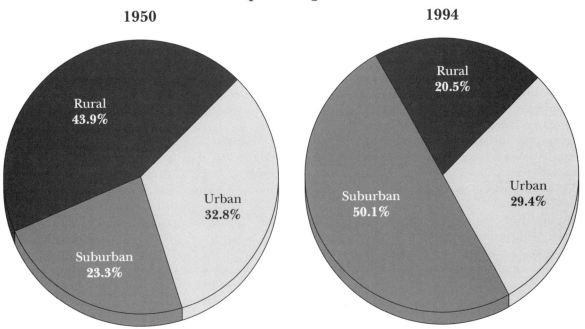

1950

Rural
43.9%

Urban
32.8%

Suburban
23.3%

1994

Rural
20.5%

Suburban
50.1%

Urban
29.4%

Source: Data are calculated from Council on Environmental Quality, *Environmental Quality 25th Anniversary Report, 1994-1995* (Washington, D.C.: U.S. Government Printing Office, 1995).

also be reflected in physical space, or in the way in which a landscape was arranged. Accordingly, it was seen as only proper that buildings, towns, and landscapes reflected particular patterns in form and ownership. In the United States, examples of kingly intervention can be found in the marking of potential mast pine trees in New England with the king's broad arrow, the designation of village squares, and the reservation of lots for the king's agents. However, the rise of a democratic society provided a shift in decision-making authority to elected representatives.

By the mid-nineteenth century, early planning laws in the United States began to regulate urban tenement housing and prohibit "obnoxious uses." In 1893 the World's Columbian Exposition in Chicago promoted the exchange of planning and design concepts in exhibits by landscape architect Frederick Law Olmsted, Sr., artist Saint Gaudens, and others. By 1895 Los Angeles had an ordinance prohibiting steam

shoddying (reprocessing of wool scrap) plants within 30.5 meters (100 feet) of a church. In addition to ordinances and the designation of parks, planned communities were initiated. Ebenezer Howard's *Tomorrow: A Peaceful Path to Real Reform* (1889) launched the Garden City movement. In an 1899 decision, the Massachusetts Supreme Court upheld a building height limitation; by 1909 zones of limits for buildings were upheld in the U.S. Supreme Court.

By the early twentieth century, organized planning efforts were underway in some large cities. Hartford, Connecticut, established a planning board in 1907. By 1909 Wisconsin had passed the first state enabling act for planning, and Los Angeles provided the first American use of zoning to direct future development in a series of multiple zoning ordinances. In 1913 Massachusetts became the first state to make planning mandatory for local governments. Newark hired the first full-time city planner in 1914. The first comprehensive zoning code in the United States was enacted in New York City in 1916. By 1925 Cincinnati, Ohio, had become the first major U.S. city to adopt a comprehensive plan, and Burlington, Vermont, had authorized a municipal planning commission.

POSTWAR POLICIES

In the 1940's planning for war and postwar public housing brought federal review to the land-use planning process. The mustering of resources for World War II demonstrated the value of planning, and many towns implemented town plans as postwar prosperity began. In 1962 the Chicago Area Transit Study showed the applicability of cost-benefit studies to planning for suitable development.

In the 1970's people became increasingly concerned with the need to control development. Performance standards were upheld in the court systems as one mechanism to manage growth through land-use policy. In 1971 the concept of transferrable development rights (TDRs) was introduced to help preserve urban landmarks. Creative land-use tools such as conservation easements, controlled access rights, planned unit development (PUD) density credits, and overlay districts were increasingly used in the 1970's as communities expanded their regulatory schemes. In their zeal to manage growth, some communities enacted land-use policies that were discriminatory. In an early decision on discriminatory land-use regulation, in 1975 the New Jersey Supreme Court struck down a restrictive Mount Laurel zoning ordinance on the basis that it did not allow a regional "fair share" of low- and middle-income housing.

By the late 1980's a sufficient number of court cases existed to reduce the likelihood of discriminatory ordinances, but a slump in the U.S. economy fueled the challenge of some land-use ordinances and policies as unauthorized "takings"—the erosion or loss of landowner rights through excessive regulation without due compensation. Yet the vast majority of land-use regulations are not generally found to be true takings because at least some economic use of land is allowed. Still, the issue of a taking is frequently raised when regulations deprive an owner of one or more desired uses. The economic climate has a strong influence on the reaction of a population to particular issues such as takings and to land-use regulation in general.

Land-use regulation is generally viewed as the control of two categories: subdivision (physical size or boundary change) and development (physical use or alteration within the boundary). Changing land ownership or subdivision is a land use because it affects the management of resources on the land and can fragment habitat. Most land that has been subdivided stays that way and seldom reverts to an original larger tract. Land that is used or developed through construction, clearing, or other alteration, including various forms of land management such as agriculture, also undergoes a change to the natural path of succession. In fact, sufficient past alteration through introduction of new species or direct physical action has so altered some landscapes as to render it almost impossible to determine the true "natural" condition. It is this perspective that, along with concern over environmental impacts from the yet-to-occur changes, is most commonly used to justify land-use policies.

CONVENTIONAL AND CONSERVATION PLANNING

Conventional land-use planning assumes the desirability of economic growth through new development and therefore tends to favor revenue generators. Tools such as zoning are the main techniques for conventional planning. A government land-use plan is implemented through a series of development regulations that differ according to zones. A series of base maps are prepared after completion of a natural, social, and infrastructural resource inventory. The maps can be viewed as opportunities and constraints for growth. Individual base maps, when combined with the community's or the region's goals and objectives for growth, result in a land-use map containing zones. Each zone reflects a particular category: commercial, industrial, residential, recreational, historical, governmental, agricultural, special, and others that are suggested by the inventory and goal processes. Each category can contain a variety of subcategories based on lot size and range of allowable uses.

The size, configuration, and pattern of the arrangement of lots can all affect growth. For example, large lots reduce the number of houses, while small lots increase the number of houses, reduce the amount of open space, and increase the fragmentation of land ownership. The community uses the planning process to agree upon the development capacity. There are tools and approaches that seek a balance between achieving appropriate densities and maintaining natural and aesthetic resources; the objective is to achieve sustainability.

Conservation planning is one approach to improving land use within conventional planning. In conservation planning, structures and uses such as septic systems are located away from valued or critical natural resources on a tract. Conventional planning, even if conservation oriented, can still lead to checkerboard or highway strip patterns of development that are harmful to open space. Clustering is a technique that goes one step further by attempting to preserve or conserve open space by treating it as a natural resource. Other cited benefits of clustering are the fostering of a sense of community, reduction of urban sprawl, and the presentation of traditional village appearances.

In clustering, dwelling units (or commercial structures) are grouped together to allow a larger uninterrupted area of open space (generally 25 to 30 percent), which is often maintained in a residential or commercial subdivision as common land under shared ownership. Some communities encourage clustering by a policy of allowing a greater density of units or square feet of construction. Thus, a 50-acre tract might be approved for construction of five houses in a traditional checkerboard subdivision of ten lots but might be allowed to contain up to fifteen units of housing if the houses (or condominiums) are clustered and if a specified percentage of the parcel is preserved as open space. The open space might be a separate 20-acre lot that is deeded to the prospective unit owners as shared or common land to be managed in a certain way.

Conservation planning can be employed on a city- or statewide level through the administration of a tiered level of permits and other development review processes coupled with a system of land-use planning in which designated areas are conserved. Oregon's greenline boundaries limiting growth at the edge of cities are one example. Trails and greenbelt corridors also reflect conservation planning but require considerable coordination and cooperation when linking more than one community or state.

ECOLOGICAL PLANNING

Ecological land-use planning expands upon conventional land-use planning by taking a more integrative perspective of ecosystem dynamics and applying it over a greater period of time than the normal five-year period. This comprehensive approach requires significant data about a wide variety of resources and also requires a fairly stable set of goals and objectives in a relatively constant sociopolitical setting. Ecological planning provides the benefit of long-range dynamic planning while attempting to prevent, rather than remedy, problems. However, it is more costly in initial expenditures. The Netherlands and some other northern European countries have a long history of ecological land-use planning, particularly in response to increased population pressures in areas of finite land resources.

The decrease of rural inhabitants and the growing number of suburbanites in the United States in the latter half of the twentieth century caused an increase in the consideration of various planning techniques and tools. Vermont adopted a statewide land-use policy law in which individual development decisions are made by a regional volunteer citizen panel at quasi-judicial hearings in which dispute resolution techniques and consensus building are encouraged. Vermont has found this case-by-case process to work quite well despite the lack of a comprehensive statewide plan and despite an expanding population. Oregon uses a similar process. Florida's Environmental Land and Water Management Act of 1972 allows the state to designate areas of critical interest that local governments must consider when enacting local policies. It also requires state review of development projects that are large or have regional impact. However, the accumulative adverse environmental effects of many small subdivision and construction projects significantly outweigh those of the larger projects that are more intensely regulated. Most states are beginning to recognize this through increased local control of land use and through statewide natural resource programs that reflect an ecological understanding of the interactions between changing land use and the need to manage wildlife and other natural resources.

FORMS OF LAND-USE REGULATION

Land use can be controlled through incentives or restrictive processes. Incentive-based land-use control includes direct funding, grants, tax abatements, tradeoffs such as credits for clustering, and other forms of positive feedback. Restrictive land-use control is the form more commonly recognized and employed, notably through the use of permits, licenses, environmental assessments, taxes, and direct prohibitions. Restrictions may be imposed by government or by the landowner via covenants or easements. Government restrictive regulations have two forms: prescriptive, in which the objectives and specifications are precisely articulated; and subscriptive, in which the outcomes are specified but the individual means of achieving them are left to the discretion of the owner or community. Critical land uses and issues, such as matters of public health, are more likely to be prescriptively regulated. Less critical matters might be handled by subscriptive means, also called performance-based planning. Even tax structuring can be performance-based when land is taxed based on actual use rather than potential use. This can reduce the pressure to commercially develop high-value or expensive properties.

Most communities and governments use a combination of the two forms of land-use control and the two forms of regulations. Much of land-use policy concerns the manner in which the combination is achieved. Issues of land-use control can become highly politicized, as in the wise-use movement or the controversy involving the northern spotted owl and logging policy in the northwestern United States. In such cases, differing cultural and social values render it difficult to achieve sufficient commu-

nity consensus to support consistent regulation.

Although the United States and Canada have contributed greatly to planning literature and the development of innovative techniques, North America is considered to have relatively weak land-use controls, as does Australia. Northern Europe and countries such as Japan are known for their comprehensive land-use controls. In the United States, as in many countries, the redistribution of population and wealth, together with the restructuring of public lands, necessitates constant reexamination of land-use policies and regulations. For developing nations, land-use policies become particularly critical in the evaluation of tradeoffs between natural and land-based resources on one hand and economic well-being on the other. Pressures exist for these countries to exploit their natural resources while striving toward the prosperity of more developed countries such as those in North America and Northern Europe. As global markets expand, the need for dialogue on international land-use issues grows.

R. M. Sanford and H. B. Stroud

SUGGESTED READINGS: Randall Arendt's *Rural by Design* (1994) is a practical handbook for planners, volunteer board members, landowners, and other members of the general public. Aldo Leopold's *A Sand County Almanac* (1949) is a classic, highly readable treatment of the relationship between humans and the land that inspired subsequent generations of planners. *Landscape Planning: Environmental Applications* (1998), by William M. Marsh, gives a structured approach to planning and resource analysis. Ian McHarg's landmark *Design with Nature* (1995) explains and promotes ecological principles in design and planning. *Managing Growth in America's Communities* (1997), by Douglas R. Porter, is a straightforward treatment of land-use development issues for the layperson. *Rural Environmental Planning for Sustainable Communities* (1991), by Frederic O. Sargent, Paul Lusk, José A. Rivera, and María Varela, is a useful overview for citizen board members interested in rural planning.

SEE ALSO: Conservation policy; Forest and range policy; Greenbelts; Open space; Planned communities; Urban planning.

Landfills

CATEGORY: Waste and waste management

Sanitary landfills are sites at which solid wastes are disposed of without creating a hazard to public health or safety. Solid waste is usually compacted and covered at the end of each day to keep out pests, air, and water and keep in gases produced within the landfill.

Prior to the 1930's most solid waste was discarded in unsightly, open areas called dumps. The open dumps attracted animal pests, and the chemicals and bacteria in the dumps could easily move into nearby water sources. In the 1930's an increasing number of communities began disposing of solid waste in sanitary landfills; by 1960 more than 1,400 cities had adopted this method of dealing with waste materials. Subsequently, government regulations were adopted that mandated the use of sanitary landfills to avoid problems posed by open dumps.

The solid waste placed in sanitary landfills in the United States is about 40 percent paper, 20 percent yard trimmings, and lesser amounts of glass, metal, plastics, wood, food, and other materials. The ratio of material is different in other countries. For example, one Asian city has a landfill composed of mostly vegetable matter.

At the end of each day, bulldozers compact the trash in the landfill, then cover it with a 15 to 30 centimeter (6 to 12 inch) layer of soil to keep out animals and air. In areas where the water table (a zone in the ground below which all the pore space contains water) is close to the surface, the waste may be placed directly on the ground in flat areas and covered with soil each day. In places with a lower water table, the waste is placed in a trench before it is covered with soil. On sloping areas, the waste is placed on the slope and covered with material taken from the

A landfill in Wayne County, Michigan. After the solid waste is dumped onto the ground, it is usually compacted and covered with a thin layer of soil. (Jim West)

Schematic of a Municipal Landfill

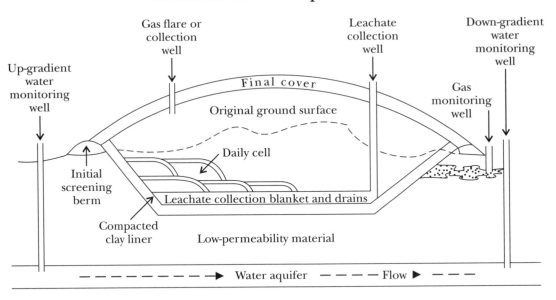

Note: Not to scale.

base of the slope. The most favorable sites for landfills are dry areas far from populated areas, although this may not be possible if the waste has to be transported too far, as this would be too expensive. Dry areas are the best locations for landfills since there will be little or no water to percolate into the solid waste.

As water moves through solid waste in a landfill, it dissolves many constituents to high concentrations. This produces a corrosive liquid called leachate. Solid waste in landfills is ideally placed in an area that contains rocks such as mudrocks or igneous rocks that are impermeable to the movement of the landfill leachate to groundwater or surface water. Burial in mudrock is especially desirable since the fine-grained clay minerals in it act as a natural filter and adsorb many of the dissolved or suspended constituents in the leachate. Ideally, a minimum of about 10 meters (33 feet) of impermeable rock is desirable between the base of the landfill and the water table. Landfills placed in permeable sediment such as sand or gravel with the water table close to the surface favor rapid movement of leachate into the groundwater system. Also, a double thickness of an impermeable liner such

as plastic is usually placed between the landfill material and the impermeable rock. A monitor system is placed between the two impermeable layers to check for leakage of the leachate. In addition, landfills must have a way to remove the leachate to minimize the amount of waste that can potentially move into the groundwater. Finally, monitor wells adjacent to the landfill must be periodically sampled to see if any contamination is moving into the groundwater.

As water slowly moves laterally through waste material, bacteria and many organic and inorganic substances are incorporated into the leachate and can reach high concentrations. The concentration of constituents in the leachate depends on the composition and grain size of the solid waste, the pH (negative log of the hydrogen ion concentration), the amount of water, contact time of the solid waste and leachate, and composition of the gas produced in the landfill. For example, most of the biodegradable solid wastes are in the organic material called carbohydrates. The carbohydrates, with the help of bacteria in an oxygen-free environment, break down into methane and carbon dioxide gas as heat is produced. A variety of organic compounds such as sugar, acetic acid, butyric acid, and proprionic acid have been detected in leachate. Also, the acidic leachate helps to dissolve many inorganic compounds out of the solid waste. Thus, fresh leachates from landfills in the United States contain relatively high concentrations of constituents such as lead (0.4 milligrams/liter), zinc (22 milligrams/liter), copper (0.1 milligrams/liter), and iron (540 milligrams/liter). Over time, the acidity of the leachate decreases, and the concentration of most constituents tends to correspondingly decrease (for example, lead drops to 0.1 milligrams/liter).

When landfills are closed, they must be covered with impermeable material, and a layer of soil up to 1 meter (3 feet) thick is placed on top of it. The layer of soil is graded and planted and may be used for parks or other recreational activity. This, for example, was done at Virginia Beach, Virginia. At Virginia Beach, the water table is close to the surface, so the landfill was placed on the surface. As waste was placed in the landfill, a hill more than 280 meters (920 feet) by 120 meters (395 feet) by 20 meters (65 feet) high was built above the land surface. When the landfill was abandoned, it was used as a recreational area. However, buildings should not be constructed on landfills because, as the trash is gradually compacted, the building may subside, resulting in structural damage. In addition, gas such as methane could potentially move into the building.

Robert L. Cullers

SUGGESTED READINGS: A detailed discussion of landfill practices and chemical changes occurring in the landfill is given in Kenneth Westlakes's *Landfill Waste Pollution and Control* (1995). *Sanitary Land Filling*, edited by Thomas H. Christensen, contains a specific discussion of gas production, chemical changes, environmental impact, and design of landfills. Debra Reinhart and Timothy Townsend, *Landfill Bioreactor Design and Operation* (1998), summarizes a specific type of landfill in which environmental impact is minimized and waste degradation is maximized.

SEE ALSO: Hazardous waste; Waste management.

Law of the Sea Treaty

DATE: passed 1982; became effective November 16, 1994
CATEGORY: Water and water pollution

The 1982 Law of the Sea Treaty, which contains 320 articles and nine additional annexes, established a broad range of international laws that regulate the use of the ocean and seabed.

The Law of the Sea Treaty is one of the most complex and comprehensive international agreements ever negotiated. Begun in 1973, the Law of the Sea Conference spent nine years formulating the treaty. It then took nearly twelve years before the resulting treaty was ratified by the required sixty countries. The treaty became effective on November 16, 1994, one year after the sixtieth nation, Guyana, signed the agreement.

Among the many law of the sea issues for which new rules were devised or customary international law was codified were innocent passage in the territorial sea, transit of passages and innocent passage in international straits, extension of coastal state jurisdiction on the continental shelf, conservation and management of the living resources of the high seas, the regime of islands, enclosed or semienclosed seas, landlocked states, protection and preservation of the marine environment, environment of ice-covered areas, marine scientific research, settlement of disputes, exploration of mineral resources in areas beyond national jurisdiction, and exclusive economic zone.

The Law of the Sea Conference and subsequent actions by many countries have effectively divided the ocean into four main zones: a territorial sea extending 12 nautical miles from a nation's shoreline, an exclusive economic zone (EEZ) that extends 200 nautical miles from the coast and includes the territorial sea, a continental shelf that may extend up to 350 nautical miles from land, and a zone enclosing the rest of the ocean called the "Area," which includes much of what has traditionally been considered the high seas.

Within its EEZ, a coastal nation has complete control over marine resources and scientific research. It has jurisdiction of all use of the ocean and can make rules concerning pollution control. Since 1982 many countries (including the United States, despite its unwillingness to sign the treaty) have declared 12-nautical-mile-wide territorial seas and 200-nautical-mile-wide EEZs. The United States' EEZ is the largest in the world (3,362,600 square nautical miles, or approximately 11,500,000 square kilometers), about 20 percent larger than the nation's land area.

Management of living resources of the ocean, especially species that migrate, was an important issue of the Law of the Sea Conference. Also, areas where many nations have traditionally fished that might come under a single country's jurisdiction under the treaty presented complex problems. The treaty provides a mechanism for binding arbitration through the International Court of Justice at The Hague, Netherlands, to resolve such disputes. In 1984 the court resolved disputed claims between the United States and Canada over control of Georges Bank in the North Atlantic Ocean by awarding a portion to each nation.

Charles E. Herdendorf

SEE ALSO: Ocean dumping; Ocean pollution; Seabed disposal.

Lead poisoning

CATEGORY: Human health and the environment

Excessive, long-term exposure to lead—which has been used in the manufacture of gasoline, paint, and water pipes, among other things—may lead to severe health problems and even death.

Lead is a relatively rare element that has been mined and used for centuries. The concentration of lead in the earth's crust is only 12.5 parts per million (ppm), and most of this lead is found in the ore mineral galena. Evidence of the use of lead dates back at least five thousand years, and the environmental impacts of smelting and using lead can be detected in ancient bog deposits of England and Spain that date back at least 2,800 years. Lead was used by ancient European and Chinese civilizations for plumbing, drinking vessels, weights, ornaments, and even storing wine. There is considerable evidence from human skeletons that lead poisoning was a significant problem in ancient Roman society.

Lead is now the fifth most commonly used metal in the world. The majority of lead is used for the manufacture of lead acid batteries, but much of the lead mined in the twentieth century was used for the production of antiknock compounds (such as tetraethyl lead) that were added to gasoline. Lead has also been used in the manufacture of some paints, ceramic glazes, ammunition, and solder, and it is the preferred material for shielding X rays. Because of the widespread burning of gasolines with tetraethyl lead and the transport of lead compounds throughout the atmosphere, lead may be the most widely distributed heavy metal. Measured levels of lead in remote areas range from three to five times that of natural background levels.

About 60 percent of the lead exposure for adults comes from food, both fresh and canned. Lead is naturally taken into plants, but it also accumulates in food because of dry deposition from the atmosphere and storage in cans. Approximately 30 percent of the lead in humans comes from the inhalation of air, although this exposure has dramatically decreased since leaded gasoline was phased out of use for most vehicles in the United States in the late 1970's. Lead is also released into the atmosphere through the burning of solid waste, coal, and oil. Tobacco smoke is an additional source of airborne lead exposure. The remaining 10 percent comes from water. Much of the exposure through water is the result of lead pipes, fixtures, or solders used in older plumbing. An additional significant exposure route for children is through ingestion of lead-based paint chips used on many homes or from the eating of dirt containing lead paint residues.

Lead is a toxic element that can cause both acute effects from short-term high dosage exposure and chronic effects that result from long-term exposures at lower levels. Children and pregnant women are at particularly high risk with regard to lead exposure. Children may ingest higher levels of lead from soil, and the effects of lead poisoning in children begin at lower blood levels. Pregnant women are a high risk because lead can substitute for calcium in bones and may be mobilized during calcium deficiencies such as pregnancy. The lead released into the blood can cross the placenta and cause damage to the fetus or even a miscarriage.

At high levels, lead poisoning can cause severe brain damage, gastrointestinal disorders, kidney damage, and even death. At lower levels, the symptoms of lead poisoning are not as severe and include constipation, vomiting, abdominal pains, and loss of muscular coordination. Because these symptoms may result from other causes, it is not always clear when lead poisoning has occurred.

The long-term chronic effects of low-level lead exposure may lead to many health problems, primarily with the circulatory and central nervous systems. Most of the lead that is absorbed ends up in the blood, and its general residence time is between two and three weeks. Long-term exposure can lead to disruption of the formation of heme in the blood and cause other enzymatic disorders. Anemia can result, with the symptoms arising at much lower blood lead levels in children.

Lead is also a neurotoxin that primarily affects workers in lead-related mining and manufacturing, as well as children. In children, exposure to lead may lead to lower mental capabilities, but it is not yet clear to what extent this can be established. It is known that lead can cause hearing disorders and even slow the growth of children. Chronic exposure in humans can cause blood-pressure problems and interference with vitamin D metabolism. Reproductive problems such as lower sperm counts and spontaneous abortions have also been linked to lead exposure. It is not yet clear whether lead is a carcinogen. Studies on workers in lead industries have been inconclusive with regard to links between cancer and lead exposure, but animal studies have shown an increase in cancers of the kidney. The Environmental Protection Agency (EPA) considers lead to be a probable human carcinogen and has classified it as a Group B2 carcinogen.

Because of its highly toxic nature, lead has been extensively studied and closely regulated in the United States and many other countries. Lead was one of the six primary pollutants regulated by the Clean Air Act, and the EPA was required to establish ambient outdoor air-quality standards for this pollutant. The removal of leaded gasoline from the United States market reduced the concentrations of lead in the atmosphere by 93 percent between 1979 and 1988. In addition, the removal of most lead-based paints has reduced the risk of accidental ingestion of lead by children. As a result of these and other regulations, the lead exposure of most adults and children has been significantly reduced. However, lead levels in the blood of many children indicate that exposure is still a great concern.

Jay R. Yett

SUGGESTED READINGS: A good source on toxics and toxicology is *Toxics A to Z* (1991), by John Harte, et al., which includes general information about toxicology and a section on each of the major toxicants, including lead. A more detailed but readable text on toxicology is *Introduction to Toxicology* (1995), by J. A. Trimbrell. *Mineral and Energy Resources* (1990), by Douglas Brookins, is an excellent source for information about the use of lead and environmental problems associated with lead and other metals.

SEE ALSO: Biomagnification; Clean Air Act and amendments; Environmental illnesses; Hazardous and toxic substance regulation; Heavy metals and heavy metal poisoning.

Leopold, Aldo

BORN: January 11, 1887; Burlington, Iowa
DIED: April 21, 1948; near Baraboo, Sauk County, Wisconsin
CATEGORY: Ecology and ecosystems

Wilderness conservationist and environmental philosopher Aldo Leopold has been called the father of modern wildlife management and ecol-

ogy. He is historically significant for applying his insightful concepts of ethics and philosophy to conservation strategies.

Aldo Leopold developed an interest in wildlife while observing waterfowl and animals living in the Mississippi River marshes near his childhood home. After completing graduate studies at the Yale School of Forestry in 1909, he worked for the U.S. Forest Service in Arizona Territory and New Mexico. With his colleague Arthur Carhart, Leopold demanded that the Forest Service preserve wilderness areas in national forests for their aesthetic and recreational value. As a result, in June, 1924, the Forest Service set aside 574,000 acres in New Mexico as the Gila Wilderness Area, initiating the protection of millions of forest acres for environmentalists to study and enjoy.

Leopold then initiated a game protection movement in the southwestern United States. An avid sportsman, he initially promoted the hunting of predators to protect game. When a deer herd on the Arizona Kaibab Plateau suffered starvation because of a high population caused by a lack of predators, Leopold realized that the predator-prey balance was crucial to a healthy, stable environment. As game consultant for the Sporting Arms and Ammunition Manufacturers' Institute, his *Report on Game Survey of the North Central States* (1931) was one of the first studies of American game population. He traveled to Europe to assess foreign game management techniques and developed a national game management policy for the American Game Protective Association.

In 1933 Leopold published *Game Management*, a popular textbook in which he discussed wildlife population dynamics and habitat protection, emphasizing the preservation of ecosystems. He considered the wilderness as a community to be shared rather than a commodity to be controlled by humans, stressing that nature should be treated ethically and not appropriated for economic gain. Leopold accepted the chair in game management, created especially for him, at the University of Wisconsin in 1933. He built a cabin by the Wisconsin River, planted trees, experimented with land restoration, and observed the

wildlife of Sauk County. Active in conservation groups, he helped organize the Wilderness Society and the Wildlife Society. President Franklin D. Roosevelt appointed Leopold to the Special Committee on Wild Life Restoration in 1934.

Leopold wrote essays about environmental ethics, which were compiled in *A Sand County Almanac and Sketches Here and There* (1949) after his death. He outlined his concept of a land ethic in which all wilderness residents had a vital role. He believed that humans were a part of nature and that individuals had a responsibility to understand, protect, and live in harmony with a healthy environment. He expected humans to respect nature and "preserve the integrity, stability, and beauty of the biotic community." Because of its literary appeal, Leopold's book introduced a broader audience to the conservation movement. When the book was reissued in the 1960's, a new generation of environmental readers embraced Leopold's passion for nature, praising his conservation initiatives and wilderness wisdom.

Elizabeth D. Schafer

SEE ALSO: Conservation Environmental ethics; Gila Wilderness Area; Kaibab Plateau deer disaster; Nature preservation policy; Predator management; Wilderness areas; Wildlife management.

Limited Test Ban Treaty

DATE: August 5, 1963
CATEGORY: Nuclear power and radiation

The 1963 Limited Test Ban Treaty was an agreement signed by the United States, Great Britain, and the Soviet Union to halt nuclear test explosions in the atmosphere, under the sea, and in outer space.

The atomic bomb was developed by the United States during World War II. The first test explosion was carried out in July, 1945, in the desert of New Mexico. Less than one month later, Hiroshima and Nagasaki, both in Japan, were destroyed by atomic bombs. Other countries soon

U.S. president Bill Clinton signs the Comprehensive Nuclear Test Ban Treaty in 1996 with the same pen used by President John F. Kennedy to sign the Limited Test Ban Treaty in 1963. (AP/Wide World Photos)

developed their own bomb technology. The Soviet Union exploded an atomic bomb in 1949, followed by Great Britain in 1952 and France in 1960. During the 1950's both the United States and the Soviet Union embarked on military programs to build intimidating nuclear arsenals.

In 1954 the United States detonated a powerful hydrogen bomb at Bikini Atoll in the South Pacific. A Japanese fishing boat, the *Lucky Dragon V*, was contaminated by radioactive fallout when the wind unexpectedly shifted. The twenty-three sailors aboard suffered radiation sickness, leading to worldwide protests against further testing. Antinuclear rallies in the United States and Europe mobilized public opinion against the escalating arms race. In 1958 U.S. president Dwight David Eisenhower and Soviet premier Nikita Khrushchev agreed to a moratorium on nuclear weapons testing in the atmosphere. It lasted for almost three years. Renewed tension between the two superpowers arose because of

the U-2 spy plane incident and the Berlin Crisis of 1961, and both countries resumed nuclear testing. The most powerful bomb in history was a 58-megaton device detonated by the Soviet Union in October, 1961.

Nuclear explosions in the atmosphere create radioactive particles that are spread around the world by prevailing winds and return to earth with precipitation. Radioactive cesium and iodine are two notably harmful materials that have contaminated grass pastures, causing grazing cows to produce radioactive milk. A study of baby teeth in the early 1960's showed the presence of radioactivity, primarily from milk consumption. The amount of radioactivity was not enough to cause radiation sickness, but even small doses of radiation have a statistical probability to increase the likelihood of cancer. A study by the Brookings Institute estimated that at least seventy thousand additional cases of cancer worldwide resulted from bomb testing.

The Cuban Missile Crisis of 1962 brought the world to the brink of nuclear war. This U.S.-Soviet confrontation led to the mutual realization that the nuclear arms race could escalate into annihilation for both sides. Subsequent negotiations resulted in the signing of the Limited Test Ban Treaty in 1963, which prohibited nuclear explosions in the atmosphere, under the ocean, and in outer space. However, this treaty still permitted underground explosions.

The agreement opened a dialogue between the United States and the Soviet Union and significantly eased Cold War tensions. It represented a fundamental step toward a global policy of arms control and disarmament. The treaty also had a positive ecological impact, particularly in the atmosphere, where it led to reduced radioactive contamination; as a result, human health problems linked with nuclear testing and radioactive fallout substantially decreased. By 1992, 125 countries had signed and become parties to the Limited Test Ban Treaty. A voluntary ban on underground explosions took effect in 1992, and negotiations for a Comprehensive Test Ban Treaty were initiated at the United Nations in the late 1990's. Nevertheless, nuclear testing continued as France exploded bombs at the Mururoa atoll in the Pacific Ocean as late as 1995, and as Pakistan and India conducted underground nuclear testing in 1998.

Hans G. Graetzer

SEE ALSO: Antinuclear movement; Bikini Atoll bombing; Nuclear testing; Nuclear weapons; Radioactive pollution and fallout.

Limits to Growth, The

DATE: published 1972
CATEGORY: Ecology and ecosystems

The Limits to Growth, published in 1972, marked the general public's first exposure to computer modeling of complex systems, such as the world ecosystem. The book ignited a great debate concerning the long-term sustainability of existing demographic, economic, and resource-use trends.

The Limits to Growth, written by Donella Meadows, Dennis Meadows, Jorgen Randers, and William Behrens, is sometimes referred to as the Club of Rome study because that organization sponsored the research effort. The authors used a computer modeling technique developed by Massachusetts Institute of Technology (MIT) professor Jay Forrester to trace the interaction of five major variables in the global ecosystem: population, pollution, nonrenewable resources, per capita food production, and industrial output. The technique used the power of the computer to track the behavior of these variables as they responded to the interaction of multiple influences through feedback loops. For example, population change is influenced by the birth rate relative to the death rate. Both of these rates are influenced by the amount of pollution, which is, in turn, influenced by the population level. The complexity of the multiple interactions quickly exceeds the ability of the human mind to anticipate the behavior of the system. Using the calculating power of a computer, the model could track the performance of the variables.

The model was based upon five major assumptions: There is a finite stock of exploitable nonrenewable resources, there is a finite amount of arable land, there are limits to the ability of the environment to absorb pollutants, there is a limit to the amount of food produced on each unit of arable land, and population, pollution, resource use, and industrial output grow exponentially, but technological change is incremental and noncontinuous.

The baseline scenario, based upon the rates of change of each of the five variables during the twentieth century, projected a rapidly growing global population until the middle of the twenty-first century. According to the model, population growth would greatly overshoot the earth's long-term carrying capacity. At some point during the twenty-first century, a massive population collapse would occur as the depletion of resources, exponential increases in pollution, and dramatic decreases in food per capita would combine to increase death rates and decrease birth rates. Altering one or more of the variables—for example, making resources unlimited—would change the timing of the collapse but would not avoid a catastrophic outcome.

While the authors took great pains to explain that the model results were projections rather than predictions, critics attacked the study as a "doomsday" report. Critics noted that the results of the simulation would radically change—there would be no population collapse—if a variable incorporating continual technological change was included in the model. Because ongoing technological change is a common feature of industrialized society, critics argued that this omission was fatal to the model. Despite the criticism, *The Limits to Growth* had a significant impact on the emergence of the notion of sustainable development.

Allan Jenkins

SEE ALSO: Club of Rome; Population growth; Sustainable development.

Logging and clear-cutting

CATEGORY: Forests and plants

Logging is the removal of timber from forestlands with the intention of using it for a specific purpose, such as lumber, fuelwood, or the production of pulp or chemicals. Clear-cutting is a harvesting technique in which all the timber is removed from a stand at the same time.

While the general public often thinks of logging and clear-cutting as practically synonymous, they are not. Similarly, while many people believe commercial logging is responsible for the loss of all forestland, particularly in tropical areas, many acres of forestland are annually cleared for other purposes. Rain forests in Amazonia, for example, are often bulldozed to create pastureland for cattle. Rather than harvesting the timber, the wood is simply pushed into piles and burned at the site.

Logging and clear-cutting, if improperly done or motivated by short-term economic goals, can pose significant threats to the environment. Logging always involves some disturbance to soil and wildlife. If performed in environmentally sensi-

tive areas, it can destroy irreplaceable habitat, contribute to problems with erosion and flooding, and worsen the threat of global warming. Heavy equipment can compact soil, leaving ruts that may persist for many years, while clear-cutting hillsides can lead to erosion, stream siltation, and devastating floods. In Asia, for example, clear-cutting in the mountains of Nepal and India has caused disastrous floods in Bangladesh. Even when logging does not inflict long-term damage on the immediate environment, the simple removal of trees can aggravate global warming. Slash (waste material) burnt at logging sites pumps greenhouse gasses into the atmosphere, while the loss of forest means that there are fewer trees to break those gasses down into oxygen and organic compounds.

Regardless of whether a logger is cutting only one tree or one thousand, logging involves four basic steps: selecting the timber to be harvested, felling the trees, trimming away waste material, and removing the desired portion of the tree from the woods. Equipment used in logging ranges from simple hand tools, such as axes and crosscut saws, to multifunction harvesting machines costing hundreds of thousands of dollars each. Mechanized feller bunchers, for example, can fell the tree, trim off the branches, cut the stem into logs of the desired length, and stack the logs to await removal from the forest. The choice of equipment utilized in harvesting any specific stand of timber will depend on factors such as the terrain, the type of timber to be logged, and whether the logger intends to harvest only selected trees or clear-cut the site. Loggers are more likely to clear-cut, or remove all the standing timber from a section of land, if the timber is plantation grown and of a uniform age and size. Clear-cutting also occurs in forests where the desired species of trees need large amounts of sunlight to regenerate. Many conifers, such as Douglas fir, are shade intolerant. Landowners will occasionally decide to change the dominant species on a tract and so will clear-cut existing timber to allow for replanting with new, more commercially desirable trees. Clear-cutting can be an acceptable practice in sustainable forestry when plantation stands are harvested in rotation.

Selective harvesting, in contrast with clear-cutting, leaves trees standing on the tract. Selective harvesting can be utilized with even-age plantation stands as a thinning technique. More commonly, it is used in mixed and uneven-age stands to harvest only trees of a desired species or size. In cutting hardwood for use as lumber, for example, 12 inches may be considered the minimum diameter of a harvestable tree. Trees smaller than that will be left in the woods to continue growing.

An individual, noncommercial woodcutter may fell only a few trees per year on small parcels of land. Commercial loggers, in contrast, annually harvest hundreds of thousands of trees and operate on large parcels of land. Nonetheless, significant acres of forestland are annually cleared by people who rely on noncommercial logging for wood for their own individual needs, such as fuel for cooking or heating their homes. Although some woodcutters may cut more than they need for their own use and then sell the surplus, fuelwood for individual households is usually gathered by members of that household. Other examples of noncommercial logging include farmers cutting trees for use as fencing or building materials on their own property.

From an environmental viewpoint, the biggest difference between commercial and noncommercial logging would seem to be one of scale, but even this is not always true. An improperly logged small parcel can have more of an impact on a watershed or ecosystem than a professionally harvested large stand. Even if no single household's logging practices pose a problem, collectively the gathering of fuelwood or other timber can be devastating. With no guidance from professional foresters, trees are logged based on convenience for the woodcutter rather than principles of sustainable forestry or watershed management. Many nations have developed programs in which professional forest-

A clear-cut area on the Olympic Peninsula in Washington State. (Jim West)

U.S. Lumber Consumption
In Millions of Board Feet

	1970	1976	1986	1991	1992
Species group					
Softwoods	32.0	NA	48.0	44.0	45.7
Hardwoods	7.9	NA	9.0	10.8	10.3
End Use					
New housing	13.3	17.0	19.3	15.0	
Residential upkeep and improvements	4.7	5.7	10.1	11.6	
New nonresidential construction	4.7	4.5	5.3	5.4	
Manufacturing	4.7	4.9	4.8	5.6	
Shipping	5.7	5.9	6.8	8.2	
Other	6.8	6.7	10.9	8.8	

Source: U.S. Department of Commerce, *Statistical Abstract of the United States, 1996,* 1996.

ers provide advice on environmentally sound harvesting practices and timber-stand improvement for small property owners, but the availability of such help varies widely from country to country.

Nancy Farm Männikkö

SUGGESTED READINGS: *Logging and Pulpwood Production* (1985), by George Stenzel et al., provides a thorough description of logging practices, including discussions of environmental regulations and the various government agencies involved with commercial forestry in the United States. Laurence C. Walker's *The Southern Forest* (1991) is an accessible and interesting description of regional forests and forestry practices. Readers concerned about logging in tropical forests and developing nations may find William W. Bevis's *Borneo Log: The Struggle for Sarawak's Forests* (1995) both disturbing and enlightening as Bevis describes the exploitation of Third World resources by more industrialized nations.

SEE ALSO: Deforestation; Forest management; National forests; Old-growth forests; Rain forests and rain forest destruction; Slash-and-burn agriculture; Sustainable forestry.

London smog disaster

DATE: December 5, 1952
CATEGORY: Atmosphere and air pollution

On December 5, 1952, a "killer smog"—a lethal combination of fog, smoke, and pollutants—settled over London, England, for several days. Transportation was severely disrupted by low visibility, and many outdoor sporting events were canceled. During the episode, London's death rate dramatically increased.

The chief culprit in the London smog disaster was deemed to be Great Britain's heavy dependence on coal, which was used by industry and burned in almost every household in London, since the country's wood supplies had long been depleted. Particular blame was placed upon the popular "nutty slack," a low-grade soft coal used in most households that burned inefficiently and gave off noxious smoke and odor. Authorities estimated that approximately one-half of the smoke emitted during the killer smog was from household hearths.

As the smog accumulated during the last three weeks of December, the death rate in London climbed dramatically. The primary causes of the deaths were circulatory or respiratory problems, including bronchitis, influenza, pneumonia, pulmonary pneumonia, and cancer of the lungs. During a five-week period ending January 3, 1957, the minister of health announced that there were 15,114 deaths registered in Greater London, compared with 9,125 during the same period one year previously. It is estimated that at least 4,000 deaths were directly attributable to the smog, the chief victims being babies under one year old and the elderly over the age of fifty-five. The death rate for this period exceeded that of previous disasters in London's history, including the worst periods of the cholera epidemic of 1866 and the great fog of December, 1873.

The British Clean Air Act of 1956 was a direct result of this disaster and an attempt to correct the smog problem. The act banned the emission of black smoke from locomotives, vessels, and chimneys; all new furnaces had to be capable, as far as practical, of not producing smoke; no grit or dust could be emitted from furnaces, new or old; and a Clean Air Council was created to advise government ministers in regard to clean air policy. One of the act's most important provisions gave cities the power to establish "smoke-control zones" in order to combat the problems of burning coal in home fireplaces and local factories. To assist in this last objective, the government decreed that 40 percent of the cost of converting home appliances to burn smokeless fuel would be paid by financial grants from the national government, while the local government authority was required to contribute 30 percent, thus requiring the owner to pay only 30 percent of the total cost of conversion.

Even then, it was estimated that it would take up to fifteen years for the full impact of the improvements to be felt. Nevertheless, the smog problem of London was eventually overcome, thanks to the Clean Air Act of 1956 and the long-term trend of factories, railroads, and households converting to oil, gas, nuclear, and electrical energy.

David C. Lukowitz

SEE ALSO: Air pollution; Environmental illnesses; Smog.

Los Angeles Aqueduct

DATE: completed 1913
CATEGORY: Preservation and wilderness issues

The Los Angeles Aqueduct, completed in 1913, was constructed to meet the growing demand for water in Los Angeles. However, the diversion of water from the Owens River led to the drying of the Owens River Valley and the subsequent collapse of the region's agricultural industry.

The Los Angeles Aqueduct extends 378 kilometers (235 miles) from the Owens River a few miles north of Independence, California, to San Fernando, California, on the north side of Los Angeles. Begun in 1908 and completed in 1913, it is a complex system of unlined, lined, and lined and covered canals, tunnels, and steel pipes. For seventy-three years the entire flow of the Owens River at the aqueduct intake point, an average of about 984 million liters (260 million gallons) per day, was diverted from the Owens Valley to Los Angeles. Beginning in 1986, a portion of the original flow was restored to the Owens River below the intake point. The aqueduct was extended north to the Mono Lake Basin in 1940, and a parallel aqueduct was completed in 1970.

In 1900 the population of Los Angeles was about 200,000 and was rapidly increasing. The city's water supply came from the Los Angeles River and a few wells and local springs. A substantial supply of water was required for the city to continue to grow, and there was no local source. To solve the problem, Fred Eaton, an engineer and former mayor of Los Angeles, conceived the Los Angeles Aqueduct and discussed his idea with William Mulholland, superintendent of the Los Angeles Department of Water. Mulholland spent forty days surveying the proposed aqueduct route and discussed the proposal with the Los Angeles Board of Water Commissioners. All this was done in secret to avoid a burst of land speculation. Eaton had already obtained options to buy most of the private land along the aqueduct route and later agreed to sell the options to Los Angeles at cost.

One obstacle remained. The U.S. Reclamation Service, created in 1902 and renamed the

Bureau of Reclamation in 1923, was in the process of planning a major irrigation project in the Owens Valley and had withdrawn the public lands in the area from claim. Part of the aqueduct route passed through these public lands. The residents of the valley were enthusiastically in favor of the irrigation project.

Plans for the aqueduct were made public, and a committee from the Los Angeles Chamber of Commerce met with President Theodore Roosevelt. To the disappointment of the Owens Valley residents, President Roosevelt reached the conclusion that the Owens River water would be much more beneficial to Los Angeles than Owens Valley. In 1906 the U.S. Congress granted the necessary right-of-way for the aqueduct. To assure its right to the Owens River water and increase the available supply, the city of Los Angeles began buying the irrigated ranches and farms above the aqueduct intake, in addition to properties below the intake that were now useless for lack of water. It was this, rather than the aqueduct itself, that angered Owens Valley residents.

The Owens River Valley lies between the Sierra Nevada to the west and the White Mountains and Inyo Range to the east. It is about 16 kilometers (10 miles) wide and 160 kilometers (100 miles) long. In its pristine state, the valley was a desert. The Owens River, flowing south from the Sierra Nevada, supported a fringe of willows and other riverside vegetation; the rest of the valley was thinly covered with cactus, chaparral, and sagebrush. The river ended at Owens Lake, an alkaline lake with no outlet. Because of evaporation over many thousands of years, the lake water was highly mineralized, primarily with sodium bicarbonate, but also with sulfates, chlorides, and other salts. The upper end of the lake was a freshwater marsh, which provided good habitat for waterfowl, as did the river

north of the lake. The river, its fringe of vegetation undoubtedly home for many native birds, was on a major migratory pathway for several species of birds. Irrigation ditches serving farms, ranches, and orchards extended as far as 8 kilometers (5 miles) from the river. Carp, an exotic fish imported from Europe, swam in the irrigation ditches.

On completion of the Los Angeles Aqueduct in 1913, the lower 85 kilometers (53 miles) of the Owens River channel became dry; partial flow was not restored until 1986. The narrow fringe of riverside vegetation died, the marshes at the head of Owens Lake dried, and, eventually, the lake itself dried. Windstorms crossing the dry lake bed carried irritating alkali dust

Large doors reroute water from Owens Lake to Southern California. Heavy winds in Owens Valley raise alkali dust from the dry lake bed, creating health problems for local residents. (AP/Wide World Photos)

through the valley and often far beyond. The irrigated farms and ranches almost entirely disappeared, and Los Angeles owned most of the valley, allowing it to return to desert. The economy of the valley changed from one based on agriculture to one based on tourism and outdoor recreation. From purely economic considerations, the income from tourism and outdoor recreation vastly exceeds what might have been expected by an expansion of irrigation-based agriculture.

By 1970 Los Angeles was using large quantities of well water to supplement aqueduct flow, lowering groundwater levels and drying local springs. In addition, the water level at Mono Lake was declining, threatening island breeding grounds of California gulls by creating land bridges that coyotes could cross.

If the aqueduct had not been built, the Reclamation Service probably would have constructed massive irrigation systems that might have been much more damaging to the environment. Indeed, a Sierra Club spokesman once said, "We recognize that Los Angeles is probably the savior of the valley. Our goal is to save the valley as it is now." By 1990, court actions had temporarily reduced Owens Valley groundwater pumpage and diversion from the Mono Lake Basin, but no permanent solution had been achieved. Restoration of flow in the lower Owens River was a positive action for the environment of the valley and may be sufficient to eventually re-create the marshes and lake at Owens Lake basin.

Robert E. Carver

SUGGESTED READINGS: A readable and informative history of the Owens Valley and the aqueduct is given in the short book *The Owens Valley and the Los Angeles Controversy: Owens Valley as I Knew It* (1973), by Richard Coke Wood. Excellent histories of the Los Angeles Aqueduct and all the other major California water projects are presented in Norris Hundley, Jr., *The Great Thirst: California and Water, 1770's-1990's* (1992), and Phillip L. Fradkin, *The Seven States of California: A Natural and Human History* (1995).

SEE ALSO: Irrigation; Prior appropriation doctrine; Roosevelt, Theodore; Water rights; Water use; Wetlands.

Love Canal

DATE: 1976
CATEGORY: Human health and the environment

In 1976 residents in the Love Canal area of Niagara Falls, New York, discovered that their homes had been built near a chemical waste burial site. Alarmed by the rising numbers of illnesses in the neighborhood, residents formed the Love Canal Homeowners Association and fought the government to win permanent relocation.

The discovery and identification of dangerous chemical wastes at Love Canal, in Niagara Falls, New York, transformed a community whose livelihood depended on the chemical companies long established in the area. Houses were boarded up and abandoned, a school was left empty and falling down, warning signs were posted, and the entire area was fenced off. The completeness of the human and ecological devastation at Love Canal made it the standard against which all chemical waste disasters are compared. The toxic terror generated by Love Canal was caused by chemical waste buried at a time when the term "pollution" was not yet part of the American vocabulary.

In 1892 entrepreneur William Love arrived in Niagara Falls with plans to construct a navigable power canal between the upper and lower portions of the Niagara River and use the 90-meter (300-foot) drop in water level to generate electric energy. Digging began in May, 1894, but the depressed state of the economy resulted in a withdrawal of investment capital, which ended the project with the canal less than one-half finished. Love Canal became the property of the Niagara Power and Development Company, which gave the Hooker Electrochemical Corporation permission in 1942 to dump wastes in the canal. The site was ideal for disposal of chemical wastes as the walls were lined with clay, which has a low level of permeability. Hooker purchased the property in 1947.

Between 1942 and 1952, Hooker disposed of 21,800 tons of chemicals at the site, burying them at depths of 6 to 8 meters (20 to 25 feet).

The foundation of a house that was moved to another section of Niagara Falls after the owners learned that their neighborhood had been built on top of the Love Canal toxic waste dump. The houses in the background are abandoned. (AP/Wide World Photos)

The majority of chemicals were contained in metal and fiber barrels, although some waste was reputedly dumped directly into the canal. The customary method for disposal of chemical wastes throughout the United States in the 1950's was to dump them directly into unlined pits, lagoons, rivers, lakes, and surface impoundments, or to burn them. Apart from disposal into bodies of water, all of these methods were legal up until 1980.

In 1953 Hooker filled the canal and topped it with a 0.5-meter (2-foot) clay cap. Beneath lay 43.6 million pounds of eighty-two different chemical residues. Included were benzene, a chemical known to cause anemia and leukemia; lindane, exposure to which results in convulsions and excess production of white blood cells; chloroform, a carcinogen that attacks respiratory, nervous, and gastrointestinal systems; trichloroethylene, a carcinogen that attacks

genes, livers, and nervous systems; and methylene chloride, which can cause recurring respiratory distress and death. The most dangerous chemical in the waste however, was dioxin, a component of the 200 tons of trichlorophenol dumped in the canal. Dioxin has been described as one of the most powerful known carcinogens.

Shortly after the canal was filled, the Niagara Falls Board of Education purchased the canal property for the token sum of one dollar on the condition that they warn future owners of the buried chemicals, use the land only as a park with the school in close proximity, and build no houses on the property. Shortly after taking possession, however, the school board removed 17,000 cubic yards of topsoils from the canal for grading at the site of the 99th street school and the surrounding area. City workmen then installed a sewer that punctured both the canal walls and the clay covering to facilitate adjacent housing tracts. Breaks in the clay cap and walls of the canal created openings through which the chemicals eventually flowed.

During the winter of 1975-1976, heavy snowfall caused the groundwater level to rise in the Love Canal area, filling the uncovered canal. Portions of the landfill subsided, and drums surfaced in a number of locations. Surface water, heavily contaminated with chemicals, was found in the backyards of houses bordering the canal. Residents complained of discomfort and illness caused by unpleasant chemical odors coming from the canal.

Lois Gibbs, president of the Love Canal Homeowners Association, united the community and became an effective and persuasive advocate for families seeking government aid. Gibbs involved herself in December, 1977, when her son, Michael, began to experience asthma and seizures just four months after he entered kindergarten at the 99th street school. She went door to door questioning residents about their health in an attempt to discover the full extent of contamination. In 1979 Gibbs traveled to Washington, D.C., where she testified on behalf of the Love Canal homeowners.

In 1977 studies conducted by city and New York state health officials showed extensive pollution affecting 57 percent of the homes at the southern end of the canal and a "moderate excess of spontaneous abortions and low birth weight infants occurring in households on 99th street bordering the landfill." In August, 1978, the New York state health commissioner closed the 99th street school and evacuated 240 families living within two blocks of the canal. In October, 1980, after the release of an alarming Environmental Protection Agency (EPA) health study, President Jimmy Carter ordered a total evacuation of the community.

The chemical disaster at Love Canal left behind a legacy of lawsuits and bitterness. In October, 1983, a tentative settlement of the billions of dollars in lawsuits was reached by lawyers of the Hooker Electrochemical Corporation, the city of Niagara Falls, the Niagara Falls Board of Education, Niagara County, and former residents of the Love Canal area. Claims against Hooker and the public agencies totaled $16 billion. On February 19, 1984, former residents of Love Canal received payments ranging from $2,000 to $400,000.

In November, 1980, Congress passed legislation to deal with the cleanup of toxic wastes known as the Comprehensive Environmental Response, Compensation, and Liability Act (CERCLA). This new legislation, commonly referred to as the Superfund law, established a $1.6 billion fund for the cleanup of hazardous substances to be administered by the EPA. The money was to be used when "no responsible party could be identified or when the responsible party refuses to or is unable to pay for such a cleanup." Of the money allocated for the Superfund, 87.5 percent came from taxes on oil and specified chemicals, while 12.5 percent was designated by Congress.

Peter Neushul

SUGGESTED READINGS: Sociologist Adelaine Gordon Levine's *Love Canal: Science, Politics and People* (1982) is a comprehensive but somewhat uncritical account of Love Canal. Eric Zuesse, "Love Canal: The Truth Seeps Out," *Reason* (February, 1981), provides the best description of agencies contributing to the disaster. Gina Bari Kolata, "Love Canal: False Alarm Caused by Botched Study," *Science* 208 (June, 1980), provides insight into scientific studies at Love Canal. Lois Gibbs's story is told in her books *Love Canal: My Story* (1982) and *Love Canal: The Story Continues . . .* (1998).

SEE ALSO: Gibbs, Lois; Hazardous Waste; Superfund.

Lovejoy, Thomas

BORN: August 21, 1941; New York, New York
CATEGORY: Philosophy and ethics

Thomas Lovejoy is a tropical biologist recognized for his contributions to conservation policy making. He is best known for developing creative solutions to issues of scientific concern, such as debt-for-nature swaps.

In 1971 Thomas Lovejoy earned a doctor of philosophy degree in biology from Yale University. His early research consisted of a long-range study of birds in the Brazilian Amazon. From 1973 to 1987 Lovejoy served as program director for the World Wildlife Fund (WWF) and was responsible for Western Hemisphere and tropical forest projects. He is credited with being the first person to use the term "biological diversity," or "biodiversity," in 1980. Lovejoy also began speaking of global extinction rates, and in 1980 his first projection of such rates appeared in *The Global 2000 Report* submitted to U.S. president Jimmy Carter. This marked the beginning of his many contributions to the policy arena.

It was during his tenure as executive vice president of the WWF from 1985 to 1987 that he originated the debt-for-nature swap, a concept that involves forgiveness of foreign debts incurred by developing nations in exchange for protection of fragile ecosystems within those nations. Debt-for-nature swaps have been used to preserve wilderness in countries such as Costa Rica, Bolivia, Ecuador, the Philippines, and Madagascar. While working for the WWF, Lovejoy increased public interest in the issue of tropical forests.

Lovejoy is recognized for his practical approach to solving environmental problems. This is illustrated by his input for the Minimum Critical Size of Ecosystems Project, also known as the Biological Dynamics of Forest Fragments Project. This was a joint research project undertaken by the Smithsonian Institution and Brazil's National Institute for Amazon Research (INPA). The project was one of the early attempts to define the minimum size for national parks and biological reserves and to formulate management strategies for such protected areas. In recognition of his conservation work in Brazil, Lovejoy became the first environmentalist to receive the Order of Rio Branco, which the Brazilian government awarded him in 1988.

In 1987 Lovejoy founded the public television series *Nature* and served as its advisor for many years. In the same year he was also appointed assistant secretary for environmental and external affairs at the Smithsonian Institution. By 1994 Lovejoy had risen to become counselor to the secretary for biodiversity and environmental affairs at the Smithsonian. While still associated with the Smithsonian Institution, he accepted a temporary assignment to the World Bank in 1998, where he served as chief biodiversity advisor. His primary responsibility at the World Bank was the environment in Latin America.

While continuing his association with the Smithsonian, Lovejoy also served in a variety of advisory positions. In 1993 he was chosen to be science advisor by the secretary of the interior and helped set up the National Biological Survey. From 1994 to 1997 Lovejoy served as scientific advisor to the executive director of the United Nations Environment Programme (UNEP). His professional contributions included serving as chair of the United States Man and the Biosphere Programme (MAB) and president of the Society for Conservation Biology. Lovejoy is the author of numerous articles and four books on environmental conservation, including a volume he edited with Robert L. Peters titled *Global Warming and Biological Diversity* (1992).

Michele Zebich-Knos

SEE ALSO: Debt-for-nature swaps; International Union for the Conservation of Nature/World Wildlife Fund.

Lovelock, James

BORN: 1919; Letchworth, Hertfordshire, England
CATEGORY: Ecology and ecosystems

Scientist and inventor James Lovelock is best known for his Gaia hypothesis, which suggests that the earth is the source of life for all things and that all life on earth has coevolved and is therefore inextricably intertwined.

James Lovelock was born in 1919. He was educated at the University of London and Manchester University and was awarded a Ph.D. in medicine. He taught in the United States at Yale, the Baylor College of Medicine, and as a Rockefeller Fellow at Harvard University. During the 1970's he served as a consultant for the National Aeronautics and Space Administration (NASA), and was also known as an independent scientist and inventor. While at NASA he developed the electron capture detector, which has been utilized to study and analyze atmospheric gases as they exhibit themselves on other planets such as Mars and Venus. Data gained from his electron capture studies strongly suggested the uniqueness of earth with respect to the fostering of life.

During his tenure as a NASA consultant, Lovelock developed the Gaia hypothesis, which suggests that as the earth's early atmosphere cooled, mixing of gases and gradual increases of water on the planet provided the necessary nutrients for the evolution of microorganisms that began to utilize the earth's toxic gases. Those that could perform photosynthesis began to contribute heavily to the oxygen levels of the earth's troposphere. Eventually, other nonphotosynthetic microorganisms evolved that returned carbon dioxide to the atmosphere. In a continuous cycle, the microscopic biota cycled and recycled the oxygen and carbon dioxide gases to provide the atmosphere as it is recognized today.

Lovelock named his hypothesis after the Greek goddess of the earth, Gaia, upon the advice of his friend and colleague, William Golding (author of the 1954 novel *Lord of the Flies*). The powerful notion that life itself contributes to and continually modifies the atmosphere was not readily accepted at the time. Skeptics argued that the sum of the biota could not purposefully maintain an entire atmosphere. Yet Gaia has remained a viable theory that continues to explain not only earth's early evolution and the importance of early life-forms but also the importance of life in maintaining atmospheric balance. Current problems of global warming may also be related in that macroscopic organisms, namely humans, may be upsetting this delicate balance.

Lovelock is the author of two major books on the Gaia hypothesis, *Gaia: A New Look at Life* (1979) and *The Ages of Gaia: A Biography of Our Living Earth* (1988), and his friend Lynn Margulis of Boston University continued work on the Gaia hypothesis as Lovelock turned to his own independent research. Working out of a barn converted to a laboratory at Coombe Mill in Cornwall, England, he made his home on the farm with his wife Helen and their pet peacocks. In 1986 he became president of the Marine Biology Association, and he was elected a fellow of the Royal Society of London in 1974.

Kathleen Rath Marr

SEE ALSO: Gaia hypothesis.

Lovins, Amory

BORN: November 13, 1947; Washington, D.C.
CATEGORY: Energy

Physicist and energy consultant Amory Lovins cofounded the Rocky Mountain Institute. He has written and lectured extensively to promote sustainable and clean energy, particularly as a means to attain global stability and security.

A native of Washington, D.C., Amory Lovins was a student at Harvard University and Magdalen College, Oxford, England, during the mid- to late 1960's. He received his master's degree from Oxford University in 1971. Thereafter he worked as a consultant physicist specializing in energy concerns and promoting the use of energy sources that are economical, efficient, diverse, sustainable, and environmentally sound. He advocated an approach that maximizes energy efficiency and minimizes environmental impact as the best way to provide for the world's long-term energy needs without incurring the security risks inherent in nuclear power or foreign oil.

Lovins lectured extensively, held several visiting academic chairs, and published hundreds of papers. Among his many books are *World Energy Strategies: Facts, Issues, and Options* (1975), *Soft Energy Paths: Toward a Durable Peace* (1977), and *Non-Nuclear Futures: The Case for an Ethical Energy Strategy* (1980; with John H. Price). Lovins was the recipient of six U.S. honorary doctorates and a 1993 MacArthur Fellowship. He consulted for utilities, industries, and governments worldwide; briefed heads of state; and served on the U.S. Department of Energy's senior advisory board.

In 1979 Lovins married L. Hunter Sheldon, one of the cofounders (and for six years the assistant director) of the urban forestry and environmental education group TreePeople. The Lovinses would coauthor several books, including *Energy/War: Breaking the Nuclear Link* (1981), *Brittle Power: Energy Strategy for National Security* (1982), *The First Nuclear World War* (1983; with Patrick O'Heffernan), *Energy Unbound: A Fable for America's Future* (1986; with Seth Zuckerman), *Least-Cost Energy: Solving the CO$_2$ Problem* (1989; with F. Krause and W. Bach), and *Factor Four: Doubling Wealth—Halving Resource Use* (1997; with Ernst von Weizacker). The Lovinses shared a number of honors, including the 1982 George and Cynthia Mitchell International Prize for Sustainable Development, the 1983 Right Livelihood Award (often called the "alternative Nobel Prize"), the 1989 Delphi Prize (one of the world's top environmental awards, presented by the Onassis Foundation), and the 1993 Nissan Prize at the International Symposium on Automotive Technology and Automation, Europe's largest car-technology conference.

In 1982 the Lovinses cofounded the Rocky Mountain Institute, a nonprofit resource policy center in Old Snowmass, Colorado, that promotes resource efficiency and global security, with Hunter as the institute's president and executive director and Amory as its vice president and director of research. The institute's stated mission is to "foster the efficient and sustainable use of resources as a path to global security." It focuses its efforts on such areas as transportation, green design and development, greenhouse gas reduction, water-efficient technologies, economic renewal, and corporate sustainability. In addition to his technical and policy-oriented writings, Lovins is known as a poet, photographer, musician, and mountaineer.

Karen N. Kähler

SEE ALSO: Alternative energy sources; Alternative fuels; Alternatively fueled vehicles.

McToxics Campaign

CATEGORY: Waste and waste management
DATE: 1989-1990

The McToxics Campaign was a consumer boy-cott that prompted McDonald's restaurants to replace polystyrene food containers with more environmentally friendly packaging.

In 1989, the Environmental Defense Fund (EDF) inaugurated a grassroots campaign against McDonald's restaurants. By convincing consumers to boycott McDonald's, the EDF hoped to persuade the fast-food giant to stop packaging its food in polystyrene "clamshells," which were cheap and efficient food insulators but which did not readily biodegrade and which were accused of emitting chlorofluorocarbons (CFCs) harmful to the earth's ozone layer. EDF members worked to educate children about the environmental consequences of polystyrene packaging, then employed the young activists in its protests. Ronald McDonald, the company's well-known advertising character designed to appeal to young consumers, was redubbed "Ronald McToxic" by the campaign's directors.

Predictably, the children's participation drew media attention, and the protest took effect as consumers began to boycott McDonald's. Some wrote angry letters; others mailed empty clamshells to the company's offices. As the corporation's profits dropped, McDonald's lawyers began urging the company to undertake the packaging changes demanded by the EDF and its supporters. On August 1, 1990, McDonald's came to a tentative agreement with the EDF, and on November 1, 1990, McDonald's announced its decision to abandon the use of polystyrene packaging.

Nevertheless, the agreement was not without detractors. Some scientists expressed doubts that the changeover from polystyrene to paper packaging would produce environmental benefits, questioning whether the use of clamshells represented a legitimate environmental hazard. In addition, some environmental extremists objected to the EDF's accommodation of corporate culture. Nevertheless, most observers agreed that the McToxics campaign and its conclusion represented a welcome example of how environmental and economic interests could be made to coexist.

Alexander Scott

SUGGESTED READINGS: E. Bruce Harrison's *Going Green* (1993) describes the McToxics affair and the unlikely partnership that evolved between McDonald's and the EDF. "McDonald's Won't Give Break to CFC Foam Blowing Agents" (*Chemical Marketing Reporter* 10, August, 1987) discusses the extent to which CFCs are incorporated in the manufacture of foam containers. Laki Tisopulos's "A Local Response to Control Air Emissions from Foam Products" (*Nation's Cities Weekly* 14, July 29, 1991) states the case against polystyrene.

SEE ALSO: Waste management.

Mad cow disease

CATEGORY: Human health and the environment

Mad cow disease, or bovine spongiform encephalopathy (BSE), is a cattle disease of the central nervous system first identified in 1986 in Great Britain.

BSE is believed to be related to new-variant Creutzfeldt-Jakob disease (nvCJD), a fatal human illness. It is marked by brain tissue deterioration and progressive degeneration of the central nervous system, which cause symptoms such as impaired physical coordination, staggering, unusual aggression, and other abnormal behavior. Since its discovery, BSE has become epidemic in Great Britain. Substantial numbers of infected cattle have also been found in Ireland, Switzerland, Portugal, and France. Scattered cases have occurred in Germany, Italy, the Netherlands, Luxembourg, Oman, Canada, Denmark, and the Falkland Islands.

BSE is a transmissible spongiform encephalopathy (TSE), a class of diseases so named because of the brain damage that characterizes them—the tissue is left riddled with small holes, like a sponge. Other species affected by TSEs include deer, elk, and antelope (chronic wasting disease); sheep and goats (scrapie); mink (transmissible mink encephalopathy); cats (feline spongiform encephalopathy, or FSE), and humans (kuru, Gertsmann-Sträussler-Scheinker syndrome, fatal familial insomnia, and Creutzfeldt-Jakob disease, or CJD). Diagnosis is confirmed only through autopsy and examination of the brain tissue. At present, all known TSEs are incurable, untreatable, and fatal.

BSE is believed to have arisen because of an infectious agent in cattle feed during the 1970's and 1980's. Meat-and-bone meal, a protein concentrate produced by rendering plants (facilities that process animal carcasses and slaughterhouse wastes into commercial products), was added to cattle feed as a nutritional supplement with increasing frequency during the 1980's. Among the rendered wastes from which the supplement was made were the carcasses of sheep that died from scrapie. Through infected protein supplement, scrapie is thought to have crossed the species barrier from sheep into cattle and become a new TSE. An increase in Great Britain's sheep population during the late 1970's and early 1980's elevated the number scrapie cases, and thus more infected material was rendered. Coincidentally, at about this time Britain's rendering industry made changes to the rendering process—eliminating the solvent extraction of fats and a subsequent steam-stripping treatment—that may have allowed more infectious agent to pass into the finished product.

Not long after it was first identified, BSE reached epidemic proportions in Great Britain.

The epidemic peaked in 1992, with 37,487 confirmed cases. Mandatory reporting of BSE cases and destruction of symptomatic animals began in 1988, as did a ban on the feeding of ruminants (cattle, sheep, goats, and deer) with ruminant-derived protein. In 1989 the British government banned the human consumption of cattle brains and other "specified offals" believed to be possible carriers of infection. In 1990 a British domestic cat was diagnosed as the first victim of FSE, causing public concern over the possible spread of BSE beyond livestock and a temporary drop in the nation's beef consumption. That year, the British government established a system for reporting and tracking CJD to monitor possible BSE effects on people.

In 1993 the British National CJD Surveillance Unit began to receive reports of unusual CJD cases. While CJD typically affects one person in one million between the ages of fifty-five and eighty, CJD was appearing in younger patients, with uncharacteristic patterns of brain damage, and with atypical frequency. By early 1996 eight cases of nvCJD in people under forty-two years of age had been identified. Global attention turned to Britain's mad cow crisis in March, 1996, when British health secretary Stephen Dorrell announced that ten cases of nvCJD had been confirmed and that they were most likely the result of exposure to BSE-contaminated beef before the 1989 specified offal ban. The announcement triggered a nationwide panic in Britain, caused a sharp drop in British beef sales, and exacerbated political and economic friction between Britain and the rest of the European Union.

Researchers have since found more conclusive evidence linking BSE and nvCJD. However, they have yet to determine the exact cause of these and other TSEs. One of the more widely accepted theories is that the illnesses are induced by tiny proteinaceous infectious particles, or prions. Prions resemble an abnormally folded version of prion protein, a harmless, naturally occurring protein found in the brains of all mammals. Prions do not break down in mammalian digestive systems and provoke no immune response. Lacking genetic material (unlike more familiar infectious agents such as bacteria, viruses, and fungi), prions are believed to spread by invading cells and converting normal prion protein into the aberrant form. As the prions accumulate, they damage brain cells and ultimately kill the organism. Whatever the agent that causes BSE, it is highly resistant to ultraviolet light, pH and temperature extremes, ionizing radiation, and many chemical disinfectants.

British efforts to contain a livestock epidemic that is only partially understood, protect human health, and allay public fears about beef consumption have created new problems. In a government-instituted culling program, more than 3.8 million animals considered at risk have been slaughtered. Neither the designated rendering plants that process the remains nor the designated incinerators that burn the rendered product are able to keep up with the slaughter rate. The resulting backlog of hundreds of thousands of tons of waste beef and rendered products has been warehoused pending disposal. Incineration plans have proved unpopular with Britons unconvinced that burning cull material is safe. Public concern has also arisen over the possible groundwater contamination risk posed by cattle carcasses landfilled and liquid rendering wastes discharged to the environment during the early years of the epidemic. Between 1993 and August, 1998, twenty-seven cases of nvCJD were diagnosed in Britain, with an additional case confirmed in France. It is not known how many other people may be incubating the disease.

Karen N. Kähler

SUGGESTED READINGS: Richard Rhodes, *Deadly Feasts: The "Prion" Controversy and the Public's Health* (1998), is a highly readable account of the discovery and study of TSEs and the social consequences of BSE and nvCJD. Sheldon Rampton and John Stauber, *Mad Cow U.S.A.: Could the Nightmare Happen Here?* (1997), investigates how corporate and government public relations efforts and public policy influenced the course of Great Britain's epidemic and explores the possibility of a similar outbreak in the United States. Howard F. Lyman and Glen Merzer, *Mad Cowboy: Plain Truth from the Cattle Rancher Who Won't Eat Meat* (1998), written from Lyman's perspective as a former Montana cattle rancher

turned vegetarian environmentalist activist, discusses BSE and other health and environmental consequences of cattle- and dairy-industry practices.

SEE ALSO: Environmental illnesses; Vegetarianism.

Malathion

CATEGORY: Pollutants and toxins

Malathion is an organophosphorus pesticide used against mosquitoes, fleas, and other insects. While no adverse health effects have been found from low-level exposure to malathion, its use in aerial spraying programs has been controversial.

Pesticides have been used to control crop loss and prevent the spread of disease for hundreds of years. Until the middle of the twentieth century, pesticides were either toxic metals (such as arsenic or lead) or natural products (such as nicotine sulfate). With the discovery of dichloro-diphenyl-trichloroethane (DDT) in 1939, new human-made pesticides became available for use. However, it was soon found that DDT and related chlorinated hydrocarbon pesticides persisted in the environment for years, causing damage to birds, fish, and other wildlife. Other pesticides, including organophosphate compounds, were therefore developed as an alternative to DDT.

Malathion is an organophosphate compound that is chemically similar to some types of nerve gas but less toxic than most other organophosphate pesticides. Unlike chlorinated hydrocarbons, malathion breaks down in the environment, transforming into carbonic and phosphoric acid over a period of days to a few weeks. Because of its relatively short lifetime, malathion does not bioaccumulate in aquatic organisms or contaminate groundwater. Following its introduction by the American Cyanamid Company in 1952, malathion quickly became a popular substitute for DDT in controlling mosquitoes, fleas, lice, and other insects. In 1956 malathion was first used to counteract fruit fly infestations.

In the following decades, the use of malathion continued to increase. In the 1980's, however, the spraying of malathion in California to control the Mediterranean fruit fly became a source of controversy. Those critical of the use of malathion in populated areas noted that commercial preparations of malathion contain trace impurities that are potentially toxic and that the initial breakdown products from malathion include malaoxin, a compound whose acute toxicity is forty times that of malathion itself. Critics also suggested that malathion might be carcinogenic or weaken the immune system, making people more susceptible to disease. However, extensive studies of human populations exposed to malathion through the spraying program have found no adverse health effects from low levels of exposure. Harmful effects from low-level exposure to malathion have likewise not been observed in laboratory studies on animals. Despite the controversy, malathion continued to be used to prevent damage to citrus crops in California, Florida, and other states, and to be available in many commercial products.

By the end of the twentieth century, controversy over the use of synthetic pesticides such as malathion convinced many people that the release of such compounds into the environment should be greatly reduced. Critics of human-made pesticides have called for more intelligent use of pesticides and alternative methods for controlling insect populations.

Jeffrey A. Joens

SEE ALSO: Medfly spraying; Pesticides and herbicides.

Manabe, Syukuro

BORN: September 21, 1931; Shingu-Mura, Uma-Gun, Ehim e-Ken, Japan
CATEGORY: Weather and climate

Malathion, a potent insecticide, is used to control populations of disease-carrying mosquitos in forested areas. (AP/Wide World Photos)

Syukuro Manabe is an award-winning meteorological scientist who specializes in climate modeling.

Syukuro Manabe is a meteorological scientist whose research area involves modeling the earth's atmosphere in order to predict global climate change. Modelers construct a system of mathematical equations that can be used, with a computer, to mimic a particular segment of nature. In Manabe's case, the equations center on the heat balance of the earth's atmosphere in order to mimic the earth's climate. The goal is to use data from previous years to accurately predict global weather patterns for future years. Even an approximate model allows its users to find the long-term effects of short-term environmental atmospheric inputs. This allows a better understanding of the causes and effects of environmental disturbances such as global warming.

Manabe completed his formal education in his home country of Japan, finishing a bachelor's degree in 1953, a master's degree in 1955, and a doctoral degree in 1958, all from Tokyo University. With a doctorate in meteorology, he moved to the United States and spent the years between 1958 and 1997 working his way up from a research meteorologist at the United States Weather Bureau to a senior scientist at the National Oceanic and Atmospheric Administration (NOAA). On January 26, 1962, he married; the marriage produced two children.

The major part of Manabe's stay in the United States was as a researcher with the NOAA, located in Princeton, New Jersey. During this tenure, Manabe contributed his expertise to many national and international committees and panels exploring the global climate. He also held an appointment as professor in the Atmospheric and Oceanic Sciences program at Princeton University. In 1997 Manabe returned to Tokyo as the director of the Global Warming Research Program of the Frontier Research System for Global Change.

Manabe's research, the details of which have appeared in more than one hundred publications, has brought him the international recognition of his peers, as is demonstrated by the approximately two dozen awards he received between 1966 to 1998. A survey of the citations accompanying the awards suggests the Manabe is without peer among scientists working to understand the global climate by using physical-mathematical models. He is considered a pioneer in the development of atmosphere-ocean circulation models and their application to the analysis of climate disturbances involving greenhouse gases. This research improves humankind's understanding of the role that the oceans play in the global climate. His work aids in the analysis of the interplay between dynamics and chemistry that controls the distribution of stratospheric ozone. The research also provides the intellectual underpinning for the increasing concern for the future of the natural environment.

Kenneth H. Brown

SEE ALSO: Climate change and global warming; Greenhouse effect; National Oceanic and Atmospheric Administration; Ozone layer and ozone depletion.

Marine Mammal Protection Act

DATE: 1972
CATEGORY: Animals and endangered species

In 1972 the United States Congress enacted legislation to provide safe environments for marine mammals.

The United States Marine Mammal Protection Act (MMPA) prohibits ownership or importation of any marine mammal or any of their products. However, it does allow a limited catch by Eskimos, American Indians, and Aleuts for purposes of material survival or cultural heritage. An amendment was added in 1994 to restructure jurisdiction and enforcement of laws and to establish guidelines for transportation of marine mammals. Controversies exist among fishing interests, environmentalists, and cultural societies as to the interpretation and effects of the MMPA.

When the MMPA became law in 1972, the Fish and Wildlife Service (FWS), which is part of the Department of the Interior, became responsible for manatees, dugongs, polar bears, walruses, and sea otters. The National Marine Fisheries Service (NMFS), a division of the Department of Commerce, was assigned management of whales, dolphins, sea lions, fur seals, elephant seals, monk seals, true northern seals, and southern fur seals.

The 1994 amendment stipulated stronger fishing regulations, especially the use of improved equipment to reduce the number of accidental killings of marine mammals and to exclude the by-catch of turtles, nontarget fish, and undersized fish of the targeted species. Before the amendment, jurisdiction over care and transport of captive marine mammals was jointly shared by the NMFS and the Department of Agriculture's Animal and Plant Health Inspection Service (APHIS). The amendment eliminated the NMFS's part of the administration and enforcement, which has caused concern in the environmental community because the Department of Agriculture officials do not have much experience working with marine mammals.

The Humane Society of the United States appealed for reinstatement of the NMFS as a joint authority, but APHIS was delegated as sole authority. Zoos and aquariums supported APHIS control since that agency made it easier for marine mammals to be captured and transported. Public facilities that already own a marine mammal need only to send a notice to APHIS after acquiring additional mammals, whereas previously it was necessary to obtain a permit. Other changes brought about by the amendment ease regulations on scientists and researchers, who no longer need permits unless there is the potential of harm to a marine mammal. Ecologists also approve of the amendment because it places emphasis on maintaining healthy ecosystems, particularly in the northwestern and

northeastern coasts of the United States where the seal and sea lion populations are declining at an alarming rate.

Many of the human activities that threatened marine mammals in the past are now violations of the MMPA. Sea otters were overhunted for their skins, but government protection has allowed their populations to recover around Prince William Sound and off the California coast. Whales are completely protected by the MMPA and the International Whaling Commission (IWC). However, the natural renewable resources on which the whales feed may be endangered since some countries are harvesting large quantities of krill, which is the mainstay of many whales' diets and is an important link in the marine food chain. As human populations increase worldwide and become more industrialized, demands on the oceans as a food source and a place to dump chemicals also increase. Noise disturbances come from sonic testing and boat traffic. Continued publicity and pressure

Pressure from environmental activist groups such as the Sea Shepherd Conservation Society encouraged the U.S. Congress to pass legislation that conserves marine environments. Among such legislation is the Marine Mammal Protection Act. (AP/Wide World Photos)

from environmental advocates such as Greenpeace and the World Wildlife Fund (WWF) have led wildlife managers, fishers, animal rights supporters, and scientists to work together under the terms of the 1994 MMPA amendment.

Dale F. Burnside with Aubyn C. Burnside
SEE ALSO: International Whaling Ban; International Whaling Commission; Whaling.

Marsh, George Perkins

BORN: March 15, 1801; Woodstock, Vermont
DIED: July 23, 1882; Vallombrosa, Italy
CATEGORY: Ecology and ecosystems

Scholar, author, statesman, and diplomat George Perkins Marsh's importance for environmental issues rests primarily upon his book Man and Nature: Or, Physical Geography as Modified by Human Action *(1864), which was the first treatise on environmental history and one of the early treatises on historical geography.*

George Perkins Marsh's childhood home in Vermont was located at the base of a mountain along the Quechee River, which originally flowed all summer. However, because of lumbering and sheep grazing, the river flow changed to flooding in the spring and little or no flow in the late summer. The river flow problem was explained to the young Marsh by his father, an eminent lawyer, and the situation eventually became a major influence upon Marsh's thinking. In the short term, however, he followed the example of his father and grandfather and trained for the law. He then entered the Vermont legislature and later represented his state in Congress.

In 1850 Marsh became the U.S. minister to Turkey. While serving there, he traveled around the eastern Mediterranean region and became impressed with the evidence of civilization's impact upon the land. He realized that deforestation and grazing by goats were important causes of desertification in arid regions. Traveling to France, he saw both the severe erosion that followed deforestation in the mountains and the value of reforestation for restoring the land.

After five years abroad, Marsh returned to the United States, and the governor of Vermont appointed him state fish commissioner. He published *Report on the Artificial Propagation of Fish* (1857), which explained the impacts of logging, livestock, farming, and industry on fish streams. In 1861 President Abraham Lincoln appointed Marsh minister to Italy, in which position he remained for the rest of his life. At his wife's urging, he had begun writing *Man and Nature* in 1860, which he completed in Italy in 1863. The 1864 edition was soon reprinted, and Marsh produced a revised edition in 1874. Italian editions were published in 1869 and 1872.

In the book, Marsh urged civilization to learn to wisely manage both natural and domesticated resources. He believed that science should provide the guidance. To him the most relevant discipline was geography; today his subject might be called applied ecology. *Man and Nature* includes chapters on wildlife, water, sand, and the side effects of engineering projects. His longest chapter is on the ecology of forests and the consequences of deforestation. The influence of *Man and Nature* helped convince the U.S. government to establish a forest policy and then the Forest Service. The book also exerted an influence upon prominent scientists in Europe, and it became one foundation for the conservation and environmental movements of the twentieth century.

Frank N. Egerton
SEE ALSO: Conservation policy; Environmental policy; Logging and clear-cutting; Sustainable forestry.

Marshall, Robert

BORN: January 2, 1901; birthplace unknown
DIED: November 11, 1939; on a train en route from New York to Washington, D.C.
CATEGORY: Forests and plants

Forester, plant physiologist, and public servant Robert Marshall published numerous writings in which he urged people to conserve the environment. He also urged the government to set aside large areas of forestland in order to protect their natural condition. In 1935 he cofounded the Wilderness Society.

Robert Marshall was born in 1901. He was the son of Florence Lowenstein Marshall and constitutional lawyer and conservationist Louis Marshall, who had been a delegate at the New York State Constitutional Convention of 1894, which placed in the state constitution the famous provision that the New York Forest Preserve shall be "kept forever wild." Marshall's extensive utilization of the family library introduced him to books and topographical surveys of the Adirondack Mountains. At the age of fourteen, he, along with his brother George and a guide, ascended a high Adirondack peak, thus cementing a lifelong love affair with wilderness exploration and celebration.

Marshall's higher education at three universities (bachelor of science in Forestry at Syracuse, Master of Forestry at Harvard, and Ph.D. in Plant Pathology at Johns Hopkins) served him well as he developed a literary and resource management career. During his early professional work in the U.S. Forest Service, he had the opportunity to observe large, unbroken wilderness conditions, which served as the catalyst and foundation for his first major article on extensive natural landscapes. Titled "The Problem of the Wilderness" and appearing in the February, 1930, issue of *The Scientific Monthly*, it was a clarion call for setting aside and protecting large tracts of land in their natural and, to the extent possible, primeval condition. The article elucidated four salient wilderness themes: its great beauty and wildness with integrated aesthetic, mental, and physical values; the rapid disappearance of wilderness; the need to look beyond commodity value of resources in wilderness as the sole arbitrator of its value; and the urgency to act for wilderness preservation.

In 1931 Marshall settled in Washington, D.C., and immediately devoted his efforts to writing assignments. Collaborating with the U.S. Forest Service on *A National Plan for American Forestry* (1932), he contributed sections on national parks, wilderness, and recreation. One year later he published *The People's Forests*, in which he ar-

ticulated the importance of conserving water, soil, and forests. Again he upheld forested areas in relation to human aesthetic needs and claimed that this was a pivotal and important value to contemporary society.

In 1933 Marshall was appointed director of forestry at the Office of Indian Affairs, where he helped develop sixteen wilderness areas on Indian reservations. Two years later, he was a leader of eight people who founded the Wilderness Society. In 1937 he became chief of the new Forest Service Division of Recreation and Lands and immediately began moving official Forest Service policy toward supporting wilderness. He also drafted new administrative regulations relating to a wilderness and wild area classification system. Approval came just months before his untimely death in 1939 at age thirty-eight. In 1964, as the Wilderness Act became law, the twentieth area to be named to the National Wilderness Preservation System was the Bob Marshall Wilderness Area in the Flathead and Lewis and Clark National Forests.

Charles Mortensen

SEE ALSO: Conservation; Forest Service, U.S.; Wilderness Society.

Mather, Stephen T.

BORN: July 4, 1867; San Francisco, California
DIED: January 22, 1930; Brookline, Massachusetts
CATEGORY: Preservation and wilderness issues

As the first director of the National Park Service, Stephen T. Mather personified the national parks movement during the early decades of the twentieth century.

A descendent of one of America's early Puritan families, Stephen Mather was born and raised in California, where he came to love the mountains and forests of the western United States. He joined John Muir's Sierra Club in 1904. He was educated at the University of California at Berkeley, where he made many life-long contacts. After graduation he went into the borax business, which gave him considerable wealth.

Mather, a Republican, followed Theodore Roosevelt, the most famous conservationist in the United States, into the Progressive Party in 1912. However, Mather had the ability to transcend partisan politics. Several earlier attempts had been made to bring order to the small number of national parks that had already been established, and in 1913 President Woodrow Wilson's secretary of the interior, Franklin K. Lane, decided to upgrade the parks by appointing Adolph C. Miller, a fellow University of California graduate, as assistant secretary. After Miller was reassigned elsewhere, Lane challenged Mather to take over the position. He did so in 1915, with still another Berkeley graduate, Horace Albright, as his chief assistant. In 1916 the Park Service Act became law, and Mather became the National Park Service's first and arguably most significant director.

Mather faced a series of interrelated issues: The parks required both financial and political support from Congress, which in turn depended upon public demand. Energetic, charismatic, and brilliant at public relations, Mather generated popular support through various avenues, including national publications such as the *Saturday Evening Post* and the *National Geographic*—1,050 articles about the parks appeared from 1917 to 1919. He organized national parks conferences, brought politicians and writers to the parks, and used his own funds in support of the park system; he personally paid the salary of Robert Sterling Yard, who became the parks' first publicist. In order to provide better access to the parks, Mather strongly supported at least some development; he worked with railroads and automobile associations in building roads and providing other public amenities. Through the years he had his differences with politicians, bureaucrats, and others, but he worked successfully with both Republicans and Democrats.

Mather retired in 1929 because of ill health and was succeeded by Albright, but he left the National Park Service as a major icon in America's consciousness. Under his tenure the park system increased from thirteen parks, eighteen national monuments, a total of 4.75 million acres of area, and 334,000 annual visitors to twenty national parks, thirty-two national monuments, 8.27 million acres, and three million visitors each year. However, development engendered criticism, and Mather was accused of being too prodevelopment. Despite the criticism, he had no doubt about his responsibility as director: "Our job in the Park Service is to keep the national parks as close to what God made them as possible." No director has had as much impact on parks as did Mather, and no one since has been able to generate the public and political support for the park system that he did.

Eugene Larson

SEE ALSO: Muir, John; National parks; Yosemite.

Medfly spraying

CATEGORY: Pollutants and toxins

Medfly spraying is the use of pesticides to eliminate the Mediterranean fruit fly, which poses a threat to a large variety of fruits and vegetables.

The small, two-winged Mediterranean fruit fly, or Medfly (*Ceratitis capitata*), belongs to a group of insects commonly called fruit flies. Considered a major agricultural pest around the world, the Medfly is a threat to more than 250 vegetables and fruits, including guava, peaches, cherries, avocados, pears, and citrus. A long-time inhabitant of Hawaii and the tropics, the Medfly has repeatedly tried to establish itself in the United States since its first unsuccessful foray into Florida in 1929. Attempts to control the fly with pesticides have so far proven successful but controversial.

The incompatibility of the Medfly with humans stems from its fondness for domestic crops. Typically, the female fly lays eggs—sometimes as many as three hundred—in the fleshy parts of fruits or vegetables. The eggs turn into larvae, which tunnel through the fruit and make it unfit for human consumption. When the damaged fruit falls to the ground, the larvae exit and burrow into the ground until they transform into flies and start the cycle over again.

Because of the Medfly's capacity for causing widespread crop destruction, governments around the world have imposed various quarantines, embargoes, and postharvest treatment requirements on any fruits and vegetables that originate in areas known to be infested with the insect. Various eradication programs were used against the Medfly in the United States during the twentieth century. Early efforts in Florida included the use of a compound of arsenate and copper carbonate that was applied with handheld equipment. Another control method was the removal of infested fruit trees.

The most common form of eradication practiced is the aerial spraying of the pesticide malathion mixed with a bait, such as syrup. A poison employed to control mosquitoes, malathion is used in weaker amounts in the war against the Medfly. Though all aerial spraying attempts must receive prior approval from the Environmental Protection Agency (EPA), the use of malathion has generated controversy in California and Florida. Opponents commonly complain about the pesticide's possible harmful effects on other insects, wildlife, livestock, aquaculture, water supplies, and human health—especially that of children. In response to these concerns, the EPA now requires various federal and state agencies involved in eradication programs to seek more environmentally friendly methods to battle the Medfly.

One alternative to malathion involves the use of domestically raised flies that have been sterilized through radiation and released to mate unsuccessfully with wild Medflies, thus reducing the regenerative capacity of the population. Another possible, but experimental, pesticide is made from a mixture of two dyes, phloxine B and uranine, which are commonly used to tint drugs and cosmetics. Once ingested by the Medfly, the dye particles absorb light, which in turn produces oxidizing agents that destroy cell tissues. As a result, most flies die within twenty-four hours. The dyes quickly lose their potency and become nontoxic. Preliminary tests in Hawaii indicate that the use of these dyes may be more effective and safer to the environment than malathion.

John M. Dunn

SEE ALSO: Malathion; Pesticides and herbicides.

Mediterranean Blue Plan

CATEGORY: Water and water pollution

The Blue Plan, a U.N.-sponsored multinational effort to curb pollution in the Mediterranean, met with mixed results.

In 1980, under the auspices of the United Nations Environment Programme (UNEP), the nations bordering the Mediterranean signed an agreement setting forth ways in which they would cooperate to reduce pollution of their common sea. The agreement, which soon came to be known as the "Blue Plan" for its efforts to clean the Mediterranean's waters, represented the culmination of several years of international efforts. In 1975, for example, UNEP had provided more than $7 million to the Mediterranean Action Plan (MAP), an earlier program designed to help Mediterranean countries fight pollution. Early in 1979, however, UNEP informed signatories to the 1975 agreement that it would cut back future financial support for MAP. UNEP thus called a February, 1979, conference in Geneva, Switzerland to prepare a new approach to budgetary demands for immediate environmental remedies and to map out a strategy for protecting the ecology of the Mediterranean basin. In Geneva, a program was drafted identifying twenty-three environmental protection projects demanding immediate attention. A budget of $6.5 million was established, one-half to come from the participating countries, one-fourth from UNEP, and the remainder from contributions of services and staff time by environmental organizations.

Staffs of international researchers formed Blue Plan Regional Activity Centers, which provided information to signatory governments to help them plan future economic development in such a way as to prevent a repetition of the environmental damages that had been done to the sea and its coastline in earlier decades. The U.N. role was to facilitate communication between these centers and to sponsor international meetings to share findings and propose solutions on a regular basis. Blue Plan researchers focused their attention on food production, industry, energy use, tourism, and transport. The jointly sponsored research suggested, for example, that some ecologically harmful industries, such as mining and metallurgical processing and petrochemical production, were overproducing in Mediterranean areas; in such cases, plan officials suggested ecologically preferable and economically logical adjustments.

The attainment of such goals, however, was complicated by political and economic factors. For example, efforts to streamline supply and production of coal and steel on a geographic basis were hindered by longstanding tensions between Turkey and Greece. Similarly, Tunisia and Algeria resisted energy market cooperation with each other, although Tunisia needed the natural gas and petroleum that neighboring Algeria produced; Tunisia thus continued to pursue, at substantial economic and ecological cost, its own limited petroleum production. Despite such setbacks, Blue Plan efforts did have an effect, and pollution levels in the Mediterranean dropped throughout the 1980's and 1990's.

Alexander Scott

SEE ALSO: Water and water pollution.

Mendes, Chico

BORN: December 15, 1944; Acre, Brazil
DIED: December 22, 1988; Xapuri, Acre, Brazil
CATEGORY: Forests and plants

Rubber tapper and trade union leader Chico Mendes spent his entire life working against the forces of environmental destruction in the Amazon forest in order to sustain a way of life for his fellow rubber tappers and other indigenous peoples of western Brazil. He earned international recognition as a defender of the Amazon ecosystem.

Francisco "Chico" Mendes Filho's early years were like those of all rubber tappers. His family lived on a *seringal* (rubber estate) owned by a master. Tappers received artificially low prices for latex, they had to purchase all their tools and foodstuffs from the estate store, and education of children was prohibited, though Mendes secretly learned to read and write from a fellow tapper.

In the 1960's the Brazilian military government embarked on a massive effort to open the rain forests in the Amazon region to development. Roads were built, and large cattle ranches were developed to make cheap meat available to the frustrated working class. The area also became a dumping ground for the surplus population in the east. By 1980 deforested land in the Amazon totaled 123,590 square kilometers (47,718 square miles), threatening indigenous people's way of life.

In 1971 Mendes began organizing rural rubber tappers, culminating in the formation of a union in Xapuri in the northern state of Acre in 1977. Unions sprang up in other towns in Acre as well. Their main tactic in challenging developers and ranchers was the *empate*, a standoff between tappers and laborers who were clearing the forest during the dry season. From the mid-1970's to 1988, the Acre rubber tappers staged over forty major *empates*, persuading the laborers to lay down their chain saws and return home. Over two million acres of forest were saved during this period.

Mendes's leadership was recognized in 1985 when the first National Rubber Tappers Congress was held in Brasília. Soon after, Mendes gained international recognition through the efforts of a maverick Brazilian agronomist and a British journalist. Although his first concern was sustaining his people rather than promoting an environmental agenda, Mendes eventually realized that environmental concerns would give his cause attention and shape world opinion in the rubber tappers' favor. In 1987 he traveled to Miami, Florida, to speak with the Inter-American Development Bank, which was financing road building in the Amazon. Mendes also traveled to Washington, D.C., to testify before a Senate committee considering funds for the bank.

Mendes's crowning achievement came in October, 1988, when the Brazilian government declared a 61,000-acre tract of rubber tapper territory near Xapuri an extractive reserve. Originated by Mendes, an extractive reserve is an area designated solely for sustainable use. However, the local ranchers' opposition to the reserves signaled death for Mendes; he was assassinated in December, 1988, by a rancher's hired gunman.

Ruth Bamberger

SEE ALSO: Deforestation; Greenhouse effect; Rain forests and rain forest destruction.

Mercury and mercury poisoning

CATEGORY: Pollutants and toxins

Mercury is a highly toxic element that can cause acute episodes of poisoning and chronic health effects. Mercury is found in the environment in elemental and combined forms; people are exposed to its various forms in the workplace, at home, and through mercury-laden foods. Numerous federal regulations have set standards of minimum exposure for humans and have helped reduce the occurrence of mercury poisoning.

Mercury, often called quicksilver, is a naturally occurring silver-white metal that is liquid at room temperature. It is a relatively rare element, making up only about 3 parts per billion (ppb) of the earth's crust. Mercury occurs in its elemental form as a liquid but is more commonly found combined with other elements to form various minerals. The most common mercury-containing mineral is cinnabar (mercury sulfide), which is found mainly in Spain, Algeria, and China.

Mercury and its inorganic compounds have been used in producing caustic soda, dry-cell batteries, scientific measuring devices, dental amalgams, and mercury vapor lamps. Workers in factories making these products may be exposed to relatively high levels of mercury vapors and compounds. Dental assistants may also be exposed to relatively high levels of mercury vapor. Mercury is also released into the environment by the waste incineration of these products and may eventually accumulate in food sources such as grains and fish. Mercury was also used as an additive to latex paints to inhibit the growth of bacteria and fungi, but this use was eliminated in the United States in 1991. Mercury compounds have also been used in agriculture as a fungicide and may cause health effects in workers who handle these compounds. Finally, mercury mines and smelters may release considerable amounts of mercury into the surrounding regions.

Some bacteria can convert inorganic mercury into organic compounds, such as methyl mercury. Methyl mercury bioaccumulates in biotic systems and biomagnifies until the highest members of the aquatic food chain have high levels of mercury in their fatty tissues. The greatest exposure to mercury is through eating mercury-rich fish; individuals who rely heavily on fish such as tuna, pike, swordfish, and shark may therefore suffer adverse health effects.

Since elemental mercury is a liquid at room temperature and easily volatilizes, it can be inhaled and cause damage to the central nervous system, lungs, and kidneys. Mercury is a strong neurotoxin that can cause a series of effects such as memory loss, tremors, and excitability at low levels; higher levels may cause more severe mental problems and even death. Inhalation of high concentrations of mercury leads to hallucinations, delirium, and even suicide in some cases. It is not clear whether elemental mercury causes cancer in humans; the Environmental Protection Agency (EPA) classifies elemental mercury in Group D (not classifiable as to human carcinogenicity).

The inorganic forms of mercury are not easily absorbed by the body and pose a relatively small health risk. Oral ingestion of inorganic mercury can cause nausea, stomach pains, and vomiting. The EPA states that the acute lethal dose of most inorganic mercury compounds for a normal adult is 1 to 4 grams (14 to 57 milligrams/kilogram for a 70 kilogram person). The EPA classifies inorganic mercury in Group C (a possible human carcinogen).

Organic mercury compounds, particularly methyl mercury, are absorbed easily and constitute the greatest health risk for most people. Methyl mercury exposure is greatest from consuming fish and fish products. Low-level poisoning by methyl mercury may cause vision problems, paresthesia (the sensation of prickly skin), speech difficulties, shyness, and a general malaise. Acute exposure causes numerous central

When mercury is released into water, it bioaccumulates in fish. From there it may be passed on to humans and other organisms that eat the tainted fish, causing adverse health effects. (Jim West)

nervous problems, including blindness, deafness, and loss of consciousness. The EPA estimates that the minimum lethal dose of methyl mercury is 20 to 60 milligrams/kilogram for a 70 kilogram person. Lower doses are lethal for more sensitive people, particularly children. Methyl mercury has been shown to have significant developmental effects on babies born to women who ingested high levels during their pregnancy. The infants from such pregnancies may exhibit mental retardation, visual problems (including blindness in some), and cerebral palsy. Methyl mercury is a Group C carcinogen.

Mercury has caused serious environmental and human health problems when released into ecological systems and when used inappropriately. The most infamous incident was the mercury poisoning at Minamata Bay in Japan. In the 1950's and early 1960's, inorganic mercury waste was released into the bay, and bacteria converted the inorganic mercury to methyl mercury. The mercury was then consumed by marine organisms and passed up the food chain, eventually resulting in human mercury poisoning. At least forty-three people died of "Minamata disease" from eating fish and other marine organisms that contained high levels of mercury. Even more people were killed in Iraq by the inappropriate used of methyl mercury fungicide. In 1972 more than 460 people died from mercury poisoning after eating bread made from wheat that had been treated with the fungicide. Many of the animals and plants within the area were also poisoned by the methyl mercury.

Jay R. Yett

SUGGESTED READINGS: The toxic effects of mercury, as well as laws governing its use, are discussed in *Toxics A to Z* (1990), by John Harte et al. The origin of mercury deposits and the uses of the element are cogently described in *Mineral and Energy Resources* (1990), by Douglas Brookings. F. Bakir et al., give the best discussion of the Iraqi mercury poisoning in their article "Methyl Mercury Poisoning in Iraq," *Science* 181 (1973).

SEE ALSO: Biomagnification; Food chains; Heavy metals and heavy metal poisoning; Minamata Bay mercury poisoning; Oak Ridge, Tennessee, mercury releases.

Methyl parathion

CATEGORY: Pollutants and toxins

Methyl parathion is a highly toxic neurotoxin used on agricultural crops to kill pests, especially boll weevils and mites. In irrigated fields, methyl parathion is used to exterminate mosquito larvae.

Methyl parathion's molecular formula is $C_8H_{10}NO_5PS$, and its chemical name is O,O-dimethyl (4-nitrophenyl) phosphorothioate. Also known as parathion-methyl and metafos, and commonly called cotton poison and roach milk, methyl parathion has been marketed under such trade names as Metacide and Penncap-M and packaged in dust, liquid, and emulsifiable concentrate forms. The Environmental Protection Agency (EPA) classifies methyl parathion as a restricted-use pesticide (RUP) that can be sold and utilized solely by certified applicators. Methyl parathion packaging is distinguished with the word "danger" to alert users of potential harmful contact. The EPA forbids agricultural workers from being in a field treated with methyl parathion for forty-eight hours after application of the pesticide.

Methyl parathion is an organophosphate that inhibits the enzyme cholinesterase, which is crucial to healthy functioning of the nervous system. If absorbed through the skin, inhaled, or ingested, the pesticide blisters tissues, retards reproductivity, and can cause death. Methyl parathion is toxic to aquatic organisms such as fish, shrimp, and crabs and harmfully affects carnivores that consume tainted organisms. Contaminated wildlife poses a danger to hunters who eat their flesh. By killing algae-eating insects and crustaceans, methyl parathion inadvertently causes algae to rapidly reproduce and use oxygen that other water inhabitants need. In addition, methyl parathion poisons small mammals

such as rats and rabbits, is harmful to birds and beneficial insects such as honey bees, and has been reported to injure alfalfa and sorghum.

Methyl parathion usually requires several months to break down in soil but does not leave dangerous residues or kill soil microorganisms. The speed of degradation increases with higher temperatures, with exposure to sunlight, and in flooded soil. While small amounts of methyl parathion do not contaminate groundwater, spills require longer to degrade. The pesticide degrades more quickly in flowing water, fresh water, and water bodies with sediments. In plants, methyl parathion is almost completely metabolized within a week.

Although it is considered safe only for outdoor use, methyl parathion is effective against roaches and rats, which often results in unlicensed exterminators spraying it inside homes. In 1996 the EPA evacuated residents of one thousand contaminated homes in Mississippi and spent millions of dollars to rebuild the structures. Many methyl parathion containers are tracked by computers to prevent misuse. Methyl parathion poses combustion hazards above 100 degrees Celsius and releases toxic fumes, including dimethyl sulfide, sulfur dioxide, and carbon monoxide. Concerns about adverse toxic effects of pesticide products have resulted in studies of methyl parathion's potential carcinogenic tendencies. In 1992 producers of methyl parathion stopped supporting its use on such agricultural crops as strawberries and tobacco and for mosquito control in forests.

Elizabeth D. Schafer

SEE ALSO: Agricultural chemicals; Pesticides and herbicides; Runoff: agricultural; Water pollution.

Mexico oil well leak

DATE: June 3, 1979
CATEGORY: Water and water pollution

In 1979 a drilling well in the Gulf of Mexico blew out and spilled an estimated three million barrels of oil into the sea over a nine-month period before it could be capped.

On June 3, 1979, an offshore oil operation in the Gulf of Mexico drilled into a high-pressure gas pocket, causing a blowout that ignited the oil well. The accident produced a massive oil slick nearly 65,000 square kilometers (25,000 square miles) in area. According to a July, 1979, issue of the *Christian Science Monitor*, at the time it was considered not only "the world's worst oil spill from an oil well, but also the worst spill ever." The runaway well, Ixtoc I, belonged to Petroleos Mexicanos (PEMEX), the Mexican state-owned oil company. The exploratory well was being drilled in the Gulf of Campeche off the Yucatán Peninsula, a low-lying limestone tableland that separates the Caribbean Sea from the Gulf of Mexico.

Natural gas escaping from the well was ignited and destroyed the $22 million drilling rig. Crude oil issuing from the damaged well, which is toxic to most marine life, polluted waters in the region and ruined much of Mexico's shrimping industry. Large numbers of Atlantic sea turtles nest along the Mexican coast. The young of this endangered species, which normally swim out to seaweed beds after they hatch, were in danger from the toxic effects of the oil slick.

The oil slick also posed a threat to certain regions along the Texas coast, including the Padre Island National Seashore area and resort beaches such as Galveston, Texas. As a preventive measure, floating booms and other types of barriers were placed as shields for bays and estuaries along the south Texas coast by the Department of Texas Water Resources. The south Texas beaches escaped extensive damage; only isolated patches of diluted oil reached the Texas coastline. Some Texas beaches were littered by nontoxic, pancakelike hydrocarbon globs that resembled chocolate mousse.

Immediately after the blowout, PEMEX began drilling relief wells, named Ixtoc 1A and 1B, in an attempt to intersect the runaway well and divert the escaping oil. Later, the famous oil field troubleshooter Paul "Red" Adair and his crew were called in to cap the well. They succeeded in temporarily shutting the well down in late June, but it burst out again shortly thereafter. Over the next several months steel balls, lead balls, and gelatin were pumped down the well in repeated attempts to check or stop the oil flow.

Although most of the early balls were expelled, a later injection of 108,000 balls reportedly reduced the flow from 20,000 to 10,000 barrels per day on October 12. Meanwhile, several different organizations, including Oil Mop Incorporated and Shell Oil, used a variety of equipment in an attempt to skim oil from the ocean surface at or near the well site.

Ixtoc 1A and 1B ultimately injected drilling mud, water, and cement into the reservoir rock and base of the well, which cut off flow. The well was finally blocked with three cement plugs on March 22, 1980, after more than three months of uncontrolled oil and gas flow.

The Ixloc I and other wells drilled in the Gulf of Campeche indicated the presence of large quantities of oil in this offshore region. However, the disaster led to questions about PEMEX's credibility and Mexico's entire oil policy.

Donald F. Reaser

SEE ALSO: Oil spills; Santa Barbara oil spill.

Minamata Bay mercury poisoning

CATEGORY: Human health and the environment

The discharge of toxic wastewater into Minimata Bay, Japan, spread mercury into the local food chain and produced an environmental health tragedy now known as Minamata disease. The episode represents one of the first identified cases of a clear cause-and-effect relationship between toxic chemical discharge and severe harm to humans and their environment.

Minamata is located on the southwest coast of Kyushu, the southernmost of Japan's four main islands, in Kumamoto Prefecture. During the 1950's, a local chemical plant owned by Chisso Corporation was engaged in the production of acetyl-aldehyde and vinyl chloride. One of the chemicals used as a catalyst in the production processes was mercury oxide (HgO). The industrial waste, including mercury, was discharged into Minamata Bay. At the time, this was an acceptable industrial practice, and the amount of discharge increased during the early 1950's. It was generally understood that mercury, a heavy metal like lead, cadmium, and arsenic, could be injurious to health if people mishandled, ingested, or inhaled it. However, since mercury is dense and quite insoluble in water, it was presumed that it would quickly sink into the bay-bottom sediment, be slowly buried, and disappear.

By 1953, unusual medical symptoms had begun to appear in area residents; most of the symptoms were neurological, such as tremors and impaired senses. In 1956, the first case of a distinct medical condition was reported. By 1959, the cause of the affliction had been identified and the effects established. The first patients had acute or near-acute levels of mercury poisoning, an ailment that became known as Minamata disease.

The residents of the area had not realized that the dense mercury (Hg) on the bay bottom was being acted on by microorganisms such as bacteria and algae and was being converted into methyl mercury and dimethyl mercury. Methyl mercury, the more injurious and toxic of the two substances, is much less dense that mercury itself and is more soluble. In the bay environment, it worked its way to the upper sediment and water, was taken up in the food chain by bottom-dwelling shellfish and fish, and was then consumed by the local people, for whom seafood was a dietary staple. Mercury can also be taken up by fish-eating birds and mammals. Heavy metals tend to accumulate in the bodies of organisms that ingest them, as in the edible flesh of fish. They concentrate up the food chain, as one species retains and accumulates the contaminant from its regular diet.

In natural environmental cycling and processes, elements such as mercury are weathered out of rock to soil and enter the hydrosphere (lakes, rivers, groundwater, and oceans). They might become more concentrated, but organic decay eventually returns them to the soil or water. This cycling can be disrupted when industrial or other human activities introduce an anomalously large amount of the element into the cycle that cannot be accommodated by natural means. Whether an element is toxic to humans or others is a function of concentration;

some elements are essential in small amounts but become toxic in large concentrations.

Mercury occurs as a trace element in average crustal rock at a level of 0.1 parts per million (ppm); in seawater, it is typically under 0.05 parts per billion (ppb). At Minamata Bay, mercury levels in fish and shellfish were measured up to 50 ppm (25 ppm represents a biomagnification factor of 500,000 over normal seawater). For comparison, the U.S. Food and Drug Administration (FDA) prohibits fish with more than 1 ppm of mercury from being commercially marketed. Residents diagnosed with the disease had mercury levels of 20 to 100 ppm in their livers and kidneys and 3 to 25 ppm in their brains. Minamata Bay sediment levels of mercury were as high as 7,000 ppm.

Medical effects of consuming mercury include progressive damage to the central nervous system and birth defects from prenatal exposure. Symptoms include shakiness (tremors), blurred vision, impaired speech, numbness in limbs, loss of memory and intelligence, nervousness, and, in extreme cases, death. By the time symptoms are apparent, damage is irreversible. Of the first 34 people diagnosed with Minamata disease, 20 died within six months. By 1960, 43 people had died and 116 were permanently affected. About one-third of the area children being born were affected by symptoms. As of 1992, 2,252 people had become ill and been officially diagnosed with the disease, and 1,043 had died. Several thousand more claimed to have the disease but had not been officially diagnosed by the examining board set up by the government.

The developing environmental health problem of the 1950's at Minamata Bay, however inadvertently it my have started, became a broader and more persistent tragedy because of the lack of corporate or governmental response. Chisso Corporation denied responsibility for the problem, arguing that dumping of industrial waste was an accepted practice and was legal at the time. Some have speculated that the company may have suspected the cause of the growing problem but concealed it. The national government declined to act on the matter. At the time, there was little governmental consciousness of environmental quality issues, and intervention might have harmed the chemical industry and slowed Japan's post-World War II industrialization and economic recovery. A similar problem occurred at Niigata in west-central Japan in the 1960's, when mercury waste was dumped into the Agano River. The national government thus did not intervene in the Minamata situation until 1968, and the dumping of mercury into the bay was not stopped until 1971, twelve years after the problem became known.

The National Institute for Minamata Disease was set up in 1978 to assist with medical studies, research, and following the progress of victims over time. The Japanese government provides medical care for diagnosed victims, along with financial compensation. Active victims' movements and support groups have used the court system to pursue responsibility and liability.

It is estimated that the total discharge of mercury from the Chisso plant was between 70 and 150 tons. Major efforts were implemented to remediate the site and restore the local environment so that it would once again be fit for sea life and human use. From 1977 to 1990, the mercury-contaminated sludge was dredged from 58 hectares (143 acres) of the bay in a project funded by the national and prefectural governments and Chisso Corporation. In 1974, a huge net was placed to isolate 380 hectares (939 acres) of the bay area to prevent fish from swimming in or out. In 1997, Minamata Bay was declared to be free of mercury contamination, and the net was removed.

Robert S. Carmichael

SUGGESTED READINGS: F. M. D'Itri's "Mercury Contamination—What We Have Learned Since Minamata" (*Environmental Monitoring and Assessment* 19, 1991) explores the effects of mercury contamination and poisoning. *Environmental Geology* (1997), by C. Montgomery, contains a discussion of environmental remediation efforts such as the Minamata cleanup.

SEE ALSO: Biomagnification; Environmental health; Environmental illnesses; Food chains; Hazardous waste; Heavy metals and heavy metal poisoning; Mercury and mercury poisoning; Oak Ridge, Tennessee, mercury releases; Reclamation; Water pollution.

Mobro barge incident

DATE: 1987
CATEGORY: Waste and waste management

The Mobro barge sparked national interest in recycling as a solution to landfill problems during its 9,600-kilometer (6,000-mile) odyssey to find a place to unload its cargo of garbage.

On March 22, 1987, the *Mobro* left New York loaded with 3,186 tons of garbage from Islip and New York City. The garbage was to be unloaded in the rural Jones County, North Carolina, town of Morehead City, where it would be converted to methane gas. The project was the brainchild of Lowell Harrelson, an Alabama entrepreneur who envisioned shipping garbage from northern states where landfills were becoming scarcer to cheaper southern landfills that had adequate space. Jones County manager Larry Meadows saw the plan to accept out-of-state garbage as a way to boost the county's tax base.

The *Mobro* arrived in Morehead City sooner than expected. North Carolina officials questioned whether the *Mobro*'s hasty arrival meant that its cargo might contain hazardous waste. North Carolina officials denied permission to dock, and the barge set forth in search of another port. The media learned of the *Mobro* and closely reported its journey. Before its journey finally ended back where it began in New York, the *Mobro* traveled to Louisiana, Alabama, Mississippi, Florida, New Jersey, the Bahamas, Mexico, and Belize, only to be turned away. Amid intense publicity, the "not in my backyard" (NIMBY) syndrome quickly took hold. Mexico tracked the *Mobro* with gunboats, while Belize threatened to dispatch its air force to prevent the barge from approaching its waters.

As a result of media coverage, the barge became a symbol of U.S. trash-disposal problems. New York environmentalist Walter Hang spoke of a garbage crisis on television. Other environmentalists referred to a nation buried in trash. Some Environmental Protection Agency (EPA) officials took note of the *Mobro* incident and publicly acknowledged a national garbage crisis. By September, 1988, EPA administrator J. Winston Porter asked each state and municipality to formulate a recycling plan that would reduce disposal needs by 25 percent. In an EPA report titled "Solid Waste Dilemma," Porter implied that the country was running out of space to bury garbage. Others in the EPA disputed Porter's claim and maintained that the United States had an average of twenty-one years of remaining landfill capacity.

The crisis ended after the *Mobro* returned to New York. Lawsuits filed in New York State Supreme Court by the borough of Queens and the New York Public Interest Group (NYPIRG) failed to prevent incineration and burial of the garbage. Queens officials feared a health hazard, while NYPIRG contended that the garbage contained high levels of cadmium and lead. All three thousand bales of trash were finally burned in September, 1987, at Brooklyn's Southwest Incinerator, and 500 tons of ash were buried at Islip's Blydenburgh landfill.

Michele Zebich-Knos

SEE ALSO: Landfills; Recycling; Solid waste management policy; Waste management.

Mario Molina, a chemical engineer, was awarded the 1995 Nobel Prize in Chemistry, along with F. Sherwood Rowland and Paul Crutzen. All three were honored for their studies of the formation and decomposition of ozone in the stratosphere. (AP/Wide World Photos)

Molina, Mario

BORN: March 19, 1943; Mexico City, Mexico
CATEGORY: Weather and climate

Mario Molina shared the 1995 Nobel Prize in Chemistry with Paul Crutzen and F. Sherwood Rowland for his pioneering work concerning the formation and catalytic decomposition of ozone in the stratosphere.

Ozone is produced naturally in the stratosphere and provides a filter to restrict the amount of harmful ultraviolet radiation that strikes the surface of the earth. It was the work of Mario Molina, Paul Crutzen, and F. Sherwood Rowland that first drew attention to the catalytic role played by chlorofluorocarbons (CFCs) in the decomposition of ozone. CFCs have been used in spray cans, electronic parts cleaning, and refrigeration systems. Their production was banned by an international agreement called the Montreal Protocol in 1987.

Molina was born in Mexico City in 1943 to Pasquel and Leonor Henriquez. He obtained a degree in chemical engineering in 1965 from the Universidad Nacional Autonoma de Mexico. His masters degree was earned at the University of Freiburg in Germany in 1967 for work done on the study of rates of polymerization reactions. Molino returned to his undergraduate institution as an assistant professor for one year, and then began working on a doctorate at the Berkeley campus of the University of California. He earned his Ph.D. in 1972 in physical chemistry and remained at Berkeley as a postdoctoral student. The following year he married Luisa Y. Tan, who is also a chemist. The couple has one son.

In 1973 Molina accepted a postdoctoral appointment with Rowland at the University of

California at Irvine. At this time Molina began to look at the question of what happens to chemicals and pollutants in the troposphere and stratosphere, where they are subject to high-intensity ultraviolet radiation. A 1974 article by Molina in *Nature* began the series of papers pointing out the connection between ozone destruction and CFCs. Molina's continuing research led to the prediction of the ozone hole that was later discovered over Antarctica in 1985. Molina then began researching the interface between the atmosphere and the biosphere; he hoped his investigations would lead to an understanding of global climate-change processes.

Since his original publication, Molina has held appointments with the Jet Propulsion Laboratory at the California Institute of Technology as a senior research scientist and with the Department of Earth, Atmospheric, and Planetary Sciences and the Department of Chemistry at the Massachusetts Institute of Technology (MIT) as a professor. Molina has received several awards besides the Nobel Prize in Chemistry. These include the American Chemical Society Esselan Award (1987), the American Association for the Advancement of Science Newcomb-Cleveland Prize (1988), the National Aeronautics and Space Administration Medal for Exceptional Scientific Advancement (1989), and the United Nations Environmental Programme Global 500 Award (1989).

Kenneth H. Brown

SEE ALSO: Chlorofluorocarbons; Freon; Ozone layer and ozone depletion.

Monkeywrenching

CATEGORY: Philosophy and ethics

Monkeywrenching is a controversial, direct-action tactic used by radical environmentalists to disrupt equipment and processes that degrade the environment.

The term "monkeywrenching" was first used in 1904 to refer to throwing a monkey wrench, or a spanner with a movable jaw, into machinery to obstruct or hinder factory work. These acts were preceded in early nineteenth century England by Luddites and machine breakers who protested the mechanization of the workplace during the Industrial Revolution. The term was appropriated later by environmentalist and author Edward Abbey in his 1975 novel *The Monkey Wrench Gang*. Abbey detailed the exploits of three men and one woman who took the law into their own hands to defend the wilderness from excavation. They performed acts of sabotage, or ecotage, on road-building equipment and entertained visions of blowing up Arizona's Glen Canyon Dam.

The radical environmental group Earth First!, founded in 1980, used direct action, civil disobedience, and monkeywrenching tactics to defend wilderness areas from developers. The general idea behind monkeywrenching is to create a stir, delay or halt projects, and gain publicity for the cause. Dave Foreman, one of the founders of Earth First!, published *Ecodefense: A Field Guide to Monkeywrenching* in 1985 and *Confessions of an Eco-Warrior* in 1991. He credits Abbey's book as a major motivation and inspiration. In 1990 Earth First! promoted the Redwood Summer, a ten-week campaign to slow logging of redwoods. Among the tactics used for such actions was tree spiking, in which long nails were driven into trees to dissuade loggers. Since the nails could potentially shatter chain saws and thus hurt loggers, Foreman, in *Ecodefense*, advised monkeywrenchers to warn loggers when an area has been spiked. Incidents occurred, however, in which no notice was given. Another tactic is tree sitting. In 1998 a California woman lived for many months in a tree to prevent logging of a particular grove. Earth First! activists have also been known to knock down billboards and destroy heavy equipment. The self-proclaimed "navy of Earth First!'s army" is the Sea Shepherd Conservation Society, directed by Paul Watson, who published *Ocean Warrior* in 1994. This group aims to prevent whaling ships and others from killing and capturing marine mammals.

Monkeywrenchers are determined and not easily reformed in jail, as many activists have been detained multiple times. The ideological viewpoint of those supporting monkeywrenching is strongly preservationist. Although monkey-wrenchers are often influenced by such preservationists as John Muir and Henry David Thoreau, they tend to be far more militant. The theory behind monkeywrenching stems from a field of thought known as deep ecology, whose core tenet is biocentrism, a belief that the human species is just one member of the biological community in which all species have equal standing. This view places humans on the same level as every other living thing. Followers of this philosophy feel that human conduct should proceed from an understanding that all life, no matter how big or small, has a form of equality by ethical pretenses. Therefore, those individuals willing to risk their lives act as foils to projects they feel are destroying something natural. They claim to fight for organisms that cannot defend themselves against industrial intrusion.

Oliver B. Pollak and Aaron S. Pollak

SEE ALSO: Abbey, Edward; Earth First!; Ecotage; Ecoterrorism; Foreman, Dave.

Monongahela River tank collapse

DATE: January 2, 1988
CATEGORY: Water and water pollution

The collapse of an oil storage tank caused the worst inland oil spill in U.S. history.

In the early evening of January 2, 1988, a rupture of a storage tank at the Ashland Oil terminal in Floreffe, Pennsylvania, twenty-five miles southeast of Pittsburgh, released 3.9 million gallons of diesel oil. The oil spilled over a containment dike and flowed across a road into a ravine, and much of the oil eventually found its way into a storm sewer leading to the Monongahela River.

Because of darkness and freezing weather, the extent of the damage was not fully recognized until the next morning, by which time nearly 750,000 gallons of oil had flowed through the storm sewer into the Monongahela.

Nevertheless, response efforts from a number of agencies, including the local volunteer fire department, borough police, and the Mt. Pleasant Hazardous Materials Team, began almost immediately. Within hours, a team from the Pennsylvania Emergency Management Agency (PEMA) was en route. The first concern of those responding was to stop the flow of fuel, but darkness and cold made the operation difficult. Moreover, a strong odor of gasoline indicated an additional gas leak of unknown origin. The mixture of gasoline and diesel fuel presented a dangerous situation made even more serious by the presence of hazardous chemicals at a nearby chemical plant, and the decision was made to evacuate twelve hundred residents. Emergency crews and firefighters worked throughout the night to contain the oil.

An Environmental Protection Agency (EPA) coordinator arrived the following morning and discovered the condition of the river to be worse than expected, as the oil had dispersed through the water volume rather than remaining on the surface. Water intakes for communities downstream on the Monongahela and the Ohio River were shut off as a preventative measure.

Containment booms placed downstream had little effect, so deflection booms were used to move the oil to collection areas. As a result of the cold temperatures, the oil formed heavy globs that could be picked up from the river edges and bottom. Over the next two months, nearly 205,000 gallons of oil, or 29 percent of the total spilled into the river, was recovered.

The damage caused to wildlife was difficult to assess, as much wildlife was hibernating or inactive in the winter weather. Researchers estimated that the spill killed eleven thousand fish and two thousand waterfowl.

Investigations into the cause of the collapse discovered a small flaw in the steel plates at the base of the tank. The tank's forty-year-old steel had been weakened by reassembly, and the temperature was low enough to cause a brittle fracture. When the tank was filled, the resulting stress caused a crack near a weld.

The Ashland company had not secured a written permit before constructing the tank, which had not been properly tested before filling. The company was held liable for a $2.25 million federal fine for violating the Clean Water Act and the Refuse Act; it also paid $11 million in cleanup costs and tens of millions more to other injured parties.

Alexander Scott

SEE ALSO: Oil spills; water and water pollution.

Montreal Protocol

DATE: 1987
CATEGORY: Atmosphere and air pollution

The Montreal Protocol was an international agreement with the specific goal of limiting atmospheric inputs of chlorofluorocarbons for the purpose of protecting the ozone layer, an essential component of the atmosphere.

The ozone layer, which is 10 to 20 kilometers (6 to 12 miles) above the earth's surface, screens out most of the sun's ultraviolet radiation. Ultraviolet radiation can lead to mutations and cancer in living things. The nations participating in the Montreal Protocol were motivated to act by four major developments: The accumulation of chlorofluorocarbons (CFCs) in the atmosphere was observed in the early 1970's; CFC decomposition in the atmosphere was demonstrated to cause ozone destruction in 1974; a hole in the ozone layer was discovered over Antarctica in the early 1980's, and evidence linking the ozone hole to CFCs was provided in 1985; and CFC substitutes were developed by important CFC producers.

On September 16, 1987, the Montreal Protocol on Substances that Deplete the Ozone Layer was signed by forty-six nations, including the United States. These nations represented two-thirds of the total global production and consumption of CFCs and other halogenated compounds. The protocol entered into force on January 1, 1989.

The Montreal Protocol was designed to control the production and consumption of CFCs and other halogenated compounds suspected of causing ozone destruction. The United States and other industrialized countries committed themselves to freezing immediately CFC produc-

tion at 1986 levels and reducing total CFCs leading into the twenty-first century. Consumption of the major CFCs was to be frozen at their 1986 levels by mid-1989, reduced to 80 percent of the 1986 levels by mid-1993, and reduced to 50 percent of the 1986 levels by 1998. Other halogenated compounds were to be frozen at 1986 consumption levels in 1992.

The protocol was amended in 1990 when new scientific evidence suggested that ozone was being depleted above Antarctica more dramatically than previously assumed. Measurements above Antarctica showed that ozone concentrations declined to between 50 and 95 percent of 1979 levels during certain times of the year. Improved atmospheric models also suggested that the goals of ozone protection could be met only by more stringent curbs on CFC production. The amended Montreal Protocol called for a total phaseout of specified CFCs, halons, and carbon tetrachlorides by 2000 and methyl chloroform by 1995. It also accelerated the rate at which the phaseout would be conducted for CFCs in general. A 50 percent reduction of CFC consumption was required by 1995, and an 85 percent reduction was required by 1997. There was to be a 100 percent reduction of CFC consumption by 2000.

The Montreal Protocol was a political innovation because it called for a gradual reduction in CFC production and allowed for adjustments in the activity of each treaty member that was flexible enough to respond to new scientific information. A total ban on CFCs would have been unworkable; without reasonably inexpensive alternatives, the distribution of temperature-sensitive medical supplies such as blood and food shipments would have been imperiled. Because CFCs were essential to most refrigerating units, many work places dependent on air conditioning would have been affected. Also, considerable amounts of industrial machinery with productive lifetimes of twenty to thirty years would have immediately become obsolete.

There is still considerable disagreement over the extent of ozone depletion and its effect. Debate continues over whether the economic cost of finding alternatives to CFCs is merited by the estimated increase in skin cancer caused by a

Milestones Leading to the Montreal Protocol

YEAR	EVENT
1930's	Chlorofluorocarbons (CFCs) are developed by Du Pont as a safe alternative to toxic refrigerants.
1970	James Lovelock's electron capture detector reveals the accumulation of CFCs in the atmosphere.
1974	Mario Molina and Frank Sherwood Rowland show that CFCs degrade through photodecomposition and release ozone-depleting chlorine molecules.
1978	Pressure from environmentalists causes the United States to ban CFCs as an aerosol propellant; however, worldwide use continues to grow.
1981	The United Nations Environment Programme forms the Ozone Group and discusses a global treaty to protect the ozone layer.
1985	Scientists observe a hole in the ozone layer over Antarctica that is linked to stratospheric chlorine.
1986	Major CFC producers advocate international efforts to limit the growth of CFC emissions.
1987	The Montreal Protocol is signed; 46 nations, including the United States, commit to a plan to reduce and eventually eliminate CFC production.
1989	The Montreal Protocol enters into force; signatory nations are required to begin the phaseout of CFCs.
1990	Revised estimates of ozone depletion lead to a call to cease CFC production by 2000.
1993	Scientists detect a measurable reduction in atmospheric CFCs.

global reduction in ozone. There is evidence that ozone depletion can lead to global warming, reduced productivity in Antarctic waters, and decreased reproduction in amphibians worldwide. Important arguments among the signers of the Montreal Protocol were the level of production cuts that were required to amend the problem (one reason for amending the protocol in 1990) and the level of support to which developing nations were entitled for complying with the Montreal Protocol and forgoing the benefits of cheap CFCs (particularly for refrigeration) that developed countries enjoyed.

Mark Coyne

SUGGESTED READINGS: The full text of the Montreal Protocol is in *International Environ-mental Law* (1991), edited by Michael R. Molitor. The industrial perspective on ozone protection is provided by Joseph P. Glas in *Technology and Environment* (1989). Lawrence Susskind, *Environmental Diplomacy* (1994), and Richard Benedick, *Ozone Diplomacy: New Directions in Safeguarding the Planet* (1991), both discuss the Montreal Protocol as an example of successful international environmental negotiation. Scientific evidence for the effects of the Montreal Protocol on ozone depletion is found in J. W. Elkins et al., "Decrease in the Growth Rates of Chlorofluorocarbons 11 and 12," *Nature* 234 (1993).

SEE ALSO: Air pollution; Chlorofluorocarbons; Freon; Greenhouse effect; Ozone layer and ozone depletion.

Mountain gorillas

CATEGORY: Animals and endangered species

Mountain gorillas live in the remote areas of the mountains of Rwanda, Zaire, and Uganda. After years of being hunted by poachers, the gorillas finally achieved a secure existence through the active involvement of international environmentalist organizations and individual scientists such as Dian Fossey.

Gorillas, the largest living primates, are closely related to humans. Males are fewer than 1.8 meters (6 feet) tall and weigh approximately 136 kilograms (300 pounds), while females can grow up to 1.5 meters (5 feet) in height and weigh about 90 kilograms (200 pounds). Both have a brownish-grey fur that turns gray with age. Three main types of gorilla are found in Africa: the western lowland (*Gorilla gorilla gorilla*), the eastern lowland (*Gorilla gorilla graueri*), and the mountain gorilla (*Gorilla gorilla beringei*).

Mountain gorillas, the most endangered of the three, live in the high-altitude (up to 3,700 meters, or 12,000 feet) forests of the Virunga Volcanoes National Park in Rwanda and the Bwindi Reserve in Uganda. They are dwelling animals that walk on their hind limbs and the knuckles of their forelimbs. They eat shrubs and fruit. They live in small groups that consist of a group-leading mature male (also known as the silverback), two to four adult females, and an equal number of younger gorillas. The males are generally peaceful and become aggressive only when they meet other group leaders. Females produce single offspring after a gestation of almost nine months; they lose about 50 percent their offspring within the first year of life. The average life span of mountain gorillas is estimated to be close to thirty-five years.

The end of World War II led to great changes in the colonization of African countries. Under mounting pressure, the governments of countries such as England, France, and Belgium realized that sustaining the pre-World War II colonial attitude toward Africa was too expensive, both financially and politically. Several of the struggles for independence from European co-lonial leadership resulted in the rise of corrupt African leaders whose interests were self-centered rather than patriotic. Poaching was occasionally banned, but bribery, greed, and reckless disregard of the law continued to thrive in a few of these nations. Gorillas were in great demand for their palms, skin, and skulls, whose price had risen to about US$500 by 1983. Mountain gorillas were also indirectly affected by the poaching of other animals, such as antelope, buffaloes, and elephants, which were driven to unusually higher altitudes and grazed at the expense of the mountain gorillas.

The forests of Zaire, Uganda, and Rwanda were overexploited with no regard for the environment, which led to a dangerously declining gorilla population. Individual environmentalists such as Dian Fossey, who died under mysterious circumstances in Africa, ensured that people around the world were aware of the imminent extinction of mountain gorillas. Since 1977 organizations such as the World Wildlife Federation, the African Wildlife Federation, the Gorilla Conservation Program, and the Fauna and Flora Preservation Society have been actively involved in protecting the endangered mountain gorilla. Activities include conservation, reforestation, law enforcement, environmental education, and controlled tourism and building development. The involvement of the Center for Remote Sensing and Spatial Analysis (CRSSA) in monitoring the gorilla population by using powerful analytical tools such as the imaging radar from the space shuttle *Endeavor*'s Space Radar Laboratory has enhanced the work being done to ensure the well-being of the mountain gorilla.

Soraya Ghayourmanesh

SEE ALSO: Endangered species; Extinctions and species loss; Fossey, Dian; Poaching; Wildlife management.

Muir, John

BORN: April 21, 1838; Dunbar, Scotland
DIED: December 24, 1914; Los Angeles, California
CATEGORY: Preservation and wilderness issues

Naturalist, environmental activist, and writer John Muir was America's most notable preservationists. His articles introduced Americans to California's Sierra Nevada, and he worked hard to protect much of the region's wilderness, including Yosemite, against development. Muir helped found the Sierra Club in 1892.

Born in Scotland, John Muir emigrated to the frontier area of Wisconsin in 1849 with his family. Muir's father was a dominant but negative influence on his life. The elder Muir was deeply religious but viewed the Christian god as one of justice rather than love. Muir turned away from his father's repressive religion, substituting it with an intense love for nature; for Muir, the divine seemed manifest in the wilderness.

While laboring on the family farm, Muir became an observer of the environment. He was also something of an inventor and worked as a craftsman. After attending the University of Wisconsin, from which he did not graduate, Muir trekked through the southern United States in 1867 hoping to journey on to South America. Instead, he boarded a ship from New York for California, arriving in 1868. Muir had no love for cities; he could cross a mountainous wilderness without maps but he was lost in any urban area. To him, cities were mentally and morally corrupting. He once wrote, "There is not a perfectly sane man in San Francisco."

Muir spent his first summer in California in the Central Valley; in the spring of 1869 he journeyed into the Sierra Nevada as a supervisor of a sheep herd. That year was an epiphany for Muir. Yosemite Valley was still almost virginal, and the surrounding cliffs and mountains drew him on as nothing before. He stayed in Yosemite the following winter, working in a lumber mill and escaping into the wilderness whenever possible. He wrote of the experience, "I am bewitched, enchanted. . . . I have run wild." With only bread, tea, and a blanket, Muir explored much of the central Sierras.

In 1870 Ralph Waldo Emerson visited Yosemite, and Muir eagerly sought him out. Parlor transcendentalist Emerson and natural man

John Muir, an environmental activist and writer who fought to preserve large tracts of land in the American West from development. He helped found the Sierra Club in 1892. (Library of Congress)

Muir had much in common, but Muir was unable to convince Emerson to camp under the trees and stars. Muir's attachment to the Sierras was not merely religious, philosophical, or emotional. He was a keen natural scientist. One of his major scientific contributions was his theory that glaciation was the key to explaining the existence of Yosemite Valley and the other canyons and mountains of the Sierras. This contravened the accepted doctrine that Yosemite Valley had been formed by a gigantic catastrophe that caused the valley floor to suddenly fall several thousand feet. Muir's beliefs were initially dismissed, but in the high country he discovered residual glaciers, validating his claim. In 1879 Muir became one of the first people to explore Alaska's Glacier Bay.

Something of a loner and an eccentric, Muir had few close friends. One of these was Jeanne

Carr, whose husband, Ezra, was a professor of geology and chemistry at Wisconsin who later moved to Berkeley, California. Jeanne was very supportive of Muir and his ventures, and it was she who convinced him to write about his experiences, first in the *Overland Monthly* and later in New York's *Harper's* and *Scribner's* magazines. Muir labored with his writing, but his articles and books are marvelously descriptive, with a visual immediacy that has kept his work in print for decades. It was through these articles that Muir and his beloved Sierras became known to many Americans.

Muir was always fearful of the impact of civilization on the wilderness. The fragile alpine environments were unsuited to heavy grazing by sheep and cattle, and human visitors to Yosemite and beyond often wished to bring all of their civilized amenities with them. One of Muir's greatest successes was in preserving Yosemite, then under state control and in danger of over-development. With the support of *Century* magazine's Robert Underwood Johnson, a successful campaign was mounted that saw Yosemite Valley and much of the surrounding territory become a national park in 1890.

Muir's greatest failure, however, also concerned Yosemite. San Francisco's water system was antiquated and inadequate, and local officials saw a solution in tapping the waters of the Tuolumne River, which ran through Hetch Hetchy Valley; Muir had long praised the beauty of the valley, which, at his urging, had been included within the boundaries of Yosemite park. The issue of damming Hetch Hetchy became a national conflict in the early twentieth century. Some noted conservationists favored it, most notably Gifford Pinchot, chief forester under Theodore Roosevelt and, briefly, William Howard Taft. Conservationists such as Pinchot believed that nature and its resources should be made available for human use. Muir, on the other hand, was a preservationist who believed that nature had a right to exist without human interference. Muir and his supporters fought the proposal for years but finally lost in 1913. Soon Hetch Hetchy Valley disappeared in the rising waters.

Muir died one year later, perhaps in part because of Hetch Hetchy. He left a stirring legacy, not least in the Sierra Club, which he helped found in 1892 and for which he served as president until the end of his life. Many later conservationists and organizations have been inspired by his preservationist ethic. No other environmental figure has been so widely honored: California alone boasts the Muir Woods near San Francisco, Muir Grove in Sequoia National Park, and the Sierra's John Muir Trail.

Eugene Larson

SUGGESTED READINGS: An excellent introduction to Muir and subsequent conservationist figures is Stephen Fox's *John Muir and His Legacy: The American Conservation Movement* (1981). Of equal significance is Frederick Turner's *John Muir in His Time and Ours* (1985). For information on Muir and the Sierra Club, see Michael P. Cohen's *The History of the Sierra Club* (1988). *The Wilderness World of John Muir* (1954), edited by Edwin Way Teale, is a brilliant sampler of Muir's own writings.

SEE ALSO: Hetch Hetchy Dam; National parks; Pinchot, Gifford; Preservation; Roosevelt, Theodore; Sierra Club; Yosemite.

Mumford, Lewis

BORN: October 19, 1895; Flushing, New York
DIED: January 26, 1990; Amenia, New York
CATEGORY: The urban environment

Social philosopher and urban critic Lewis Mumford was one of twentieth century America's most original thinkers. His writings, which include some thirty books and countless essays, cover a vast range of disciplines, including sociology, technology, philosophy, history, and literature. He is perhaps best known for his architectural criticism and works on urban history and planning.

Lewis Mumford did not believe in the efficacy of political action and consequently did not participate in any environmental organizations or ecological crusades in a formal manner. Yet he did speak eloquently and forcefully on a range of topics that had, as their natural corollary, beneficial consequences for the human environment.

Mumford was a strong advocate of regional planning, which was a response to the overpopulation and decline of cities in the United States. He was concerned about urban noise, congestion, senseless growth, and the way cities sprawled into the surrounding countryside, despoiling nature in the process. He proposed the creation of medium-sized garden cities located near a metropolis. These garden cities would be true cities rather than suburbs, with a balance of housing, industry, commerce, and necessary cultural and civic amenities; they would be surrounded by greenbelts made up of parks and agricultural land. The cities would exist in harmony with nature and the surrounding countryside, and relate to the soil, climate, and topography of the region. Unlike the central city, they would be built to human scale and would not be dominated by the automobile, of which Mumford was a severe critic because of its potential to create noise, congestion, and pollution.

Mumford was also a strenuous critic of the course of modern technology, as evidenced in two major works, *The Myth of the Machine* (1967) and *The Pentagon of Power* (1970). He blamed technology for encouraging the blind pursuit of power and productivity, often with dehumanizing and debasing consequences for the human personality. In its quest for profits, technology ignored human wants and desires while it strove to dominate and conquer nature. Mumford's solution was visionary, calling for a complete reordering of human values—an internal transformation—in order to curb the disastrous trend. He was misunderstood by many critics, who complained that he was acting like a petulant Old Testament prophet who hated science and scientists.

A final area that stamped Mumford as a friend of the environment was his early opposition to the atomic bomb. He criticized this weapon on moral, political, and environmental grounds. He was among the first to question the peaceful application of atomic energy. Society

Lewis Mumford was an urban theorist who believed that urban planning was necessary to overcome the environmental impacts of overpopulation. (Archive Photos)

still did not know enough about fissionable material to justify the cavalier risks that were taken, particularly because no one knew how to dispose of nuclear waste material. Mumford feared for both humans and nature if the use of nuclear energy was not constrained or eliminated. Throughout his career, Mumford praised the virtues of moderation, self-restraint, and balance in human activity, and he showed respect and love for the countryside and nature. For these reasons, he deserves an honored place among those concerned about the future of the environment.

David C. Lukowitz

SEE ALSO: Antinuclear movement; Greenbelts; Open space; Planned communities; Urban planning; Urbanization and urban sprawl.

N

Nader, Ralph

BORN: February 27, 1934; Winsted, Connecticut
CATEGORY: Human health and the environment

Ralph Nader may perhaps best be described as a consumer advocate, the champion of the underdog against corporate and political power structures in the United States. His legal advocacy work covers minorities, the poor, the elderly, consumers, environmentalists, and other groups struggling for their rights.

Ralph Nader's interest in environmental issues began long before the movement emerged in the late 1960's. When Nader was a child, his parents instilled in him a respect for the duties of citizenship in an industrial democracy. While a college student at Princeton University, Nader fought to prevent campus trees from being sprayed with dichloro-diphenyl-trichloroethane (DDT). At Harvard, where he attended law school, he was an editor of the *Harvard Law Record*, which he attempted to change from a dry academic publication to a forum for ideas on social reform.

Nader took his energy and vision for change to Washington, D.C., in the mid-1960's. He was hired by the Department of Labor's policy planning division as a consultant on highway safety. This job enabled him to amass extensive information on the automobile industry, which he used in writing *Unsafe at Any Speed* (1965), a blatant indictment of the Detroit automobile industry, specifically General Motors. The publication of this book moved Nader to center stage as a consumer advocate.

Beginning in 1969, Nader founded a number of public interest organizations, including the Center for Study of Responsive Law, the Corporate Accountability Research Group, the Public Interest Research Group, and Public Citizen, Inc. These groups focused on a wide range of consumer issues, and environmental concerns were given high priority: safety standards in the workplace, clean and renewable energy resources, automobile emissions standards, clean water standards, and food safety. Nader and his organizations were instrumental in securing passage of major environmental legislation during the 1970's. He also played a key role in the establishment of the Environmental Protection

Consumer advocate Ralph Nader lobbied for stricter environmental laws during the 1970's. In 1996 he unsuccessfully ran for president of the United States as the candidate for the Green Party. (Library of Congress)

Agency, the Occupational Safety and Health Administration, and the Consumer Product Safety Commission.

Even though the consumer and environmental agendas lost center stage in the 1980's, Nader's efforts for reform did not decline. The organizations he founded drew supporters who feared that President Ronald Reagan's administration would undo earlier legislative victories. Nader adapted his agenda to address new threats to consumers, such as proposed limits on liability for corporate negligence, restrictions on regulatory agencies protecting consumers, and trade agreements threatening workers and weakening environmental standards. In 1996 Nader ran for president of the United States on the Green Party ticket, giving him a new forum to articulate his environmental agenda and expose corporate and political leaders' priorities on protecting their narrow interests over the public welfare.

Nader has coauthored and coedited several books on environmental topics, including *The Menace of Atomic Energy* (1977), with John Abbotts; *Who's Poisoning America* (1981), with Ronald Brownstein and John Richard; and *Eating Clean: Food Safety and the Chemical Harvest* (1982), with Michael Fortun.

Ruth Bamberger

SEE ALSO: Automobile emissions; Alternative energy resources; Environmental health; Right-to-know legislation.

Naess, Arne

BORN: January 27, 1912; Oslo, Norway
CATEGORY: Philosophy and ethics

In 1972 Norwegian philosopher Arne Naess presented a paper titled "The Shallow and the Deep, Long-range Ecology Movements" in which he described deep ecology's radical orientation to the environment. Since then, his ideas have become a major force in environmental philosophy and activism.

Arne Naess is a wide-ranging philosopher whose academic thought has been enriched by exten-

sive mountaineering. Early interests were the philosophy of science, empirical semantics, and the critique of dogmatism. In the late 1960's he began to develop an ecological philosophy that he called Ecosophy T, which is heavily influenced by the metaphysics of Baruch Spinoza, the social activism of Mohandas Gandhi, and spiritual insights of Hinduism and Buddhism. At the core of Ecosophy T is an intuitive identification with all of life and a spontaneous feeling of unqualified value for the flourishing of all things. This intuition involves a wider sense of self that includes all of existence without dissolving individuality. From this comes an inclination toward maximizing biodiversity, pluralism in philosophy and religion, and decentralization in economics and politics.

A central but controversial principle that emerges is biospherical egalitarianism: All things have an equal right to live and bloom. Considered by itself as an absolute, this principle leads to crippling ethical paradoxes. However, Naess combines it with a recognition of the inevitability of harming other life and the need to develop a hierarchy of obligations. Deep ecology's holism has been criticized as devaluing individuals, but Naess does not present a simple monism that places value only on a single whole. Instead, his nondualistic holism includes an affirmation of the reality and value of individuals and relationships.

Central to Naess's method is an insistent questioning of conventional ideas and values, as well as a probing into the root causes of environmental problems. Such an inquiry leads beyond mere policy debates to a critical engagement in worldviews. Naess holds that worldviews are rooted in foundational intuitions. While reason cannot establish these intuitions, it can help people deduce general principles and particular beliefs. Similarly, his ethics reject conventional notions of duty and altruism, which are based on an atomistic view of independent selves. For him, morality is grounded in the intuitive experience of identification with others and a spontaneous sense of caring for the well-being of all things. Naess is a phenomenologist, emphasizing a direct experience of reality that goes beyond the subject-object duality. One result of this emphasis on the direct intuition of nature is a pervasive sense of joy.

Naess has not been concerned simply with articulating his own ecosophy. True to his anti-dogmatism, he presents deep ecology as an open-ended philosophical and social movement. Along with George Sessions, Naess created the Deep Ecology Platform of eight principles upon which radical environmentalists with different philosophies can agree and use as the basis for activism. He champions a diversity of worldviews and beliefs while arguing for the need for solidarity in engaging environmental problems.

David Landis Barnhill

SEE ALSO: Biodiversity; Deep ecology; Ecofeminism; Environmental ethics; Social ecology; Speciesism.

National Environmental Policy Act

DATE: 1970
CATEGORY: Ecology and ecosystems

The National Environmental Policy Act of 1970 established national policies and objectives for the protection and maintenance of the environment in the United States.

Sometimes referred to as the environmental Magna Carta of the United States, the National Environmental Policy Act (NEPA) was the first law of its kind to require a comprehensive and coordinated national environmental policy that embraced public review of environmental impacts associated with the actions of federal agencies. Oversight of NEPA compliance was facilitated through the Council on Environmental Quality (CEQ), created through a provision of the act.

In response to widespread public interest in environmental quality during the late 1960's, the U.S. Congress held separate House and Senate committee hearings to identify the best method for legislating a national policy on environmental protection. Among the problems discussed was the tendency of mission-oriented fed-

eral agencies involved in development to overlook environmentally preferable alternatives in their decision-making processes. Draft versions of House and Senate environmental bills were later integrated and approved by Congress as NEPA. Signed into law by President Richard M. Nixon on January 1, 1970, NEPA (PL 91-190) recognized the profound impact of human activity on the interrelations of all components of the natural environment, particularly the influences of population growth, high-density urbanization, resource exploration, and new and expanding technological advances. Language used in the bill embraced many of the philosophies of the conservation movement of the early twentieth century and later environmentalism of the 1960's.

Although other U.S. environmental statutes provide robust protection for the environment, such laws focus only on specific categories of resources. In contrast, NEPA serves as an umbrella statute that outlines a set of procedures and embraces the importance of public participation in federal decision making when the quality of the environment is at stake. NEPA does not demand explicit results such as limits on pollution emissions or specific actions to protect endangered species, nor does it serve as a substitute for other federal planning activities or regulatory processes. Rather, the act instructs that decisions be made on the basis of thoughtful analysis of the direct, indirect, and cumulative environmental impacts of proposed actions.

NEPA prescribes the completion of a series of steps for all actions involving federal participation that may affect the environment, emphasizing in spirit and intent the importance of public participation in safeguarding the environment. The first step in federal project review is completion of an environmental assessment (EA) with input from local governments, American Indians, the public, and other federal agencies. An EA documents influences on the environment associated with a proposed federal action, including the type and level of significant impacts. Following completion of the EA, NEPA requires that a second document be prepared for all actions going forward. A finding of no significant impact (FONSI) provides documentation in cases were actions have been determined to have no significant effect on the quality of the human environment. Federal actions that may significantly alter the quality of the human environment, including possible degradation to threatened or endangered species or their habitats, must be evaluated in greater detail. An environmental impact statement (EIS) involves additional analysis of pertinent social, demographic, economic, and ecological information and consideration of alternative courses of action. NEPA requires that the EIS process be carried out using a framework involving public input through a variety of mechanisms, including individual or group responses to proposed alternatives.

Initially, NEPA implementation was difficult within agencies struggling to establish guidelines. Within the act's first two years, federal agencies had completed more than 3,600 EISs but were also involved in nearly 150 associated lawsuits. Although the number of EISs submitted each year has fallen, more than 26,000 have been written since 1970. These reports document actions ranging from the construction of highways to the development of facilities for holding toxic wastes. About 80 percent of EISs have been produced by a small number of federal agencies, including the Forest Service, the Bureau of Land Management, the Department of Housing and Urban Development, the Federal Highway Administration, and the Army Corps of Engineers. Although there is no formal tracking process, a much large number of EAs have been prepared for other federal projects and activities.

Critics argue that the EIS process is not cost effective in terms of human resources or dollars spent and that NEPA guidelines are often inconsistently applied. As an overarching policy, NEPA cannot prevent agencies from implementing unwise actions or even from concluding that other values are more important than environmental considerations. Despite ambiguity in its language, however, the act has been credited with significantly modifying the actions of both government agencies and private industry by preventing hundreds of activities with potentially severe environmental effects.

NEPA has raised awareness of the concept of environmental impact, highlighting the need for governments and citizens to be aware of the unintended consequences of federally supported actions that affect the human environment. NEPA has also assisted in creating pathways for other federal statutes that consider environmental issues in decision making, such as the 1980 Comprehensive Environmental Response, Compensation, and Liability Act (CERCLA), also known as Superfund. Having withstood consecutive regulatory reform commissions of the Jimmy Carter, Ronald Reagan, and George Bush presidential administrations, NEPA stands out for its brevity and simplicity. Perhaps the best evidence of the act's success can be seen by the fact that it has been emulated by one-half of the state governments in the United States and by more than eighty national governments throughout the world.

Thomas A. Wikle

SUGGESTED READINGS: Ray Clark and Larry Canter, *Environmental Policy and NEPA* (1990), contains a series of essays that provide an excellent overview of NEPA's history and an assessment of its future. Other reviews of NEPA and its effectiveness are presented in Richard Andrews's *Environmental Policy and Administrative Change* (1976) and Lynton Caldwell's *Science and the National Environmental Policy Act* (1982).

SEE ALSO: Environmental impact statements and assessments; Environmental policy and lobbying; Environmental Protection Agency.

National forests

CATEGORY: Forests and plants

The national forest system was one of the first and most successful resource conservation programs initiated by the United States government. Similar programs have been developed in most of the nations of the world.

National forests are distinctly different from national parks. National parks are aesthetically pleasing, culturally or historically significant, or ecologically important tracts of land protected

Value of Timber Cut in U.S. National Forests

Millions of dollars

Year	Value
1970	~310
1980	~740
1985	~725
1988	~1,240
1989	~1,310
1990	~1,190
1991	~1,005
1992	~930
1993	~910

Source: U.S. Department of Commerce, *Statistical Abstract of the United States, 1996,* 1996.

in perpetuity for the public benefit. National forests are publicly owned resource management areas operated to maximize the cash value of timber and forest products, benefit the domestic economy, and ensure perpetual productivity of the land.

The nineteenth century saw increased efforts to preserve timber resources in the United States at the same time that unrestrained national growth encouraged increased timber harvesting. In 1827 U.S. president John Quincy Adams expressed interest in developing a plan for sustainable forestry to ensure the availability of masts for ships. The American Association for the Advancement of Science discussed the need for sustained-yield forestry in the 1860's and in 1873 asked Congress to preserve and manage the nation's forests.

The first congressional actions to preserve forests focused on areas of significant natural beauty, such as Yellowstone National Park and Yosemite Valley. The Division of Forestry was established within the Department of Agriculture in 1876 to promote harvestable timber develop-

ment. Protections for harvestable stands of timber were first approved in the Creative Act of 1891, which authorized the U.S. president to withdraw public lands previously open to preemption and homesteading rights and to establish forest reserves.

President Benjamin Harrison established the first national forest, now known as Shoshone National Forest in Wyoming, in 1891. However, confusion over the process for including land within the national forest system led Congress to suspend the law, debate the purpose of the reserves, and eventually authorize selective cutting and marketing of timber from the existing reserves. Controversy concerning the vesting of the forest resources in the Department of Interior or the Department of Agriculture ended in 1905 when the forest reserve system was established. President Theodore Roosevelt vested responsibility for national forest resources in the Bureau of Forestry of the Department of Agriculture under Gifford Pinchot, chief of the Division of Forestry. The reserve system became the national forest system in 1907.

From its inception, the forest system was intended to serve multiple purposes. The purposes of the original legislation were to improve water flows and furnish a continuous supply of timber. This evolved into the triple purposes of resource protection (especially fire protection), "wise use" of timber resources, and multiple use of system lands. The forest service lands are to be protected, harvested, and open to multiple other public uses. The Multiple Use-Sustained Yield Act of 1960 defined multiple use as a combination of outdoor recreation, fish and wildlife management, and timber production intended to best meet the needs of the American people but not necessarily giving the maximum dollar value return.

The National Forest Management Act of 1976 increased citizen involvement in decisions concerning timber harvesting, resource conservation, and multiple uses. The 1976 act also limited the technique of timber harvesting by clear-cutting the forest, a practice condemned by most environmentalists. The act also limited logging on fragile lands and encouraged actions to maintain the diversity of plants and animals and to conserve plants, animals, soils, and watersheds in the forests. However, the act also emphasized the importance of multiple uses such as mining, oil and gas exploration, grazing, farming, hunting, recreation, and logging. The conflicts among conservation, harvesting, and

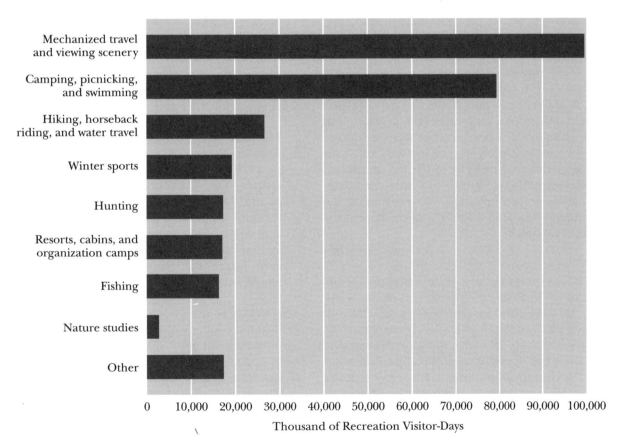

Recreational Uses of U.S. National Forests, 1993

Source: U.S. Department of Commerce, *Statistical Abstract of the United States, 1996,* 1996.
Note: One "recreation visitor-day" is the recreation use of national forest land or water that aggregates 12 visitor-hours. This may entail 1 person for 12 hours, 12 persons for 1 hour, or any equivalent combination of individual or group use.

multiple uses continue to plague forest policy decision makers. Emerging new forest policies include sustained-yield forestry; substitution of alternative harvestable crops for harvestable timber, including crops such as nuts, fruits, gums, extracts, syrups, tars, and oils; and ecosystem maintenance.

The federal government owns and manages more than 657 million acres of public land, about 30 percent of the land area of the United States. About one-third of federal lands are in the national forest system. By the 1990's the national forest system included more than 191 million acres of land in 159 individual forests in forty-four states, Puerto Rico, and the Virgin Islands. Two-thirds of the tracts are located in the West, Southeast, and Alaska. The twenty-three eastern states share about fifty forests. About 32 million acres of forest system lands are also placed within the National Wilderness Preservation System, representing about 30 percent of the land in the wilderness system.

Global forests decline at a rate of about 2 percent per decade, with tropical forests declining at about 4 percent, Asian forests at 11 percent, Latin American forests at 8 percent, and African forests at 7 percent. The number of tree plantations is rapidly increasing but still only replaces about 20 percent of the loss in forests. Environmental concerns, ecotourism, and debt-for-nature-swaps encourage the continuing development of national forest systems in most nations of the world.

Gordon Neal Diem

SUGGESTED READINGS: Harold Steen, *The Origins of the National Forest* (1992), provides a history of American forests. Jeanne Clarke and Daniel McCool, *Staking Out the Terrain: Power Differentials Among Natural Resource Management Agencies* (1985), evaluates and compares the history, professionalism, political acumen, power resources, and decision-making styles of government agencies. Paul J. Culhane, *Public Lands Politics: Interest Group Influence on the Forest Service and the Bureau of Land Management* (1981), discusses conflicting pressures on the Forest Service for the use of public forests. Gary C. Bryner, *U.S. Land and Natural Resources Policy: A Public Issues Handbook* (1998), discusses contemporary conflicting pressures. Sterling Brubaker, editor, *Rethinking the Federal Land* (1984), discusses Reagan-era criticisms of public lands policy and suggests transferring public lands, including forests, to state or private ownership. J. A. McNeely and K. R. Miller, editors, *National Parks, Conservation and Development: The Role of Protected Areas in Sustaining Society* (1984), includes an article, "Categories, Objectives and Criteria for Protected Areas," describing the International Union for the Conservation of Nature's classification for various categories of protected lands.

SEE ALSO: Conservation; Forest and range policy; Forest management; Forest Service, U.S.; Logging and clear-cutting; Renewable resources; Sustainable forestry.

National Oceanic and Atmospheric Administration

DATE: established October 3, 1970
CATEGORY: Weather and climate

The National Oceanic and Atmospheric Administration (NOAA) and the NOAA Corps, formed in 1970, bring together many related science agencies. The NOAA is chartered to measure the physical, biological, and chemical properties of the earth. Its primary focus is the collection of data on the United States and its coastal waters, and on the oceans and the atmosphere worldwide. Land-based monitoring and studies are also performed.

The NOAA is a branch of the U.S. Department of Commerce. This multifaceted federal agency evolved from an 1807 agency tasked to perform both coastal and interior geodetic surveys. The NOAA forecasts the nation's weather, warns the public of impending severe weather and flooding, predicts climate change, protects endangered ocean species, and conducts scientific research to understand and preserve the environment. Examples of NOAA strategic missions include sustaining healthy coasts, promoting safe navigation, recovering protected species, building sustainable fisheries, forecasting seasonal and long-term climate, and assessing and predicting environmental effects. Information gathered provides an important knowledge base that enables informed environmental policy decisions. National concerns include the greenhouse effect, ozone depletion, biodiversity maintenance, atmospheric and marine pollution, fishery depletion, natural resources exploitation, navigation safety, and coastal zone management. The NOAA Corps is a commissioned officer service. These officers operate the NOAA's fleet of research and survey vessels, as well as its array of aircraft.

The NOAA has five primary offices: the National Weather Service (NWS); the National Marine Fisheries Service (NMFS); the Office of Oceanic and Atmospheric Research (OAR); the National Environmental Satellite, Data, and Information Service (NESDIS); and the National Ocean Service (NOS). The NWS is the primary source of weather forecast and warning information for the United States. Weather forecast and observation data is issued several times daily. NOAA Weather Radio broadcasts around-the-clock weather reports and warnings to fixed and mobile receivers.

The NMFS consists of five Fisheries Science Centers, which provide resource managers with scientific knowledge as a basis for marine fisheries management decisions. Their programs include collection of harvesting statistics, economic data, resource surveys, biological and ecological studies, and technology development.

The OAR office runs the National Undersea Research Program, which performs oceanographic research using a scientific research support ship, a deep-diving submersible, and remotely operated diving vehicles. The OAR operates observatories that gather data on greenhouse gases, ozone, aerosols, and water vapor, while buoys measure sea and air data. The OAR also measures precipitation and other meteorological parameters and monitors solar radiation.

The NESDIS operates the Geostationary Operational Environmental Satellites program, which collects oceanographic data primarily used for daily weather forecasts and storm monitoring. The NESDIS oversees management of remote-sensing satellites and information databases for meteorology, oceanography, solid-earth geophysics, and solar-terrestrial sciences.

Finally, the NOS operates the National Earth Orientation Service and the Continuously Operating Reference Stations, which provide polar motion, universal time, and earth axis data. It implements the Physical Oceanographic Real-Time Systems, which monitor water levels, currents, and wind. It operates a network of permanent tide gauges and gathers data from drifting buoys for weather and ice forecasts and climate research. The NOS also provides Navigational Safety Aerial charts for pilots in the United States.

Robert J. Wells

SEE ALSO: Climate change and global warming; El Niño; Greenhouse effect; Ozone layer and ozone depletion.

National parks

CATEGORY: Preservation and wilderness issues

The United States Congress established Yellowstone, the first national park in the world, in 1872 as a "pleasuring ground" for Americans. Congress was willing to protect and preserve the geologic wonders primarily because politicians were convinced that the lands were economically useless. Over the years more national parks were formed in recognition of the value of preserving natural surroundings.

National parks are places where preservation for future generations must be balanced with present-day enjoyment. This has proved a difficult task, but one that has resulted in more than 370 park units, including national parks, historic sites, historical parks, memorials, memorial parks, battlefields, battlefield parks, battlefield sites, lakeshores, seashores, monuments, parkways, scenic trails, scenic rivers, scenic riverways, rivers, capital parks, memorial parks, and recreation areas. Each of these was established to preserve and protect geologic wonders, spectacular scenery,

National Park System Statistics

	1985	1990	1994
Expenditures: millions of dollars	848.1	986.1	1,404.0
Revenue from operations: millions of dollars	50.6	78.6	97.0
Recreational visitors: millions of visits	263.4	258.7	267.6
Overnight stays: millions of stays	15.8	17.6	18.3
Park system lands: millions of acres	75.7	76.4	74.9

Source: U.S. Department of Commerce, *Statistical Abstract of the United States, 1996,* 1996.

Note: Includes visitor data for national parks, monuments, recreation areas, seashores, and miscellaneous other areas.

wildlife, or a particular aspect of American history or culture.

Other countries have also established national parks, frequently using the United States as a model. Starting in the final decades of the twentieth century, the United Nations worked with countries to protect these areas. On the occasion of the one hundredth anniversary of the founding of Yellowstone National Park, the United Nations Educational, Scientific, and Cultural Organization (UNESCO) formed the World Heritage Committee. Since then, more than 150 countries have ratified the World Heritage Convention. The committee, at its first meeting in 1977, formulated the World Heritage List of cultural and natural sites considered to be of "outstanding universal value," adding or inscribing around thirty new sites each year. By the late 1990's, 550 cultural and natural sites had been so named. The list contains more cultural than natural sites, and not all the sites are parks. Yet many parks have received monetary support and advice from the United Nations to help ameliorate environmental problems, which abound in the United States and the rest of the world; the causes of these problems include the difficulties of balancing use and preservation, poaching, mineral exploitation, and too many visitors.

NATIONAL COORDINATING OFFICES

Initially, park operations in the United States were complicated because there was no one central federal coordinating office. After much controversy, Congress passed the National Parks Act, or the Organic Act, of 1916. The act established a central authority called the National Park Service and stated its responsibilities, which include conserving and providing for the enjoyment of the scenery, natural and historic objects, and wildlife in the parks while leaving them unimpaired for the enjoyment of future generations.

This was not the first national office of national parks. The Canadian Parliament had passed the Dominion Forest Reserves and Parks Act of 1911. It provided for the administration of forest reserves and dominion parks, and allowed dominion parks to be established from forest reserves. Thus, the Dominion Parks Service, created as a new branch in the Department of the Interior, became the first distinct bureau of national parks in the world. John Bernard Harkin served as commissioner from inception in 1911 to 1936. Harkin worked not only to separate the administration of the parks from that of the forests but also to emphasize resource preservation.

The number of national parks, as well as the number of visitors to each park, continued to grow in both the United States and Canada. After World War II, as the automobile became ubiquitous, visits to the national parks rapidly grew. However, the facilities were old, and the staffs were minimal. In 1956 the director of the National Park Service, Conrad L. Wirth, decided that instead of asking Congress for annual appropriations, he would package the national parks' needs into Mission 66. This one-billion-dollar, ten-year restoration program was designed to end in 1966, the fiftieth anniversary of the National Park Service. Canada was influenced by Mission 66. The Parks Policy of 1964, under the leadership of John I. Nicol, director of the National and Historic Parks Branch, emphasized the importance of protecting natural resources in the parks.

The number of visits to parks in both countries steadily increased between the 1960's and 1990's. With larger crowds has come overuse, which creates its own serious problems, including congestion, pollution, and in some cases destruction. Some popular parks have replaced or are planning to replace private automobiles with shuttle buses. Among them are Zion (Utah), Grand Canyon (Arizona), and Rocky Mountain (Colorado) National Parks. The park services of both Canada and the United States are repeatedly faced with the problem of encouraging use while protecting valuable national resources and leaving them unimpaired for future generations, as the Organic Act required.

In the United States, most of the services, such as restaurants, hotels, and souvenir shops, have traditionally been provided by private concessionaires. The contracts are usually long-term, with a small proportion of the profits returned to the federal government. In recent years, with declining public funding and increasing numbers of visitors, the National Park Service has intensified its ties with private agencies to provide public services. Some of the proposed projects have become controversial and have been stopped, including a giant theater at Gettysburg. The National Park Service has recognized the importance of co-operation with the immediate community. Partnerships and resulting plans have been developed around several national parks. Several of Canada's national parks, such as Banff and Jasper, contain towns. Animals are thus able to roam free among automobiles, homes, and stores.

MANAGEMENT OF PLANTS AND WILDLIFE

In Canada and the United States the park service must balance use with preservation of wildlife and plants. In general, the aim is to allow native species to flourish within the natural balances of the park and protect the park from exotic species. However, conflicts in remedial approaches frequently develop. Some parks, such as Shenandoah and Great Smoky Mountains National Parks, have been invaded by exotic insects that are destroying vast forests

Yellowstone, the first national park in the world, was established in 1872. (McCrea Adams)

and the component parts of the ecosystem.

One of the most controversial decisions was the reintroduction of the gray wolf into the Yellowstone ecosystem in 1995. The wolves had been exterminated by rangers in 1915 because they were seen as a menace to such animals as elk, deer, and mountain sheep. After many years of debate, the National Park Service was given permission and funds (around $6.7 million) to reintroduce wolves to the Yellowstone ecosystem, which includes the park and neighboring parts of Montana and Idaho. Many park visitors had responded to surveys by indicating that they wanted the wolf brought back. Local livestock owners were concerned about the safety of their animals. To balance the desires of park visitors, the need to restore the natural ecology of the park, and the economic concerns of landowners—and to remove the wolf from the list of endangered species—the park service carefully introduced wolves with plans to protect neighboring livestock.

In January, 1995, the first set of gray wolves from Hinton, Alberta, Canada, were brought to Yellowstone. The introduction was deemed successful, and the ecology of the park returned to a more natural state as the wolves became significant predators of elk, moose, and deer. In late 1997 a court ruled that the introduction of the wolf was illegal, but allowed a stay.

A wildlife problem that affects almost all the national parks is human-animal interaction. Feeding wildlife in U.S. and Canadian national parks is against the law. Human food is not part of the animals' natural diet and can cause all sorts of dietary problems and sometimes even kill the animal. Even so, visitors break the law, and animals come to campgrounds and cars in hopes of finding something to eat. Since they are wild animals, they occasionally harm someone. In extreme cases, the rangers may transport or shoot an animal to prevent it from hurting someone else.

ECONOMIC ACTIVITIES AND ENVIRONMENTAL
    PROBLEMS

Economic activities outside park boundaries can cause environmental problems inside the parks. Sometimes neighboring mining or logging operations contaminate waterways or hasten soil erosion. These problems have been lessened by specific compromises, as occurred with a proposed gold mine outside Yellowstone. The federal government agreed to exchange other federal lands for the site northeast of Yellowstone. In South Africa, disruption by irrigation and dam building of the Orange River interferes with the wildlife in Augrabies Falls National Park.

In other cases, distant power plants, factories, or even urban areas as a whole contribute to air pollution within the park. The combination of human-caused pollution with the natural geology and meteorology has led to diminished views, more rapid weathering of the natural wonders, and harm to wildlife in many parks in the United States. In South Africa, the industrial district of Saldanha Bay has contributed air and water pollution to the West Coast National Park.

National parks in parts of Africa and Central America face serious problems of poaching because of inadequate park administration and law enforcement. In areas such as Benin, disease-carrying livestock may invade parks and kill wildlife. Parts of national parks in the Ivory Coast and Senegal have been cultivated by farmers. However, the governments of both countries have developed successful resettlement programs, thus lessening the impact of the parks.

Individual countries often look for help from the World Heritage Committee. It has established a list of World Heritage in Danger sites that are so threatened by human interference or natural problems that they need to be brought to the attention of the world's leaders and the public in general. Approximately twenty-five sites are on the list; sometimes, however, sites are removed from the list, as occurred in 1998 with Plitvice Lakes National Park in Croatia. It had been overdeveloped and overused before the fighting in the former Yugoslavia. After its inscription on the list in 1992, the underground water supply was protected, and a new road lessened truck traffic through the park.

Margaret F. Boorstein

SUGGESTED READINGS: John Ise, *Our National Park Policy: A Critical History* (1961), presents a detailed history of national parks in the United States. Alfred Runte, *National Parks: The American Experience* (1997), is another examination of the history of U.S. national parks. *International Handbook of National Parks and Nature Reserves* (1990), edited by Craig W. Allin, provides several essays about national parks around the world. James Ridenour discusses politics and national parks in *The National Parks Compromised: Pork Barrel Politics and America's Treasures* (1994). Richard Sellars, *Preserving Nature in National Parks* (1997), discusses neglect of natural resource management by the National Park Service. The Web sites of the National Park Service (http://www.nps.gov) and the World Heritage Committee (http://www.unesco.org/whc) provide tremendous amounts of information about national parks in the United States and the world.

SEE ALSO: Grand Canyon; Kings Canyon and Sequoia National Parks; Nature preservation policy; Preservation; Serengeti National Park; Yosemite.

Nature preservation policy

CATEGORY: Preservation and wilderness issues

Nature preservation policy is legislation by any level of government that is enacted for the purpose of protecting natural resources.

The Industrial Revolution, which diminished traditional agriculture while encouraging urbanization and technology, began straining the relationship between humanity and natural resources during the early nineteenth century. As European technological advances rapidly spread West, environmental damage and natural resource depletion escalated in the United States, inspiring the American conservation movement. Conservation and environmental movements have continued to exert tremendous influence on policy making.

Artist George Catlin first proposed setting aside land for wildlife and Native Americans during the nineteenth century, while in 1864 geographer George Perkins Marsh published *Man and Nature*, the first influential manuscript on the human impact upon nature. The Homestead Act of 1867 greatly encouraged expansion in the western United States by giving over one billion acres of land to settlers, a policy that often resulted in barren landscapes. Destructive logging methods were employed, land was rapidly cleared for agriculture, and large-scale fires raged. Some western grasslands experienced such excessive grazing that many regions had not recovered their full productivity by the end of the twentieth century.

As farmer's journals described "wearing out" several homesteads during westward journeys, naturalist Henry David Thoreau and essayist Ralph Waldo Emerson countered with writings that fueled increasing nature conservation activity. Considerable public outcry was expressed against the near elimination of several wildlife species that previously existed in massive numbers, such as bison, deer, elk, and beaver. This led to legislation that created the world's first national park in 1872 at Yellowstone, Wyoming, followed by an 1873 petition to Congress by the American Association for the Advancement of Science to curtail the inefficient use of natural resources such as water, soil, forests, and minerals.

The 1891 Forest Reserve Act began the establishment of natural forests, and the 1900 Lacey Act initiated wildlife protection by regulating commercial hunting. The 1894 Buffalo Protection Act provided recognition that a previously abundant natural resource could rapidly become an endangered species. As naturalist John Muir championed numerous wilderness preservation events and became first president of the Sierra Club in 1892, legislation began classifying natural resources as renewable or nonrenewable. Renewable resources can be regenerated and even improve under proper management but can be depleted or completely eliminated if misused. Nonrenewable resources are present only in fixed amounts and will not regenerate regardless of human efforts. Examples of renewable resources include plants, animals, soils, and inland waters, whereas nonrenewable resources include minerals and fossil fuels. The founding of private conservation organizations such as the American Forestry Association in 1875, the

American Ornithologist's Union in 1883, the Boone and Crockett Club in 1887, and the New York Zoological Society in 1895 increased public influence in conservation legislation at all governmental levels.

THEODORE ROOSEVELT AND FRANKLIN ROOSEVELT

President Theodore Roosevelt initiated habitat protection for wildlife in 1903 when he set aside Pelican Island in Florida's Indian River as a federal bird sanctuary. Through such initiatives as the 1908 White House Governors' Conference on Conservation, Roosevelt's politics and personality helped establish more than fifty wildlife refuges, five national parks, and eighteen national monuments, and increased the area of national forests by more than 150 million acres. Roosevelt's policies required that certain public lands be held in trust for the "good of the country" and separated many public domain regions from commercial interests.

During and immediately following Roosevelt's legacy, Forest Service chief Gifford Pinchot and Interior Secretary James Garfield implemented more unified policies governing natural resource planning that relied upon scientific principles, leading to development of the discipline of conservation biology. Many nonrenewable resources were then protected from exploitation by private industry by the 1920 Mineral Leasing Act.

Political debates and administration changes during the Great Depression of the 1930's shelved more environmental legislation until President Franklin Roosevelt signed the Taylor Grazing Act in 1934, whereby all public domain lands would be managed as part of the public trust. The dry Dust Bowl years of the plains states during the early 1930's severely depleted migratory bird populations, motivating a renewed surge of public conservation activity and passage of the 1934 Duck Stamp Act, which tacked a conservation fee for the acquisition of wetlands onto waterfowl hunting licenses.

In 1933 the Soil Erosion Service (later called the Soil Conservation Service), the Civilian Conservation Corps, and Tennessee Valley Authority were established to provide water and soil conservation assistance to landowners as Midwest-

ern farmland continued to deteriorate under improper agricultural practices. Franklin Roosevelt's Civilian Conservation Corp provided over two million jobs planting trees and building irrigation systems and dams. More federal involvement was initiated after dust clouds from the dry soil of Midwestern farmland blew east all the way to Washington, D.C. In 1940 Congress enacted the Bald Eagle Act to protect the national bird.

POST-WORLD WAR II CONSERVATION

Advanced technology and economic development in the turbulent post-World War II era, combined with the postwar baby boom, put additional stressors upon environmental resources. President Harry Truman began a national water pollution control program, with later legislation requiring states to set and enforce standards for natural rivers. In attempts to reduce insect-borne disease and increase food production, dichloro-diphenyl-trichloroethane (DDT) and other synthetic pesticides were developed with considerable initial success, causing the near-complete disappearance of malaria and the production of bumper crops.

The 1947 Forest Pest Control Act provided for the detection and chemical destruction of insect-borne diseases, but the numerous new and powerful experimental substances being invented caused other severe environmental problems, which in many cases caused more damage than those that the pesticides were created to prevent. Grassroots public outcry stimulated several federal restrictions on chemicals such as DDT, with many citizens alerted to these dangers by former U.S. Fish and Wildlife Service biologist Rachel Carson's 1962 book *Silent Spring*. Carson is credited with warning mainstream America about health and environmental hazards posed by pesticides and other toxic chemicals. She stimulated further writings that described human threats to the environment, including *The Population Bomb* (1968), by Paul Ehrlich, and *The Limits to Growth* (1972), by Donella Meadows and others.

All forms of environmental pollution greatly increased during the 1950's and 1960's. Television beamed graphic examples of environmental problems into public view, notably the mercury poisoning at Minamata Bay, Japan; killer smog episodes in London, England, and Los Angeles, California; and the 1967 *Torrey Canyon* oil spill in the English Channel. As land and water rights prices skyrocketed, the Land and Water Conservation Fund, which was set up during the 1960's to increase outdoor recreation space, generated revenues from offshore drilling leases. Several catastrophic environmental events occurred in 1969, including the toxic waste fires on the Cuyahoga River in Cleveland, Ohio, and a coastal oil spill near Santa Barbara, California. Public pressure regarding these and other concerns led to passage of the National Environmental Policy Act in 1969. During development of this precedent-setting act, Congress discovered over eighty governmental units that had activities directly affecting the environment with no government policies to coordinate and review such actions.

Private individuals and organizations such as the Sierra Club, the Nature Conservancy, the Wilderness Society, the National Wildlife Federation, and National Audubon Society began lobbying for more laws to establish nature preservation areas for both renewable and nonrenewable natural resources. Examples of highly visible social programs that influenced public opinion by the Nature Conservancy, founded in 1951, included Oklahoma's Tall Grass Prairie Preserve "Adopt a Bison" program and Montana's Pine Butte Swamp Preserve, where dinosaur fossils were discovered in 1978. The Endangered Species Preservation Act of 1966 and the Endangered Species Conservation Act of 1969 did not directly protect any species but led to later legislation that did.

The national park system was designed to preserve the natural wonders of the United States in pristine condition. Among the protected areas is this canyon of the Virgin River in Zion National Park, Utah. (Jim West)

THE 1970'S AND 1980'S

Following unanimous passage of the National Environmental Policy Act by Congress over President Richard Nixon's objection, the 1970's saw the passage and often complicated enforcement of several laws regulating nature preservation. The Environmental Protection Agency (EPA) was established in 1970, followed later that year by passage of the Clean Air Act. Important pollution control measures were then implemented by the 1972 Water Pollution Control Act, the 1973 Endangered Species Act, the 1976 Toxic Substances Control Act, the 1976 National Forest Management Act, and the 1977 Clean Air Act amendments. The Endangered Species Act is considered the most effective and wide-reaching act ever passed by Congress to protect natural ecosystems. Authorization for this monumental legislation expired in 1992, at which time it was not renewed, although funding continued for its activities. Public support for nature preservation during the 1970's was enhanced by "Keep American Beautiful" advertisements featuring Iron Eyes Cody, a Native American shown crying while observing a littered landscape.

Key legislation during the following decade included the 1980 National Acid Precipitation Act, the 1980 Alaska National Interest Lands Conservation Act (which enabled the size of the refuge and park systems to double), and the 1987 Clean Water Act. Surveys conducted during the late 1980's revealed that over 60 percent of wildlife areas permit activities proven harmful to wildlife, with the most destructive practices such as military activities and drilling not falling under legal jurisdiction of the U.S. Fish and Wildlife Service. This era also saw increased public interest in nature preservation following events such as the 1986 Chernobyl nuclear catastrophe, the 1989 *Exxon Valdez* oil spill, acid rain and tropical deforestation controversies, harvesting of old-growth timber, and the discovery of a trend toward global warming.

Private corporations that previously sacrificed important wildlife habitat began to realize that the environment was an important issue to American consumers. In response to pressure by consumers, employees, and stockholders, many businesses implemented stewardship programs that protected natural resources and allowed public enjoyment of their underdeveloped land.

THE 1990'S

Many nature preservation goals first proposed during the Industrial Revolution finally began to be realized with the systematic creation and maintenance of healthy forests; prevention of timber depletion and the siltation of streams; provision of food, cover, and protection for wildlife; and the creation of recreation places available to escape growing urbanization. Water reservoirs could now control flooding, provide clean water for humans and livestock, keep the soil fertile for agriculture, provide irrigation, and generate power. Water treatment plants were now effective in keeping rivers clean by processing wastes from urban sewage, while fish

hatcheries provided supplemental stocks to natural and human-made reservoirs, streams, and lakes. Scenic easements along riverbanks aided antipollution efforts and reduced erosion, while green spaces established by city zoning held soil and became available for community use while maintaining the environment's natural beauty. Mass interurban transportation systems moved people efficiently, and footpaths and bicycle trails offered outdoor recreation and cultural opportunities. Continual management of resources was more successful in keeping delicate ecosystems in balance, with ongoing environmental efforts including seeding wildlife foods, controlled burning to destroy unwanted vegetation, and the closing of wildlife habitat during mating and birthing seasons.

Conservationists-turned-environmentalists greatly influenced nature preservation policies during the 1990's as President George Bush passed legislation in 1990 that amended the 1970 Clean Air Act to focus more on reducing acid rain and emissions from fossil fuels and nitrogen oxide. However, the Defenders of Wildlife organized a 1992 activist citizen's commission, which revealed that the United States was "falling far short" of meeting the urgent needs of nature preservation. This led to the 1997 National Wildlife Refuge System Improvement Act, which shifted the priorities of the nature preservation systems toward the formation of multiple-use environments. This key legislation redefined the mission statement regarding conservation habitat for fish, wildlife, and plants; designated priority public uses such as hunting, fishing, wildlife observation and photography, and environmental education and interpretation; and required that "environmental health" be maintained on public lands. The principle of multiple use, however, continues to allow mining, drilling, grazing, logging, and motorized recreation, as well as military training such as bombing, tank, and troop exercises, on lands designated for nature preservation.

INTERNATIONAL EFFORTS

Cooperative international nature preservation efforts began with the 1916 Migratory Bird Treaty signed by the United States, Great Britain, Canada, and later Mexico. The International Union for Conservation of Nature and Natural Resources, founded in 1948, represented the interests of 116 countries toward protecting endangered and threatened "living resources." The United Nations Environmental Conference of 1972, hosted by Sweden, was instrumental in establishing nature preservation as an international concern. Utilizing concepts from this conference, the U.S. Congress passed the 1983 International Environmental Protection Act, which included landmark legislation incorporating wildlife and plant conservation and biological diversity as objectives when the United States provides assistance to developing countries.

The 1973 Convention on International Trade in Endangered Species (CITES) involved the cooperation of more than one hundred nations to regulate the import and export of natural resources. Earth Day, first held on April 22, 1970, as a campus-based event encompassing an estimated 20 million people across the United States, began to be combined with other concerns and later demonstrated massive support for conservation issues. The international Greenpeace Foundation was formed in 1971 to protect against radioactive and toxic waste dumping, nuclear weapon use, and acid rain.

Nature preservation sentiment began taking political form in Europe during the early 1980's, notably with the formation of the Green Party in Germany. International collaboration on environmental preservation issues that influenced later legislation included the 1987 Montreal Protocol to protect the ozone layer, the 1992 United Nations Conference on Environment and Development (also called the Earth Summit) in Brazil, and the 1994 United Nations Population Conference in Egypt. The 1992 Earth Summit in Rio de Janeiro, Brazil, was the largest-ever international meeting (178 nations attended) and emphasized a sustainable-growth, utilitarian approach to nature conservation. A paper that resulted from the Earth Summit entitled "World Scientists' Warning to Humanity" warned that if current consumption rates continued, the earth's resources would be reduced to the point where the world would be "unable to sustain life in the manner we now know."

This monumental event was followed by another summit in 1997, after which many analysts predicted that future local and global nature preservation and environmental problems would become much more complex and irreversible, making natural resource preservation one of the most important challenges to ever face humankind.

Daniel G. Graetzer

SUGGESTED READINGS: Oliver S. Owen and Daniel D. Chiras, *Natural Resource Conservation: Management for a Sustainable Future* (1995), is an excellent text describing ecological methods employed to conserve and manage natural resources. For descriptions of environmental degradation and actions taken by individuals and societies active in nature preservation, environmentalism, and ecology, see Joseph Edward de Steiguer, *Age of Environmentalism* (1997); Alan Axelrod and Charles Phillips, *The Environmentalists: A Biographical Dictionary from the 17th Century to the Present* (1993); and John O'Neill, *Ecology, Policy, and Politics: Human Well-Being and the Natural World* (1993). Books that examine methods employed in attempts to maintain environmental balance include Benjamin Kline's *First Along the River: A Brief History of the United States Environmental Movement* (1997), Gale Bittinger's *Learning and Caring About Our World* (1990), and Mark Dowie's *Losing Ground: American Environmentalism at the Close of the 20th Century* (1995).

SEE ALSO: Biodiversity; Ecosystems; Endangered species and animal protection policy; Nature reserves; Preservation.

Nature reserves

CATEGORY: Preservation and wilderness issues

Nature reserves are areas set aside to protect natural resources, usually animal and plant species. They may also preserve geologic formations and other landscape features. Most are created and managed by governments, although some nature reserves, such as nature conservancies, are run under private auspices.

The nature reserve concept is often traced back to 1872, when the United States Congress established Yellowstone as the world's first national park. In fact, the basic idea is much older. In medieval England, the New Forest was a royal preserve. It had special protected status so the king and his nobles could enjoy fine hunting in a land where farming was rapidly encroaching upon woodlands. Despite the hunting and poaching detailed in history and legend, the New Forest remained a viable natural area for centuries.

As the vast western regions of North America were explored, spectacular places such Banff in Canada (1885) and Yosemite in California (1890) received protection. The number of protected areas grew over the following decades. North America now has several hundred natural sites and parks, ranging from almost pristine wilderness to multiple-use areas such as national forests, where some commercial logging is allowed.

The movement developed differently on other continents, but by the late twentieth century almost every nation had established nature reserves. By 1980 more than three thousand protected sites existed worldwide. Treaties give special recognition to three types of reserve: Biosphere reserves shield strictly protected areas with zones of limited, sustainable activity around them; World Heritage Sites contain extraordinary natural features; and Ramsar Sites are protected wetland systems.

Nature reserves serve as reservoirs of native species and genetic diversity, viable ecosystems that support biological and climatic cycles even outside their bounds, and places for people to observe and enjoy the natural world. These goals are not innately incompatible. In managing reserves, however, decisions may have to be made that favor one aspect over the others.

Few people oppose nature reserves in the abstract, but when it comes to real-life choices, many conflicts arise. It seems to be easier to "think globally" than to "act locally." Clashes between biological or scenic integrity and human economic needs occur everywhere. In developed countries, these often take the form of duels between industries seeking to use specific natu-

ral resources, and government bureaus and conservation groups trying to prevent such activities. The push for oil drilling or pipeline construction in remote natural areas is one example. The dispute over allowing lumbering and mining in national forests is another. These struggles become more heated when a unique species is threatened, as in the northern spotted owl controversy in the Pacific Northwest.

Developing nations face added problems. Restricted natural areas often deprive their own citizens of customary food sources and trade items. Banned natural products may bring an even higher price on the black market, encouraging poaching. This has happened in Africa with elephant ivory and rhinoceros horns. Deciding how much of their total resources to devote to nature reserves is also a sharp dilemma for poor nations.

Tensions occur between wilderness values and demands for human recreational use. Tourists and campers bring cash and public support to parks, but they can disturb delicate natural balances. Automobile traffic brings pollution and noise problems inside the reserves' boundaries. In spite of strong warnings about behavior, visitors may start forest fires or introduce alien organisms to the area. On the other hand, when such lands are public trusts, can their reasonable use be denied to the citizens who support them?

Other issues arise about how to manage nature reserves. Fires, epidemics, and unusual weather patterns may threaten nature's balances. When disasters occur, should humans intervene? A consistent environmental ethic might say no: Nature will eventually restore itself. However, sometimes entire species are lost before this happens. Many such disasters are at least partly or indirectly caused by human actions. Science does not provide clear answers. Furthermore, measures taken to prevent or relieve such disasters do not always work.

There are also technical issues that require much further consideration. These include determining the efficacy of small reserves in protecting specific species, the role of reserves in stabilizing the mix of gases in earth's atmosphere and slowing global warming, and the tipping point in the recovery of a species or a damaged ecosystem.

Several promising ideas are enriching the reserve movement. The first is international cooperation and joint action. The World Heritage Sites program is a major achievement of this approach. A second renewed approach is the establishment or recognition of natural areas sponsored by nongovernmental groups. Two different but equally positive examples in the United States are the nature conservancy movement and the new respect given to Native American sacred sites.

Emily Alward

SUGGESTED READINGS: *Nature's Last Strongholds* (1991), edited by Robert Burton, is an excellent survey of types of natural areas, giving worldwide examples. It also includes a useful glossary. *Gaia: State of the Ark Atlas* (1986), by Lee Durrell, has dozens of capsule summaries on the situations of species and biomes, along with maps, charts, diagrams, and photos. "Our National Forests," *National Geographic* (March, 1997), by John G. Mitchell, calls for less timber cutting in these multipurpose reserves and depicts the onsite conflicts between environmentalist and commercial interests.

SEE ALSO: Ecosystems; National parks; Wilderness areas; Wildlife refuges.

Neutron bombs

CATEGORY: Nuclear power and radiation

"Neutron bomb" is the popular name for a nuclear device called an enhanced radiation weapon (ERW). It is designed to put more of its energy into radiation (particularly high-energy neutrons) and less into blast, heat, and fallout than standard nuclear weapons.

During the Cold War, Warsaw Pact armaments and soldiers outnumbered those of the North Atlantic Treaty Organization (NATO) in Europe by more than two to one. Had the Warsaw Pact launched a massive tank invasion into Germany,

the NATO plan was to halt the invasion with nuclear weapons if necessary. A major flaw in this plan was that Germany might not be any better off if its land were pummeled by allied nuclear weapons instead of Warsaw Pact tanks and planes. The neutron bomb was proposed as a way to stop Warsaw Pact tanks without devastating Germany.

Since tanks are resistant to damage by blast and thermal radiation, conventional nuclear warheads with explosive yields of 10 or more kilotons would be required to destroy them in large numbers. A 10-kiloton-yield conventional nuclear warhead could destroy or incapacitate tank crews within a 1.5-square-kilometer (0.6-square-mile) area but would destroy or damage buildings in an area of nearly 5 square kilometers (2 square miles). A 1-kiloton-yield neutron bomb could deliver a prompt neutron dose that would incapacitate any tank crew within the same 1.5 square kilometer area, but it would destroy or moderately damage urban structures only within a 1-square-kilometer (0.4-square-mile) area. Fallout would be only one-half of that of the larger weapon, and within a few hours radiation levels would be low enough for people to pass safely through the area.

Clearly, neutron bombs should be able to halt an invasion with less environmental damage than would be caused by conventional nuclear weapons. Advocates argued that since the neutron bomb's use would be less costly to Germany's cities and countryside, the Warsaw Pact should believe that NATO was more likely to use it. This should deter the Warsaw Pact from attacking in the first place.

Opponents argued that the bomb's effectiveness could be reduced by spacing the tanks farther apart and adding neutron shielding. They agreed that the neutron bomb's environmental advantages made it more likely to be used, but they feared that once nuclear weapons were used, the side that was losing would resort to larger nuclear weapons. The conflict could quickly escalate to full-scale nuclear war, possibly the ultimate human-made environmental disaster.

Beginning in 1981, the United States equipped 40 howitzer projectiles and 350 Lance missiles with neutron bombs. These warheads had yields of 1 kiloton or less. All were withdrawn from service and dismantled at the end of the Cold War. The former Soviet Union, France, and China have tested enhanced radiation weapons and may have produced them.

Charles W. Rogers

SEE ALSO: Antinuclear movement; Nuclear weapons; Radioactive pollution and fallout; Union of Concerned Scientists.

NIMBY

CATEGORY: Philosophy and ethics

An acronym for "not in my backyard," NIMBY refers to organized opposition by community residents to the siting of unwanted development in the local area. The syndrome has manifested itself in particular in response to the siting of solid waste and hazardous waste facilities.

The unregulated disposal of solid waste over the years has resulted in environmental problems such as groundwater contamination. Hazardous waste disposal practices have had even more significant impacts upon the environment, as well as adverse effects upon human health. The enactment of more stringent legislation has been successful in decreasing the amounts of hazardous wastes produced and stimulating the development of better disposal practices. However, many existing facilities have closed or are scheduled to close because they are unable to comply with the newer regulations. Consequently, there is a pressing demand for new facilities that meet the upgraded standards. The proposed siting of these facilities frequently generates a NIMBY response.

The controversy develops as a manifestation of the "collective goods" problem. In other words, it is difficult to provide goods that benefit everyone and for which the costs are equally shared. Therefore, the system fosters a situation that provides an incentive to be a "free rider." A perception of inequity develops when decisions about site facilities are made without any sense of input from the local community. The basis

then for the NIMBY response is basic democratic rights, in which people are asking for participation in the decision-making process. Equity issues may further accentuate a negative public response. For example, the sites selected for such facilities are frequently located in lower income, minority neighborhoods. Therefore, the economic and social aspects often become important considerations. The question is then raised, if the facility will provide for the public good, why does it have to be here as opposed to some other place?

These are all valid concerns that must be considered in order to reach a consensus on equitable siting of solid waste facilities. Public participation in a legitimate forum on health risks, as well as economic and social implications, is a vital part of gaining community support. If the safety aspects can be demonstrated and the quality of the facility can be shown to enhance property values and not convey a negative image on the community, residents are apt to be more receptive to the local siting of a proposed facility.

The NIMBY syndrome initially had a generally negative connotation. It was characterized as an emotional, almost irrational, selfish response that interfered with the larger public good. However, it has begun to be perceived in a more neutral, if not a positive, light. The NIMBY protest, in challenging government decisions, is sometimes effective in drawing public attention to valid environmental issues as well as forcing consideration of alternatives for more effective and equitable waste management facilities.

Steven B. McBride

SEE ALSO: Environmental ethics; Environmental justice and environmental racism; Hazardous waste; Landfills; Waste management.

Noise pollution

CATEGORY: Pollutants and toxins

It is widely accepted that noise constitutes a critical problem in environmental protection. The problem is particularly acute since noise pollution increases with population density and thus affects a disproportionately large sector of the human population. The control of noise requires scientific, social, and sometimes political actions.

Music, speech, and noise are the three basic categories of sound. Noise is simply defined as any unwanted sound. The degree of being unwanted is, however, a psychological question and may range from moderate annoyance to various forms of hearing loss. Furthermore, the interpretation of noise must be subjective, since both music and conversation may be regarded as noise in places such as an office or a library.

NOISE IN THE ENVIRONMENT

With few exceptions, technological advances in the past few decades have resulted in a steady increase in the amount of unwanted sound. Examples may include jet airplanes, automobiles, and ventilation fans. It was thought at one time that noise should be tolerated in order to gain the benefits of industrial advances. Yet problems associated with exposure to noise have mani-

fested themselves often enough that there is now serious concern about noise pollution in the environment. Noise can be generated in a great variety of ways. There are, however, only a few prominent sources of noise emission in daily life. Noise in the environment can be greatly reduced if these sources of pollution can be controlled.

Noise from airplanes poses a major problem in urban areas. Airplanes produce noise through the efflux of the jet engine and the high-pitched whine of the engine fan. Since 1969 the Federal Aviation Administration (FAA) has legislated acceptable noise levels for commercial airplanes. Over the years, remarkable engineering innovations have been made in the reduction of jet noise to satisfy the relatively stringent FAA requirements. In the meantime, however, air travel has become increasingly popular. Airlines in the United States have now captured more than 80 percent of all intercity passenger traffic, and the percentage is still on the rise. It is anticipated that the number of takeoffs and landings near major cities will continue to grow in the early

Airplanes are a major source of noise pollution in urban areas. (AP/Wide World Photos)

decades of the twenty-first century. Furthermore, as land prices rise, residential dwellings will encroach on noise buffer zones near airports in greater numbers. The control of aircraft noise will certainly become more pressing in the future.

The deafening din of high-powered trucks and motorcycles is familiar to nearly everyone. Social surveys in cities consistently rank road traffic noise as one of the primary sources of annoyance. Traffic noise may be controlled by rerouting heavy traffic and smoothing the flow of traffic to avoid unnecessary starts, stops, and acceleration. Construction of sound-barrier walls near highways is an effective but passive means of noise reduction. Traffic noise can also be reduced if motor vehicles are properly maintained. At low speeds, most of the noise from a vehicle is radiated from the exhaust system, and a good muffler is very effective in controlling the acoustical emission. The Environmental Protection Agency (EPA) has made several recommendations for reducing noise from motor vehicles. Any future legislative control of traffic noise must be carefully planned, since there are more than 100 million cars in the United States alone.

INDOOR NOISE

A high proportion of the workforce is employed in interior environments in which the workers are subject to long periods of exposure to noise. Indoor noise also affects people living in apartments and homes of relatively light construction. Pollution by indoor noise may be reduced by lining walls and ceilings with acoustic panels and absorbing materials. In addition, the operation of most home appliances and factory machinery produces noise. Noise from washing machines, drills, and air-conditioning systems arises from friction, unbalanced rotating parts, and air turbulence created by fans. Such noise can be substantially reduced through proper lubrication, balance of rotating parts, and acoustic insulation. The American National Standards Institute has published guidelines for manufacturers to rate noise emission in their products. Nevertheless, these guidelines are not always adhered to, and they have not served to encourage the design of quiet products. Perhaps this

will change if consumers and employers express a willingness to pay higher prices for quieter appliances and machinery.

There are other sources of noise pollution, some of which cannot even be heard. Ultrasonic and infrasonic noise possesses frequencies above and below the audible range, respectively. Such noise is emitted, for instance, through the background hum of high-voltage transmission cables. Although not heard, ultrasonic and infrasonic noise affects people in ways similar to audible noise.

The adverse physiological and psychological effects of noise on people have been the subject of considerable study in recent years. Noise interferes with work, sleep, and recreation. It also causes strain, fatigue, loss of appetite, indigestion, irritation, and headaches. High-intensity noise has adverse cumulative effects on the human hearing mechanism that may produce temporary or permanent deafness. In fact, noise-induced hearing loss has been identified as a major health hazard by the Occupational Safety and Health Administration (OSHA). Noise decreases the efficiency of workers and increases their liability to error.

Fai Ma

SUGGESTED READINGS: A readable account of the principles of sound and noise reduction is provided by Thomas D. Rossing's *The Science of Sound* (1990). Advanced treatment of noise control is offered in Leo L. Beranek's *Noise and Vibration Control* (1971). For a technical exposition, the reader is referred to *Fundamentals of Acoustics* (1982), by Lawrence E. Kinsler et al. A lucid discussion of noise and hearing is given in *The Physics of Sound* (1995), by Richard E. Berg and David G. Stork.

SEE ALSO: Environmental health; Environmental illnesses.

North Sea radioactive cargo sinking

DATE: August 25, 1984
CATEGORY: Nuclear power and radiation

During the night of August 25, 1984, a small French vessel, the Mont Louis, *collided with the West German vessel* Olau Britannia. *The Mont Louis capsized and sank, along with its cargo of thirty containers holding 225 tons of uranium hexafluoride, a radioactive material. The containers remained intact, and no radioactive material was released into the environment.*

Uranium hexafluoride is created in the refining process of uranium. Naturally occurring uranium consists of two isotopes: uranium 235 and uranium 238. Only uranium 235 can be split apart in a nuclear reactor to produce energy. In order to separate the two isotopes, the uranium is converted to uranium hexafluoride and passed through a gaseous diffusion facility. The uranium hexafluoride on the *Mont Louis* had been created in France, and it was en route to a gaseous diffusion facility in Riga, Latvia. Most of the uranium hexafluoride had been made from freshly mined uranium, but a small amount had been made from used reactor fuel. While uranium itself is mildly radioactive, uranium hexafluoride made from used reactor fuel is contaminated with highly radioactive materials produced in the reactor.

On August 25, 1984, the *Mont Louis* was running eastward, parallel to the coast of Belgium. The *Olau Britannia* departed from Flushing, Holland, bound for Sheerness, England. In this situation, the *Olau Britannia* had the right-of-way. There was heavy fog, and the sailor who should have been the *Mont Louis* lookout had been assigned other duties.

The *Olau Britannia* struck the *Mont Louis* in the starboard, or right, side. The two ships remained locked together until pulled apart by salvage tugs, at which point the *Mont Louis* capsized and sank in 14 meters (45 feet) of water. Although five members of the ship's crew fell into the sea, they were quickly rescued. There were no serious injuries on the *Olau Britannia.* Some of the fuel aboard the *Mont Louis* leaked into the sea and fouled a Belgian beach near Ostend.

The *Mont Louis* lay with its starboard side above the water and the containers of uranium hexafluoride still in its hold. Smit Tak International, a salvage company, was hired to recover the uranium hexafluoride and then raise the sunken ship. They cut a large hole in the exposed side of the ship and used a large floating crane to lift out the containers of uranium hexafluoride. Throughout this process, radioactivity detectors were used, but no significant radioactivity was found. The last container of uranium hexafluoride was recovered on October 4, 1984. The wrecked ship was raised on September 29, 1985. Estimated cost of the salvage operation was US$4.6 million.

Rules published by the International Maritime Organization state that ships larger than a certain size are requested, but not required, to notify authorities whose coasts they pass if they carry dangerous cargo. Since the *Mont Louis* was under the limit, this rule did not apply. The crew had received no special training in carrying radioactive cargo; they may not even have known that such cargo was on board. No uranium hexafluoride was released into the environment. If a release had occurred, it would have posed a danger to personnel near the scene but not to people onshore.

Edwin G. Wiggins

SEE ALSO: Nuclear accidents; Ocean pollution.

Northern spotted owl

CATEGORY: Animals and endangered species

The northern spotted owl is a medium-sized owl native to the coniferous forests of the Pacific Northwest of the United States. Destruction of its habitat by timber companies has led to intense conflict between loggers trying to earn a living and environmentalists who wish to protect the endangered owl.

The northern spotted owl requires an extensive habitat of old-growth trees for forage and nesting, and the survival of the species has been threatened by the commercial harvesting of old-growth timber in California, Oregon, and Washington. During the 1880's the habitat for the owl totaled about 15 million acres, but one century later the habitat was reduced to fewer than 5 million acres, almost entirely on federal lands.

In 1984 four conservation groups appealed the U.S. Forest Service's regional plans for timber harvesting within the spotted owl's range, and four years later federal regulators proposed setting aside up to 2,200 acres of suitable habitat for each nesting pair of owls. While timber companies argued that such a policy would lead to dire economic consequences in the region, environmentalists insisted that more habitat was needed. In 1990 the spotted owl was listed as an endangered species. The next year, environmental groups sued the Interior Department for failing to protect the owl, and a federal district court in Seattle, Washington, ordered the cessation of logging operations in the region's old-growth federal forests. During the presidential election of 1992, candidate Bill Clinton promised, if elected, to convene a Forest Summit to consider the issue.

On April 2, 1993, President Clinton and members of his cabinet met at the resulting summit in Portland, Oregon, which included representatives from environmental groups, the timber industry, and the salmon fishery industry. Thousands of timber workers gathered in Portland to demonstrate their concerns about the recent loss of between thirteen thousand and twenty thousand jobs in the industry. On July 1, Clinton announced a compromise plan that allowed the logging industry to cut about 1.2 billion board feet of timber each year, down from about 5 billion board feet each year during the preceding decade. The administration estimated that only six thousand jobs would be lost from the logging cutback. To compensate for the loss, the president asked Congress for a Northwest Economic Adjustment Fund that would provide $1.2 billion in aid over the next five years. The district court of Seattle upheld the legality of the Clinton plan.

The timber industry, nevertheless, went to court challenging the Clinton administration's broad interpretation of the Endangered Species Act. The issue was whether the government might restrict activities, such as logging, that only indirectly limited the habitat of the spotted owl. In *Babbitt v. Sweet Home Chapter* (1995), the Supreme Court ruled that the government had the discretionary authority to protect any habitat required by an endangered species.

No affected group was entirely happy with the Clinton compromise. In 1997 the U.S. Forest Service released a study that reported that the spotted owl's population was dropping by 4.5 percent per year and that the rate was accelerating. While environmentalists worried about the fate of the owl, the timber industry claimed that the settlement would lead to a total of about eighty-five thousand job losses. Numerous lumber mills were forced to close, and many workers either moved elsewhere or retrained for other jobs, which often paid minimum wage. In 1996 the timber industry of Linn County, Oregon, generated only $2.1 million, about 17 percent of what it had generated eight years earlier.

Thomas T. Lewis

SEE ALSO: Endangered species; Endangered Species Act; Forest Summit; Logging and clear-cutting; Old-growth forests.

Nuclear accidents

CATEGORY: Nuclear power and radiation

Public concern about reactor safety and the risk of accidents remains an important reason why the expansion of nuclear power has stalled in the United States and many other nations. Although the industry's safety record has generally been good, the public remains disquieted by recurring reports of safety lapses and near-accidents and an awareness that a severe accident could have widespread consequences.

The accident risk in nuclear power plants lies in the possibility that a malfunction caused by equipment failure, operator error, or external events will disrupt the flow of vital cooling water to the intensely hot, dangerously radioactive reactor core, which could result in the fuel melting. In the unlikely event that both the huge steel vessel holding the fuel and the massive concrete containment structure surrounding the reactor were breached, large quantities of dangerous radioactive materials could be dispersed to the environment.

The U.S. Nuclear Regulatory Commission (NRC) is the federal agency that sets the safety requirements that plant owners must meet in the construction and operation of nuclear power reactors. Regulations and technical specifications cover a wide range of topics such as design requirements, safety systems, equipment quality, record keeping, and operator training.

The worst U.S. accident occurred in 1979 at the Three Mile Island (TMI) Unit 2 plant in Pennsylvania, resulting in more than one-half of the fuel melting and the total loss of the reactor. Recurring incidents and near-accidents over the years have also disquieted the public and regulators. Among these have been the 1975 fire at the Brown's Ferry plant that burned for seven hours, during which it took plant personnel several hours to shut down the reactor; the 1983 unexpected failure of the Salem reactor to shut down after a safety system was activated; and the 1985 incident at the Davis-Besse plant when multiple equipment failures caused a loss of cooling water that could have initiated a core meltdown if plant operators had not responded quickly. At some plants it has been discovered that important safety equipment was inoperable for years. The NRC has kept a number of U.S. reactors shut down for long periods until safety-related equipment problems or operator lapses were corrected.

Nuclear proponents argue that, despite recurrent mishaps, the U.S. industry's overall safety record has been excellent. They assert that the TMI accident showed that systems designed to contain an accident worked. They claim that an accident similar to the one that occurred at Chernobyl could not occur in the United States because the design of the older, Soviet-designed Chernobyl-style plants is flawed and permits a severe accident to occur far more easily than in U.S.-designed plants. They note that the number of annual safety-related occurrences at U.S. reactors has declined since the TMI accident. Finally, they point to government-sponsored studies that concluded that there is a very low probability of a severe accident occurring with significant off-site property damage and injuries to the public.

Antinuclear activists point to numerous recurring safety-related incidents and problems at U.S. reactors, including some for which serious accidents were narrowly averted. They argue that the TMI accident showed that a serious accident is not as improbable as the industry claims. While critics acknowledge the design superiority of U.S. reactors as compared to the Chernobyl plant, they note that an NRC commissioner testified before Congress in 1986 that an accident at a U.S. reactor with off-site releases equal to or worse than Chernobyl could occur under conditions regarded as improbable but not impossible. Critics have also attacked the assumptions and methodology of the government's major accident risk studies. Finally, opponents observe that those same studies indicate that if a low-probability catastrophic accident should nonetheless occur, it could result in tens of thousands of deaths and injuries, tens of billions of dollars of property damage, and widespread, long-lived radioactive contamination.

The Price-Anderson Act limits the liability of nuclear power plant owners and equipment vendors in the event of a severe accident. It was adopted in 1957 after private companies made it clear that they would not participate in the development of nuclear energy without liability protection. Critics argue that the nuclear industry's continuing insistence that such protection is required is inconsistent with its denial of the possibility of a catastrophic accident.

Much less information is generally available about accidents and incidents with serious safety implications in countries other than the United States because many nations have public disclosure requirements less stringent than those in the United States. Worrisome accidents and equipment malfunctions are known to have occurred at reactors in many nations, including Japan, India, and the countries of the former Soviet Union. In Europe, grave concern remains about twenty-six older Soviet-designed reactors, including fifteen Chernobyl-style plants, in the countries of Eastern Europe and the former Soviet Union. Almost all were still operating in the late 1990's. Western nations have provided some funding for safety improvements and have sought eventual permanent closure of many of the plants.

A number of nuclear accidents have occurred at noncommercial facilities, some of which resulted in serious environmental consequences. In 1952 a fuel melt and explosion severely damaged an experimental Canadian reactor at Chalk River, Ontario. In 1957 an explosion at a waste site at the then-secret Chelyabinsk complex near the city of Kyshtym in Russia contaminated hundreds of square miles of the surrounding countryside. Also in 1957, a fire occurred in England at the Windscale reactor (an atypical plant used to produce plutonium for the United Kingdom's weapons program); the surrounding countryside was contaminated, and radioactive fallout drifted into several neighboring countries. In 1966 the experimental Fermi breeder reactor near Detroit suffered a disastrous fuel melt that seriously damaged the reactor.

Phillip A. Greenberg

SUGGESTED READINGS: A brief but excellent introduction to the topic of nuclear accidents can be found in Richard Wolfson, *Nuclear Choices: A Citizen's Guide to Nuclear Technology* (1993). A good account of the Chernobyl accident is Grigori Medvedev, *The Truth About Chernobyl* (1989). A good summary is Yuri Shcherbak, "Ten Years of the Chernobyl Era," in *Scientific American* (April, 1996). Several books on TMI were published in the early 1980's; none stand out. The most comprehensive report was issued by the President's Commission on the Accident at Three Mile Island (1979).

SEE ALSO: Brazilian radioactive powder release; Chalk River nuclear reactor explosion; Chelyabinsk nuclear waste explosion; Chernobyl nuclear accident; North Sea radioactive cargo sinking; Nuclear regulatory policy; Three Mile Island nuclear accident; Windscale radiation release.

Nuclear and radioactive waste

CATEGORY: Waste and waste management

Nuclear and radioactive wastes encompass a broad range of radioactive by-products resulting from the use of nuclear materials for energy generation, weapons production, research, medicine, and industry.

Highly radioactive waste includes spent nuclear fuel rods from commercial nuclear energy facilities and high-level radioactive waste by-products from nuclear weapons production at defense processing plants. Low-level nuclear waste consists of items that have been contaminated with radioactivity, such as clothing, rags, equipment, tools, medical instruments and waste, and other such materials, as well as the waste generated in the processing of uranium ore. Radioactive waste deteriorates according to its half-life (reducing its level of radioactivity by one-half). While some low-level radioactive waste may be safe to handle in a relatively short time, some of the more highly radioactive waste can take thousands of years before it can be handled safely.

Regulatory control of nuclear and radioactive waste is vested in a number of government agencies. At the federal level, the Nuclear Regulatory Commission (NRC) and the Department of Energy (DOE) are the primary regulators of highly radioactive waste in the United States. Other federal agencies such as the Environmental Protection Agency (EPA), the Department of Transportation, and the Department of Health and Human Services can also play a role in the regulation of radioactive material. Individual states are responsible for the regulation, management, storage, and disposal of commercial low-level radioactive waste materials generated within their boundaries.

After the development of nuclear fission, little concern was initially exhibited over how to safely dispose of nuclear waste. Policymakers perceived nuclear waste disposal as a long-term issue that would need to be addressed sometime in the future. However, as the amount of highly radioactive and low-level nuclear waste continued to grow, the issue became increasingly contentious and politicized. Eventually, the problem of radioactive and nuclear waste disposal became one of the most controversial and environmentally challenging aspects of the nuclear era.

As of the late 1990's, there were no permanent repositories for highly radioactive waste in the United States. While Yucca Mountain, Nevada, was designated as a deep underground national repository for spent nuclear fuel and high-level radioactive waste, the selection of this site has been highly controversial because of a number of environmental and geologic concerns that could make it unsuitable for the safe storage of nuclear waste. For example, environmentalists have expressed concern about Yucca Mountain's volcanic history. While the explosive type of volcano is extinct, scientists are studying seven small, dormant volcanoes in the area. The Yucca Mountain area also contains more than thirty known earthquake fault lines, some of which have been active in recent years. There is also considerable concern about the movement of groundwater that could transport any radioactive material that may leak from the disposal facility. Scientists also have discovered renewed evidence of hydrothermal activity at Yucca Mountain. Perched water (geothermal water that has risen to the surface and drained back down) has been found in several boreholes made in the area. Similar to groundwater movement, perched water can act as a transport mechanism for any nuclear waste material that may leak from the disposal facility. DOE scientists have claimed that such potential dangers to the integrity of a deep underground storage facility at Yucca Mountain are minimal.

A container holding 10 tons of high-level nuclear waste is unloaded from a ship in northern Japan. The storage of radioactive waste materials is a controversial issue in many nations. (AP/Wide World Photos)

Yucca Mountain was not expected to be ready to accept highly radioactive waste before 2010 at the earliest, even if it received NRC approval. The delay in establishing a permanent national repository for the disposal of highly radioactive waste has prompted policymakers to search for alternative solutions to the problem of how to store nuclear waste safely. Commercial utilities and defense processing plants have begun storing spent nuclear fuel on-site (either in cooling ponds or in above-ground canisters). However, since suitable on-site storage space for highly radioactive waste is running out, maintaining the status quo clearly is not an acceptable long-term alternative. One proposal that seeks a more immediate solution is to establish an interim storage facility for highly radioactive waste. This interim facility—also known as a monitored retrievable storage (MRS) facility—would store nuclear waste above ground until a permanent repository is ready to accept highly radioactive waste. However, environmentalists have expressed concern that any MRS facility could become a permanent aboveground storage facility, which clearly would not be as safe as a deep underground repository.

Environmentalists and opponents of centralized storage facilities have also expressed concern over the potential for an accident during the transport of nuclear waste to either a national permanent repository or an MRS facility. Transportation routes would carry nuclear waste through a number of highly populated areas. Proponents of regional and national storage facilities counter that a lack of centralization enhances management and monitoring problems and increases the possibility of an accident. They also point to the successful existing safety record for the transport of radioactive material.

Existing disposal facilities for low-level radioactive waste are quite limited, and progress has been slow to develop any new facilities. The "not in my backyard," or NIMBY, movement has been effective in preventing the approval of any new disposal sites. Thus, much of the low-level radioactive waste continues to be stored at the location where it was produced pending the construction of new disposal facilities.

Kenneth A. Rogers and Donna L. Rogers

SUGGESTED READINGS: The Congressional Research Service (CRS) publishes the *CRS Issue Brief,* which contains civilian nuclear waste disposal updates. Their Web site (http://www. cnie.org/nle/waste-2.html) provides up-to-date information on nuclear and radioactive waste issues. Kenneth A. Rogers and Donna L. Rogers, "The Politics of Long-term Highly Radioactive Waste Disposal," *Midsouth Political Science Review* 2 (1998), explains the political ramifications of nuclear and radioactive waste disposal. The Department of Energy publishes two series of studies—*Yucca Mountain Project Studies* and *DOE's Yucca Mountain Studies*—and maintains a Web site (http://www.ymp.gov) that provide thorough overviews of the Yucca Mountain Project. The Office of Civilian Radioactive Waste Management (OCRWM) provides information on the Yucca Mountain project through their publi-

cation (*OCRWM Bulletin*) and their Web site (http://www.rw.doe.gov).

SEE ALSO: Antinuclear movement; Nuclear power; Nuclear regulatory policy; Yucca Mountain, Nevada, repository.

Nuclear power

CATEGORY: Nuclear power and radiation

Nuclear power has long been the most controversial source of electricity. By the mid-1990's, the growth of the nuclear power industry had slowed worldwide and stopped in the United States and most other industrialized nations. In the United States, four issues dominate: waste disposal, cost, safety, and public mistrust.

Many of the environmental impacts of nuclear power plants are common to all large-scale electric generating facilities, regardless of their fuel. The most important are land use and related impacts on plants, animals, and ecosystems; nonradioactive water effluent and water quality; thermal pollution of adjacent waters; and social impacts on nearby communities. Unique to nuclear power plants is the hazardous radiation emitted by radioactive materials present in all stages of the nuclear fuel cycle. Such radiation is contained in uranium ore when it is mined and processed, fabricated uranium reactor fuel, spent fuel that has been fissioned, contaminated reactor components, and low-level radioactive waste—such as contaminated tools, protective clothing, and replaced reactor parts—generated by routine plant operation and maintenance.

Under normal operating circumstances, commercial nuclear power plants release negligible radioactive emissions. The principal safety concern is that a severe accident could release large quantities of dangerous radioactive materials into the environment, as happened in 1986 at the Chernobyl reactor in Ukraine. Spent fuel is also highly radioactive and must be isolated from the environment for at least tens of thousands of years. Finally, at the end of a plant's useful life, the contaminated reactor must be dismantled and the radioactive and nonradioactive components disposed of in a process known as "decommissioning."

HISTORY

The nuclear power industry arose out of the technology developed during World War II to produce the atomic bomb. Enthusiasm for new technology and pressure to demonstrate peaceful uses for the expensive and fearsome development of nuclear weapons led to a strong government effort, beginning in the early 1950's, to induce industry to develop nuclear energy. Large government subsidies and preferential treatment assisted the industry from its early days. Between 1948 and 1972, nuclear power received 82 percent of all federal energy research and development funds, totaling $22 billion in 1998 constant dollars. (During the longer 1948-1997 period, nuclear power's share was 61 percent totaling $66 billion.) One of the industry's unique subsidies was the passage of the 1957 Price-Anderson Act, which limits the liability of nuclear power plant owners and equipment vendors in the event of a reactor accident.

The basic design of the current generation of U.S. nuclear power plants was adapted from early naval propulsion reactors. The choice of this "light water" reactor design was largely driven by expediency and political considerations rather than an explicit effort to seek safe or reliable design features. The first U.S. commercial nuclear plant, which began operating at Shippingport, Pennsylvania, in 1957, was a conversion of a naval reactor project for which funding had been canceled. Some observers believe this early technology decision was largely responsible for the industry's problems in later years. Government, industry, and public opinion were all largely positive about nuclear energy up through the late 1960's. Almost all U.S. reactors operating in 1997 were or-

dered between 1966 and 1973, a time of great optimism about the technology.

During this same era, the 1969 National Environmental Policy Act was enacted, which forced prospective reactor owners to address environmental impacts in plant proposals; the environmental movement arose, beginning with Earth Day in 1970; and there was a widespread increase in citizen activism. By the mid-1970's, several widely publicized safety hearings, plant incidents, and government studies had begun to focus public and media attention on regulatory and safety lapses, accident risks, and the problem of nuclear waste disposal. These forces com-

Although nuclear power plants produce minimal amounts of air pollution, public concern about radioactive waste disposal and safety issues during the 1970's led to a decline in the use of nuclear power in the United States. (Jim West)

bined to create a sizable antinuclear movement in the United States, which was bolstered by the 1979 accident at the Three Mile Island (TMI) plant in Pennsylvania. Although proponents frequently blamed licensing interventions by antinuclear activists for numerous cost overruns and plant delays, most analyses concluded that, in the majority of cases, other factors, such as capital availability and shifting regulatory requirements, were primarily responsible. By the late 1980's, a decline in the number of plants under construction and a shift of public concern to the threat of the nuclear arms race had begun to reduce the antinuclear movement's ranks.

Between 1987 and 1994 nuclear power proponents won long-sought changes in regulations governing emergency planning requirements, the licensing of new reactors, siting, and reactor design certification. The essence of these changes was to facilitate the process for approving new reactors while sharply reducing opportunities for public participation in the regulatory process. Nuclear opponents adamantly opposed many of these changes, especially a 1992 congressional decision authorizing the U.S. Nuclear Regulatory Commission (NRC) to forego a long history of issuing separate licenses for construction and operation, each of which allowed for hearings. Instead, Congress mandated the issuance of a single combined license for both activities that permits few opportunities for safety challenges after construction has been completed.

STATUS AND PROJECTIONS

By the end of 1997, there were 433 nuclear power plants licensed for operation in thirty-two countries, with thirty-six more reactors under active construction (although some of the latter may not be completed). However, four or more reactors were under construction in only five nations. In eighteen countries, nuclear power supplied at least 25 percent of total electricity. Almost 80 percent of the world's reactors (including all U.S. plants) are U.S.-designed light water reactors.

In the United States in mid-1998, there were 103 licensed, operable reactors at sixty-five sites in thirty states. In 1997 nuclear power generated 20 percent of electricity and accounted for al-

most 8 percent of total energy consumed. No new reactors were ordered after 1978, and every order placed after 1973 was subsequently canceled. As of 1997, no plants were on order or under construction in the United States.

France and Japan rank second and third in total operating reactors and, along with South Korea, are regarded as having the most active nuclear programs. By the late 1990's, a de facto moratorium on new reactors had existed for some years in Western Europe—with the exception of France—and there were few prospects for new orders. Asia and scattered developing nations were widely regarded as the only active markets for possible new plant orders.

The NRC is the federal agency responsible for nuclear power licensing and safety regulation in the United States. All U.S. reactors were originally granted forty-year operating licenses. By 1998 more than forty U.S. reactor licenses were scheduled to expire by the year 2015. Under NRC regulations adopted in 1992, plant owners may seek twenty-year license extensions. In some cases, competitive economic pressures or the costs of expensive equipment upgrades are expected to make retirement a preferable alternative. By 1998 eighteen commercial reactors had already been permanently shut down before their scheduled license expiration dates because competing power sources or a need for expensive repairs made continued operation economically undesirable.

In 1998 the U.S. Department of Energy (DOE) forecast that U.S. nuclear capacity would decline by the year 2020, owing to a lack of new orders and expected retirements of existing units. The decline could range from a 20 percent to a 98 percent drop from 1996 capacity, depending primarily on assumptions about whether reactors would be retired early or at their license expiration dates, or would be granted ten-year license extensions. Worldwide over the same period, the DOE forecast a 49 percent decrease to a 20 percent increase, depending upon the rate of reactor retirements and the number of new orders. These projections explicitly excluded consideration of whether the 1997 Kyoto Accords on global climate change might generate new interest in nuclear power.

World Nuclear Power, 1997

REGION	OPERABLE REACTORS	REACTORS UNDER CONSTRUCTION	NUCLEAR PERCENTAGE OF TOTAL ELECTRICAL ENERGY	NUCLEAR PERCENTAGE OF TOTAL ENERGY CONSUMED
United States	103	0	20.1	7.7
France	59	1	78.2	39.4
Japan	54	1	35.2	13.5
United Kingdom	35	0	27.5	10.1
Russia	29	4	13.6	4.4
Germany	20	0	31.8	10.8
Canada	16	0	14.2	8.2
Ukraine	16	4	46.8	13.3
Sweden	12	0	46.2	31.7
South Korea	12	6	34.1	9.8
India	10	4	2.3	0.8
Spain	9	0	29.3	12.0
Belgium	7	0	60.1	17.1
Bulgaria	6	0	45.4	19.4
Switzerland	5	0	40.6	20.7
Czech Republic	4	2	19.3	6.4
Finland	4	0	30.4	17.9
Hungary	4	0	39.9	12.3
Slovak Republic	4	4	44.0	21.2
Latin America	5	2	—	—
Other Western Europe	1	0	—	—
Other Eastern Europe	6	1	—	—
Other	12	7	—	—
Total	433	36	17.6	6.4

Source: International Atomic Energy Agency.

ARGUMENTS PRO AND CON

Nuclear proponents cite several points in the technology's favor: a good safety record, improved operating performance since the TMI accident, studies concluding that the risk of a severe accident is low, and the fact that nuclear power plants emit no significant amounts of common air pollutants or gases that contribute to global warming.

Critics of nuclear power point to the long-standing failure of any country as of 1998 to develop a viable technology and site for disposing of spent fuel and high-level wastes; flaws, uncertainties, and omissions in accident probability studies; the catastrophic consequences that could result from a severe reactor accident; a large number of safety-related incidents and near-accidents; and the high cost of nuclear plants compared to competing generating technologies and energy efficiency improvements. In addition, for nations about which there may be weapons proliferation concerns, there are inherent proliferation risks associated with nuclear power technology and fuels, which, if misused, can provide the expertise, infrastructure, and basic materials for a program to develop nuclear weapons.

Public opinion surveys in the United States after the TMI accident have shown a growing majority opposing the construction of new nuclear power plants, with a 2-1 margin against since the late 1980's. Numerous influential policy assessments have concluded that it is unlikely that new nuclear power plants will be built in the United States unless the key issues of waste disposal, costs, safety, and public acceptance can be satisfactorily resolved. Although the situation differs from country to country, some combination of these factors generally explains why the growth of nuclear power has stalled in most countries. Additional factors that contributed to nuclear power's decline in the United States included unexpectedly plentiful supplies of natural gas and slower-than-anticipated growth in electricity demand.

With regard to nuclear wastes, the DOE is performing site suitability studies for a spent fuel and high-level radioactive waste repository at Yucca Mountain, Nevada. Persistent geological, seismic, hydrological, and related scientific questions must be resolved favorably for the site to be approved. Opponents, including the state of Nevada, charge that the DOE's efforts have been geared more toward preparing the site for operation than for objectively assessing its suitability. The DOE has stated that even if the site is found suitable, it could not accept waste for disposal until the year 2010 at the earliest, twelve years after the original 1998 target date for opening.

High costs remain another obstacle to new plant orders. U.S. reactors completed in the 1990's cost $2 billion to $6 billion each, averaging more than $3 billion. Although proponents argue that new plant designs could be built more cheaply, the industry's history of large cost overruns discomforts prospective plant owners, capital markets, and state regulators. Many analysts believe that unpredictably high capital costs have been among the most important reasons for the industry's decline. The picture is complicated further by the ongoing deregulation of the electric utility industry, where uncertainty surrounding the future economic and regulatory climate raises questions about the competitiveness of both new and many existing reactors.

Concern about accidents continues to trouble the public. Precise accident probability estimates are impossible to derive given the complexity of nuclear reactor systems. Industry proponents point to several government-sponsored studies that concluded that the probability of an accident with off-site consequences was extremely low. Critics charged that the studies were methodologically flawed, omitted important factors, and underestimated the true risks. They also emphasized the catastrophic consequences that could result if a low-probability accident should nonetheless occur.

The public's mistrust of nuclear power is generally acknowledged as a roadblock to new plant orders. Risk perception studies have shown that the public's lack of trust is deeply rooted, widely felt, and resistant to change. NRC officials and others have noted that, regardless of one's explanation of the public's views, it is unlikely that new plants will be built in the United States without increased public acceptance.

Phillip A. Greenberg

SUGGESTED READINGS: An excellent summary of current policy, controversies, and legislation can be found in the regularly updated issue briefs on nuclear power and nuclear waste produced by the Congressional Research Service, available via the Internet at http://www.cnie.org or by request from the offices of members of Congress. A good introduction to nuclear technology and issues is Richard Wolfson, *Nuclear Choices: A Citizen's Guide to Nuclear Technology* (1993). An excellent short history of the nuclear power industry up to the 1990's can be found in the chapter by Christoph Hohenemser, Robert Goble, and Paul Slovic in Jack Hollander, editor, *The Energy-Environment Connection* (1992). A technical book with understandable policy commentary and extensive data is David Bodansky, *Nuclear Energy: Principles, Practices, and Prospects* (1996). The Energy Information Administration of the U.S. Department of Energy (DOE/EIA) periodically publishes a review of the status of nuclear energy in the United States and around the world, with forecasts and extensive data tables, entitled *Nuclear Power Generation and Fuel Cycle Report*. A readable history of the nuclear industry and its problems in the United States is Joseph G. Morone and Edward J. Woodhouse, *The Demise of Nuclear Energy? Lessons for Democratic Control of Technology* (1989).

SEE ALSO: Alternative energy sources; Nuclear regulatory policy; Superphénix; Yucca Mountain, Nevada, repository.

Nuclear regulatory policy

CATEGORY: Nuclear power and radiation

The Nuclear Regulatory Commission (NRC) is the federal agency responsible for licensing nuclear power plants and regulating the civilian nuclear power industry in the United States. The federal government has broadly preempted state authority. The NRC's mission is to protect the public health and safety, safeguard national security, and ensure environmental protection.

The NRC licenses plant owners to construct and operate nuclear power reactors and issues a wide variety of technical and procedural rules governing reactor operation. Through the late 1990's, electric utilities historically owned and operated nuclear power plants (although restructuring of the utility industry may lead to ownership by other kinds of companies).

The nuclear regulatory system is largely based on self-regulation by plant owners, who are responsible for meeting the NRC's requirements. Utility personnel maintain and inspect their plants and submit regular reports to the NRC.

Plant owners must notify the NRC of important safety lapses and equipment malfunctions. As with many industries, federal regulators set forth specific rules, but federal inspectors only spot-check for compliance. After the 1979 accident at the Three Mile Island (TMI) plant in Pennsylvania, the NRC stationed full-time inspectors at each nuclear power plant to conduct inspections and monitor compliance with NRC rules. Special NRC inspection teams are called in when warranted by serious equipment or operator failures.

Utilities found in violation of the NRC's rules may be subject to fines and penalties. When violations are sufficiently serious to raise questions about the safety of plant operation, operators customarily shut down a reactor before being ordered to do so. Although the NRC has the authority to order an operating plant to shut down, as of 1998 it had done so only once. However, the NRC may withhold permission for a plant to restart—sometimes for years—until the utility can demonstrate it has corrected the problems.

STATE REGULATION AND ANTINUCLEAR ACTIVISM
States are largely preempted by federal law from health and safety regulation of the nuclear aspects of nuclear power plants. States retain authority to regulate other aspects of nuclear plants that are common to all electric generating facilities. The most important of these are the ability of state public service commissions to set utility rates and financial returns on utility investments, and the broad powers of state agencies to regulate land use, facility siting, and nonnuclear environmental impacts.

States and members of the public have long been able to apply to the NRC to be granted status as "intervenors" in NRC licensing proceedings. This aspect of the regulatory system has been fundamentally adversarial. NRC licensing hearings often resembled judicial proceedings in which parties enjoyed certain rights with regard to reviewing documents, cross-examination, the right of appeal to the courts, and so forth.

Beginning in the early 1970's, antinuclear activism and the number of interventions in construction permit and operating license proceedings increased. Industry proponents often claimed that most interventions were mere delaying tactics intended primarily to lengthen licensing times and drive up costs. Nuclear critics argued that most interventions were warranted, and many brought to light safety-related problems that would not otherwise have been identified. Studies by government agencies and others concluded that most—but not all—plant delays resulted from factors other than interventions, such as changing regulatory requirements and construction problems.

With regard to operating plants, the public has little access to the regulatory process. Although NRC regulations permit members of the public to submit a petition to seek a hearing regarding modification or revocation of a utility's license on safety or environmental grounds, the NRC has granted hearings for only a handful of the hundreds of such petitions received.

Many other countries have based aspects of their regulatory system on the U.S. model, particularly with regard to technical oversight, since roughly 80 percent of the world's nuclear plants utilize U.S. reactor designs. However, organizational structure, staffing levels, practices, and requirements differ from country to country. In many nations, such as France and Canada, the relationship between the regulatory agency and plant licensees is much closer than in the United States. In most developing countries, the public has little access to the regulatory and policy processes. In industrialized nations, a wide range exists with regard to opportunities for public participation and the related ability to seek judicial review. In France and Japan, which operate the world's second-and third-largest nuclear programs, the regulatory policies have been highly centralized and largely closed to public participation.

U.S. REGULATORY HISTORY

The effort to develop the atomic bomb during World War II was accompanied by a tight government monopoly over nuclear materials and technology. The 1946 Atomic Energy Act created the civilian Atomic Energy Commission (AEC), which oversaw all aspects of atomic en-

ergy, and the powerful congressional Joint Committee on Atomic Energy (JCAE), which was given authority over all nuclear-related legislation and oversight. Amendments in 1954 loosened the government's monopoly in an effort to spur the development of a private nuclear power industry. The AEC was given the additional charges of actively promoting the development of commercial nuclear energy and regulating the new industry. Government and private companies cooperated in what was regarded as a joint effort.

Between 1963 and 1974, forty-four U.S. commercial nuclear power plants began operation. Because of the cooperative nature of the endeavor and the relatively few operating facilities, federal regulation was initially modest. However, reactor safety issues soon began to draw media and public attention, and the AEC was widely perceived as too friendly with the industry it was supposed to regulate. In 1974 Congress, responding to criticism that the AEC's dual missions of promotion and regulation were conflicting, replaced the AEC with two new agencies. The task of regulation was given to the newly created NRC. Promotion of nuclear energy was assigned to the Energy Research and Development Administration, which was superseded in 1977 by the Department of Energy. In 1976 the JCAE was dissolved for largely the same reasons as the AEC, and congressional authority was divided among various committees.

Critics, including the U.S. General Accounting Office, noted that the staff of the new NRC was drawn largely from the AEC and thus retained essentially the same mentality. However, the years following the NRC's creation were also marked by increased regulatory scrutiny. Between 1975 and the mid-1980's, NRC safety research increased; by 1981 the inspection and enforcement staff had grown to become the agency's largest division. The 1979 TMI accident created a new atmosphere that was initially marked by a brief licensing moratorium for new plants and several independent investigations into the accident's causes. An industry group was formed to try to improve safety performance. The NRC responded to public and congressional safety concerns by setting new construction and operating requirements and monitoring industry compliance more closely. However, these actions were counterbalanced by President Ronald Reagan's administration and its commission appointees, who began initiatives to reduce federal regulation, an effort from which the NRC was insulated but not immune.

CHANGES IN NRC REGULATIONS

Beginning in the late 1980's, the NRC enacted a number of new regulations long sought by the nuclear industry. The most important changes addressed licensing of new reactors, emergency planning, siting, reactor design certification, and operating license extensions.

For all reactors licensed through 1996, the NRC followed a two-step licensing process mandated by Congress in 1957. First, utilities applied for a permit to begin construction. When the plant was completed, the utility applied for an operating license. In numerous cases, intervenors opposed the granting of operating licenses, usually on grounds of safety-related allegations relating to design, construction, or infeasible emergency planning. In 1992 the industry achieved a long-sought goal when Congress approved a 1989 NRC regulation change that implemented one-step licensing for new reactors, permitting the NRC to issue a single combined construction and operating license prior to the onset of construction. The rule sharply limits the ability of intervenors to raise environmental and safety issues after the license has been granted, both during construction and before operation begins. Other rule changes in 1989 narrowed intervenors' procedural rights during licensing hearings.

Following the TMI accident, the NRC adopted regulations requiring the agency to approve state and local emergency response plans before a plant could be granted an operating license to ensure that the public could be protected or evacuated if a serious accident occurred. In several celebrated cases (the Diablo Canyon, Seabrook, and Shoreham reactors), states refused to cooperate on emergency planning and opposed the issuance of an operating license on the grounds that no feasible emergency planning was possible because of siting

error. In 1987 the NRC adopted new rules permitting the agency to grant an operating license based on emergency plans drafted by the plant owner alone; in 1988 President Reagan issued an executive order authorizing federal agencies to draft emergency plans. These actions effectively bypassed state opposition.

In 1989 the NRC adopted regulations that permitted utilities to seek approval for reactor sites before any license was granted and to retain the ability to use approved sites for up to forty years. Under these rules, utilities are permitted to seek approval of emergency plans for a site before filing any application to build a plant. In 1989 the NRC also adopted new rules allowing reactor vendors to seek approval for "standardized" reactor designs, which would be certified by the agency for fifteen years. The rule prohibits critics from raising design-related safety and environmental issues when a utility seeks to build a new plant using the certified designs. As of 1999, several vendors had submitted designs, and approval was pending.

All U.S. reactors operating by 1996 were issued forty-year operating licenses, 40 percent of which were scheduled to expire by the year 2015. In 1992 the NRC adopted a rule that permits plant owners to seek license extensions of up to twenty additional years. By mid-1998, three plant owners had submitted the first applications. Extensive documentation and equipment inspections and upgrades will be required; the NRC projected that the review process would take approximately three years. The net results of these regulation changes, which were adamantly opposed by nuclear critics, have been to facilitate the licensing process for new and existing reactors while significantly restricting public access to the regulatory process.

CRITICISM OF THE NRC

The nuclear industry and antinuclear activists alike have criticized the NRC's regulation, although for different reasons. In the late 1990's the nuclear industry argued that the NRC's requirements unnecessarily raised plant operating costs. The industry also complained that the NRC failed to set objective criteria by which adequate levels of safety and licensees' compliance could be judged and that the agency did not rank the relative safety significance of compliance requirements and adjust regulatory priorities and enforcement emphasis accordingly. Meanwhile, antinuclear activists have asserted that the NRC has generally been allied with the industry and has often failed to maintain an arms-length relationship with its licensees. They also argue that the NRC has retreated from tough regulation, has failed to pursue important safety questions, and has been either unresponsive or obstructive to substantive participation by individuals and citizen groups.

These allegations are not mutually exclusive, and there is evidence to support arguments on both sides. However, it is difficult to deny that there has been a fundamental alignment of interest between the NRC and the nuclear industry. Such relationships between federal regulatory agencies and the industries they regulate are widely acknowledged, although the degree of cooperation or tension between the two sides may fluctuate over time. By the late 1990's, assisted by improved safety performance, momentum had shifted toward lighter regulation under the influence of presidential administrations, appointees, and a Congress more sympathetic to the industry's views. Meanwhile, nuclear critics continued to argue that easing regulatory vigilance was a mistake in the light of repeated plant shut-downs for safety violations, unexpected aging of plant equipment, and recurring discoveries of safety-related problems in both individual plants and across the industry.

Finally, many U.S. nuclear energy proponents have advocated the amendment of corresponding aspects of the U.S. licensing and regulatory system to resemble the French and Japanese models, which permit little public participation. The key regulatory changes adopted in the United States from 1987 to 1992 indicate movement in that direction. Nuclear power critics oppose such changes, arguing that opportunities for public participation in the regulatory and policy arenas, along with the availability of judicial review, are fundamental to preserving the democratic process in making national decisions about energy and technology policy.

Phillip A. Greenberg

SUGGESTED READINGS: A summary of current regulatory issues can be found in the Congressional Research Service issue brief "Nuclear Energy Policy," updated regularly and available on the Internet at http://www.cnie.org or by request from the offices of members of Congress. The nuclear industry's initiative to improve self-regulation is described in Joseph V. Rees, *Hostages of Each Other: The Transformation of Nuclear Safety Since Three Mile Island* (1994). An excellent critical treatment of the 1987-1992 regulatory changes can be found in David O'Very, Christopher Paine, and Dan Reicher, *Controlling the Atom in the Twenty-first Century* (1994). A critical but thorough account of nuclear regulation through the mid-1990's is Robert J. Duffy, *Nuclear Politics in America: A History and Theory of Government Regulation* (1997). Two authoritative histories of U.S. regulation are George T. Mazuzan and J. Samuel Walker, *Controlling the Atom: The Beginnings of Nuclear Regulation, 1946-1962* (1984), and J. Samuel Walker, *Containing the Atom: Nuclear Regulation in a Changing Environment, 1963-1971* (1992).

SEE ALSO: Antinuclear movement; Nuclear accidents; Nuclear power; Union of Concerned Scientists.

Nuclear testing

CATEGORY: Nuclear power and radiation

Since 1945 hundreds of nuclear devices have been exploded in the atmosphere and underground to test nuclear weapons systems. Such testing has raised concerns about nuclear fallout and its effects on human health and the environment.

The United States began nuclear testing with the shot code-named "Trinity" on July 16, 1945, at Alamogordo, New Mexico. The United States has tested 219 devices in the atmosphere, with the last atmospheric test occurring in 1963. Most of these tests occurred at the Nevada Test Site, but the most powerful tests were done in the South Pacific at Christmas Island and in the

Marshall Islands at Bikini Atoll and Enewetak Atoll.

The former Soviet Union conducted 219 atmospheric tests, the last one in 1962. Nearly all these tests were at the Semipalatinsk Test Site in Kazakhstan or at the Northern Test Site, Novaya Zemlya. The higher-yield weapons were tested at the Northern Test Site, including a 50-megaton monster on October 30, 1961, the largest weapon ever exploded. The last of China's twenty-three atmospheric tests occurred in 1980. These were conducted at the Lop Nur Test Site in northwest China. Britain conducted twenty-one atmospheric tests off Christmas Island, near Monte Bello Island off Australia, and at sites in South Australia. Britain ended atmospheric testing in 1965. France continued atmospheric testing into 1974. France conducted fifty-six such tests at Reganne in Algeria and at Fangataufa and Moruroa in the South Pacific.

The United States ended underground testing in 1992, the former Soviet Union in 1990,

Britain in 1991, France in 1996, and China in 1996. Pakistan and India both conducted underground tests in 1998. With some notable exceptions, underground tests by the United States, Britain, and France have released little radioactivity into the atmosphere. The chief adverse environmental effect of most of these underground tests is that the radioactive subterranean rubble might someday pollute groundwater. The former Soviet Union was less successful in containing the radioactivity of underground tests.

TESTING AT BIKINI

The global effect of atmospheric tests is judged to be small, since the radiation dose that individuals received from global fallout was less than they received from natural background radiation. In contrast, local fallout from atmospheric tests has had serious consequences. The most notorious case of injury from radioactive fallout occurred with the Bravo shot of March 1, 1954, at Bikini Atoll in the South Pacific. The

The site of a nuclear test blast conducted by India in early 1998. Concerns about the effects of radiation on humans and other organisms led to several international treaties designed to regulate or ban nuclear testing. (Reuters/Indian Government/Archive Photos)

yield of about 15 megatons was three to four times what was expected. Twelve hours before the test, the wind direction changed so that it would no longer carry the fallout safely out to sea. The decision to proceed despite the wind change shows that problems from fallout were vastly underestimated.

The nine-man firing team crouched in a bunker only 32 kilometers (20 miles) from ground zero. As their radiation meters began to show unexpectedly high amounts of fallout, they sealed their bunker. Navy ships going in to pick up the firing team and recover scientific data encountered such heavy fallout 50 kilometers (31 miles) out that they retreated at top speed. The fallout ceased after three hours. The ships were washed and returned to rescue the firing party. Covered with bed sheets sealed with masking tape to keep out fallout particles, the firing party drove 1 kilometer (0.6 miles) to the landing site and were taken to the rescue ship by helicopter.

Radiation doses are measured in grays (Gy). (The old unit for an absorbed dose was the rad; 1 gray equals 100 rads.) Normal background radiation results in a dose of around 0.002 Gy per year. Acute doses are those doses received over several days instead of over months or years. Acute doses of 0.25 Gy or less generally result in no obvious injuries. Radiation sickness and some obvious injuries begin to occur for acute doses between 0.5 and 1.0 Gy. An acute dose of 6.0 Gy is generally fatal. The men of the firing party received only 0.005 Gy. Had they been outside their bunker, they would have received more than a fatal dose during the first few hours.

Groups farther from Bikini were evacuated as the magnitude of their peril became clear. It was not until two days after the blast that eighty-two people were taken from Rongelap Atoll 200 kilometers (124 miles) east of Bikini. They received an estimated 1.0 Gy, and some were already showing signs of radiation sickness. A survey flight failed to detect significant radiation over Rongerik Atoll 240 kilometers (149 miles) east of Bikini, but sensors on the ground did. Again the magnitude of the problem was unclear so that the twenty-eight men at the weather station on Rongerik were not evacuated before they re-

ceived an estimated 0.3 to 0.5 Gy. The final evacuation was a group of 158 inhabitants on Utirik Atoll 500 kilometers (310 miles) from Bikini. They received an estimated 0.15 Gy.

The greatest radiation dose was received by twenty-three Japanese fishermen who unwittingly sailed their vessel, the *Lucky Dragon*, into the fallout cloud only 160 kilometers (100 miles) from Bikini. Not realizing that they should wash the fallout from the ship, their dose continued to increase during their two-week voyage to port. By then, their estimated dose was between 1.3 and 4.5 Gy. The ship's radioman, Aikichi Kuboyama, died seven months later of hepatitis, a complication brought on by his treatment for radiation sickness.

AFTEREFFECTS OF THE BIKINI TESTS

Further weapons tests were conducted at Bikini Atoll in 1956 and 1958. Most of the radioactivity had been swept away by nature or had decayed by 1967, and pressure mounted to return the refugees to their home. Radioactive debris from the tests, some plants, and some topsoil were removed from Bikini. New trees were planted, and new houses were built. The first group was resettled on Bikini in 1969. However, some water wells and some local foods were mildly radioactive. Two radioactive elements, strontium 90 and cesium 137, accumulated in the inhabitants' bodies to worrisome levels. They were evacuated again in 1978.

By the late 1990's, the lush vegetation made Bikini seem like paradise, and its former inhabitants were again anxious to return. It has been shown that the coconut trees will take potassium from the soil instead of the radioactive cesium; therefore, if the food trees are given a heavy dose of potassium fertilizer, and if a few centimeters of topsoil around the living areas are scraped away, it should be safe for the inhabitants to return. Responding to considerable pressure, the U.S. Congress eventually set aside $275 million for health care and other needs of the Marshall Islanders affected by nuclear tests.

Following World War II, five nuclear tests were carried out at the Bikini and Enewetak Atolls. However, logistic and security problems made it attractive to seek a test site in the continental United States. The Nevada Test Site was chosen because the government already owned the land, its surroundings were only sparsely populated, and it was close to the Los Alamos weapons laboratory. The best evidence available showed that off-site fallout from devices with yields less than 50 kilotons would be minimal. That proved to be too optimistic. Although the contamination was light, monitoring teams occasionally closed sections of state highways for a few hours or one day and had vehicles washed off. More seriously, on May 19, 1953, fallout near St. George, Utah, amounted to 0.05 Gy for people in the open. A dose that low causes no perceptible damage, but it may produce a slightly increased risk of cancer.

IODINE 131

Years of careful study show that iodine 131 is probably the most dangerous element of fallout. It accounts for 30 percent of the radioactivity produced by a nuclear bomb, but it rapidly decays. Its activity decreases by one-half every eight days, so that after one or two months, it has virtually disappeared. Two factors make it dangerous: Cows and goats grazing on plants dusted with fallout concentrate iodine in their milk, and iodine ingested by humans concentrates in their thyroids. This two-step concentration process means that significant doses to the thyroid might result from fallout levels previously considered harmless. U.S. scientists also learned that rainstorms or snowstorms could wash enough fallout from the air to create fallout hot spots as far away from the Nevada Test Site as the East Coast.

Scientists at the National Cancer Institute estimated per capita thyroid dose in the United States caused by iodine 131 from all the nuclear tests. They accomplished this by using available fallout measurements, the predictions of mathematical models, weather data, and information on patterns of milk consumption. During testing periods, children between the ages of three months and five years received thyroid doses three to seven times the average per capita because they drank more milk and because their thyroids are smaller. The estimated twenty thousand individuals who drank goats' milk were at greater risk, since goats' milk concentrates io-

dine ten to twenty times more than cows' milk. A much smaller dose of radioactivity may have come from eating other dairy products or leafy vegetables.

Thyroid cancer is rare, occurring in only three or four people out of 100,000. Officials at the National Cancer Institute estimate that nationwide, fallout might have caused 10,000 to 75,000 cases of thyroid abnormalities. It would be prudent for individuals who were born in the 1950's, drank a lot of milk, and lived in areas that received the most fallout to have their thyroid examined by a doctor.

This fallout was probably allowed to occur because of ignorance of its effects and because of the Cold War mentality. It was believed that some risk to the public was justified by the needs of the national defense. It also took awhile to collect convincing data that dangerous levels of iodine 131 from fallout would concentrate in the thyroid. Even those doses were no higher than those given by fluoroscopy examinations and some X-ray procedures in the early 1950's.

Scientists at the University of Exeter in the United Kingdom have pioneered an ingenious use of radioactive tracers. They estimate the rate of soil erosion and deposition on agricultural land by measuring the amounts of radioactive cesium at different locations. Tiny amounts of radioactive cesium were deposited worldwide during the years when nuclear weapons were tested in the atmosphere. In undisturbed permanent pasture and rangeland, the cesium remains near the surface. Scientists need only compare the cesium concentrations from various locations with that of an undisturbed site to determine the amount of erosion.

Charles W. Rogers

SUGGESTED READINGS: The essential and authoritative source of information on nuclear testing is the two-volume set sponsored by the Nevada Field Office of the U.S. Department of Energy written by Barton C. Hacker: *The Dragon's Tail: Radiation Safety in the Manhattan Project, 1942-1946* (1987) and *Elements of Controversy: The Atomic Energy Commission and Radiation Safety in Nuclear Weapons Testing, 1947-1974* (1994). The treatment is open, honest, and extensively documented. *The Day We Bombed Utah: America's Most Lethal Secret* (1984), by John G. Fuller, is a popular account that links fallout, cancer in humans, and the deaths of sheep in southern Utah. *Killing Our Own: The Disaster of America's Experience with Atomic Radiation* (1982), by Harvey Wasserman and Norman Solomon, has some good information but also cites the conclusions of controversial studies as facts. *Day of Two Suns: U.S. Testing and the Pacific Islanders* (1990), by Jane Dibblin, exaggerates the effects of fallout but vividly portrays the concerns of the Marshallese people.

SEE ALSO: Bikini Atoll bombing; Limited Test Ban Treaty; Nuclear weapons; Radioactive pollution and fallout.

Nuclear weapons

CATEGORY: Nuclear power and radiation

Nuclear weapons, first developed during World War II, have the potential to cause massive amounts of destruction and spread dangerous radiation around the world. The manufacture of such weapons also produces nuclear waste, which creates disposal problems. Several international agreements have been drafted in an attempt to limit the production, testing, and possession of nuclear weapons.

In 1942 a research group headed by Enrico Fermi built an atomic pile—literally a pile of graphite blocks interspersed with uranium, uranium oxide, and cadmium control rods. On December 2 they achieved a self-sustaining chain reaction in which neutrons from fissioning uranium nuclei caused a constant number of other uranium nuclei to also fission. The atomic pile was the direct ancestor of the nuclear reactor. Its successful operation opened the door to the production of plutonium and the plutonium route to nuclear weapons. Since the pile's constituents became radioactive, it also began a disposal problem.

The development of the atomic bomb was a feverish race driven by the fear that Nazi dictator

Adolf Hitler would reach the goal first. For the most part, reasonable precautions were taken to protect scientists and workers from dangerous exposure to radiation, but insufficient attention was given to environmental problems. The urgency of war dictated that environmental concerns be postponed until more time and resources were available. With the Cold War following on the heels of World War II, this attitude continued for a generation.

The Environmental Management program under the Department of Energy has the assignment of dealing with the legacy of nuclear weapons production. The *1996 Baseline Environmental Management Report* is a description of the management activities they believe will be necessary to restore the environment where possible, deal with nuclear waste, shut down facilities, and safeguard special nuclear material. There are thousands of contaminated areas and buildings, along with 500,000 cubic meters (17.6 million cubic feet) of various levels of radioactive waste. Since permanent disposal sites for high-level waste are not yet available, the worst sites will only be stabilized, not cleaned up. The estimated cost for this is $230 billion spent over seventy-five years. For comparison, the estimated cost of repairs after the 1995 earthquake in Kobe, Japan, was $100 billion.

Pollution problems begin with uranium ore. It is mined, then pulverized and leached to extract the uranium. After drying, the depleted ore, called tailings, consists of particles the size of sand grains or smaller. There are more than 242 million tons of tailings in the United States and more than 1.9 billion tons worldwide. The two main hazards from tailings are radon gas produced by the decay of radium and the contamination of groundwater with heavy metals by runoff. Proper control requires placing tailings piles away from population centers, covering them with earth to trap the radon, and holding the earth cover in place with grasses and other plants. Care must also be taken to prevent runoff from entering the water supply.

Uranium leaves the mill as a compound called yellowcake. For the weapons program, the yellowcake was converted to the gas uranium hexafluoride, and then the isotope uranium 235 was separated from the more common uranium 238 at factories in Oak Ridge, Tennessee. Metal highly enriched in uranium 235 was fabricated into weapons parts at Oak Ridge. Reactor-grade uranium was shipped to the Hanford Nuclear Reservation in the state of Washington, where large nuclear reactors converted some of it into plutonium.

At Hanford, spent reactor fuel, now highly radioactive, was dissolved and its plutonium extracted. Plutonium was shipped to the Rocky Flats plant in Colorado, where it was fabricated into weapons parts. These parts were shipped to the Pantex plant at Amarillo, Texas, where the weapons were assembled. Each plant produced radioactive waste and must now deal with contaminated areas and equipment.

There are 232 million liters (61 million gallons) of waste liquids and sludges stored at Hanford. The material is highly radioactive and includes chemicals such as nitrates, mercury, and cyanide. An estimated four million liters (one million gallons) have leaked into the ground from the oldest storage tanks. Believing that the pollution would not migrate beyond the large Hanford site, low-level waste, including 1,700 billion liters (449 billion gallons) of processing water, was poured directly onto the ground or into injection wells. The Hanford site will be the most difficult to clean.

Uranium and thorium were processed near Weldon Spring, Missouri. Forty-four buildings and several waste-disposal pits were contaminated. The buildings have been demolished, and 170,000 cubic meters (six million cubic feet) of radioactive sludge is being stabilized. It will be retained on-site in a special disposal cell. A similar site near Fernald, Ohio, had two hundred structures and 2,400,000 cubic meters (85 million cubic feet) of waste. Cleanup of Fernald will involve vitrification (making it an ingredient of

A mushroom cloud forms over the Japanese city of Nagasaki on August 9, 1945, moments after the United States dropped the second atomic bomb ever used in warfare. (National Archives)

glass) of the high-level waste, shipping some waste elsewhere, and storing slightly contaminated soil on-site in a special facility. This is expected to take twenty-five years and cost $4.2 billion.

Full-scale nuclear war would do massive amounts of damage to the environment in a short time. Dust equaling that of the 1883 eruption of Krakatoa volcano would be lofted into the stratosphere. This dust, along with smoke from forest fires and noxious smoke from burning cities, would dim the sunlight reaching the ground, causing a nuclear winter. Nitrous oxide formed in nuclear fireballs would produce acid rain and, if it reached the stratosphere, would reduce the ozone layer for a few years. If targeted, a large fraction of farmland could be made too radioactive for use for weeks or months. As of 1999, the worldwide supply of nuclear weapons was less than 6,000 MT (million tons of high explosive). This is probably not enough to produce nuclear winter. Except radioactive hot spots, the worst environmental effects of nuclear war would be over in one year or perhaps a few years. Most of the casualties would not occur during the nuclear exchange, but afterward from starvation and disease. One-half of humankind could die, and it might take generations to recover.

Charles W. Rogers

SUGGESTED READINGS: *Life After Nuclear War: The Economic and Social Impacts of Nuclear Attacks on the United States* (1982), by Arthur M. Katz, is a popular assessment of the real problems that would follow nuclear war. *Nuclear Wastelands: A Global Guide to Nuclear Weapons Production and Its Health and Environmental Effects* (1995), edited by Howard Hu, Arjun Makhijani, and Katherine Yih, is a useful, although biased, source. For those who wonder how the Hanford site became such a problem, Michele Gerber's *On the Home Front: The Cold War Legacy of the Hanford Nuclear Site* (1997) tells the story.

SEE ALSO: Bikini Atoll bombing; Hanford Nuclear Reservation; Neutron bombs; Nuclear testing; Nuclear winter; Rocky Mountain Arsenal.

Nuclear winter

CATEGORY: Nuclear power and radiation

A nuclear winter is a severe change in global climatic patterns resulting from either a nuclear war or a cosmic impact event.

The nuclear winter theory predicts that a global nuclear war would cause huge amounts of dust and debris to be carried high into the atmosphere, where it could remain for as long as three years. This debris would obstruct incoming sunlight and disrupt photosynthesis, causing the extinction of many plant species. The loss of considerable amounts of plant life would eventually affect the entire food chain. Once the atmosphere cleared, only those species that could survive without sunlight would be left to repopulate the planet.

The theory of nuclear winter began in the early 1980's as a result of the ever-present fear of global nuclear war. Several prominent scientists, notably Carl Sagan of Cornell University, began to examine the possible effects of total nuclear war on the environment. Using computer simulations, they predicted that an enormous amount of dust and particulate matter from the nuclear explosions and resulting fires would be lifted high into the atmosphere. This debris would be quickly distributed around the globe by the earth's prevailing winds and would remain in the atmosphere for a minimum of three months to a maximum of three years. It was predicted that the majority of nuclear impacts would take place in the Northern Hemisphere, thereby giving the Southern Hemisphere a six-month "grace period" before it felt the full effects of the atmospheric contamination.

The loss of sunlight would affect plant and animal life. Within a matter of months, the photosynthetic cycle and the natural food chain would be totally disrupted. This would be accompanied by a worldwide drop in temperature, followed by a dramatic increase in the need for heating fuel, which would already be in short supply from the destruction of energy industries, oil fields, and forests. Many writers who studied the results of the computer simulations pointed out that the people who died from the exchange of nuclear weapons would be the lucky ones. Those who survived the bombs would have to face starvation and freezing temperatures.

A test of the nuclear winter concept resulted from the Persian Gulf War of 1991. Huge clouds of black smoke from thousands of burning oil wells entered the atmosphere. From space, astronauts clearly observed distribution patterns as the smoke slowly made its way around the world. Locally, the people of the Middle East experienced a small-scale nuclear winter, which gave scientists a first-hand opportunity to study Sagan's theory.

Smoke rises from Kuwaiti oil wells that were ignited by retreating Iraqi troops in 1991. As a result of the air pollution, the Persian Gulf region experienced the equivalent of a small-scale nuclear winter. (Reuters/Pat Benic/Archive Photos)

As the concept of nuclear winter was being discussed, a natural example of the same effect became apparent from asteroid impact. A large impact structure was found on the Yucatan Peninsula in Mexico dating back 65 million years to the extinction period of the dinosaurs. Scientists quickly employed the scenario of a nuclear winter to dramatize the last days of the dinosaurs.

They also pointed out that several other mass extinction events had previously occurred and that it was likely to happen again.

Paul P. Sipiera

SEE ALSO: Antinuclear movement; Climate change and global warming; Gulf War oil burning; Nuclear weapons.

Oak Ridge, Tennessee, mercury releases

DATE: 1950's; discovered December, 1981
CATEGORY: Human health and the environment

During the 1980's researchers discovered that mercury from a nuclear weapons plant had contaminated air and water in Oak Ridge, Tennessee. Subsequent study of the contamination raised questions about the methods used to determine the risk of toxic pollution to humans.

In December, 1981, Stephen Gough, a biologist employed at Oak Ridge National Laboratory, tested the amount of mercury in plants growing near East Fork Poplar Creek and found it to be high. Since the creek was near the Department of Energy's (DOE) Y-12 plant, where nuclear bombs had been built in the 1950's, the possibility of a large amount of mercury having been released into the environment during this time was raised.

During 1982 the DOE commissioned a study to determine how much mercury had been released. They discovered that an estimated 215,000 kilograms (475,000 pounds) had been released into the creek and that another 907,000 kilograms (two million pounds) were unaccounted for. The DOE released their report to

The Oak Ridge National Laboratory was built in 1943 as part of the national effort to build nuclear weapons. During the 1950's large amounts of mercury leaked from the facility into the surrounding environment. (Oak Ridge National Laboratory)

the public on May 17, 1983, after a small local paper and the Tennessee Department of Health and Environment requested it under the Freedom of Information Act. After a public hearing on June 11, 1983, the DOE was reprimanded by the House Science and Technology Committee on November 3, 1983. It was also ordered to assess human health risks and accurately estimate how much mercury was released into the creek and the atmosphere.

The greatest risk for human health is mercury in the methylated form. A leak of this type occurred in Minamata, Japan, in the 1960's, where many people who drank contaminated water became paralyzed. The DOE claimed that the mercury at Oak Ridge was not methylated, but in elemental form, so the risk was minimal. The DOE also claimed that the mercury was released below the public drinking water intake and that the mercury traveled downstream before sinking into the earth, where it was trapped by a layer of shale before entering the groundwater. Another danger is the consumption of contaminated fish, which were determined to have twice the level of mercury deemed safe by the Food and Drug Administration (FDA); therefore, fishing in Poplar Creek was banned.

A report released by the DOE in 1997 stated that 127,000 kilograms (280,000 pounds) of mercury had entered the East Fork Poplar Creek and that 33,000 kilograms (73,000 pounds) of mercury had entered the atmosphere over several years, with peak years being 1957 and 1958. The study also estimated vegetable and milk production and consumption near the creek during critical years, since the airborne mercury could have contaminated gardens, crops, and grass eaten by cattle. The mercury in the soil was determined to be mercuric sulfide, a stable compound that was not likely to migrate to centers of population.

The DOE's report concluded that small doses of mercury had been ingested by local residents during the decades following 1950 but that such low levels did not pose a health risk. Controversy followed the report, as health experts claimed that individuals react differently to mercury ingestion and that local immune disorders and diseases could be the result of the mercury release.

Rose Secrest

SEE ALSO: Heavy metals and heavy metal poisoning; Mercury and mercury poisoning; Minamata Bay mercury poisoning.

Ocean dumping

CATEGORY: Waste and waste management

The disposal of sewage, dredging sludge, garbage, chemicals, and other waste into the ocean disrupts individual ecosystems and kills fish and marine life. Though marine pollution treaties and laws regulate dumping, such legislation is difficult to enforce.

The ocean is the final stop for much of the garbage generated on land. Sewage processing centers release treated wastewater into it. Waste management officials at overflowing landfills look to the ocean as an alternative for burying rubbish and debris. Oil tankers rinse and flush holding compartments in the open ocean. Fishing vessels dump scrap parts and unwanted or spoiled fish into the water. All these sources of pollution are thought to have little effect on the ocean because it is so large. Whether in large or small amounts, however, this dumping upsets the balance of the ocean ecosystem. The environment to which fish, plants, and other marine organisms are accustomed is altered. Some survive the change, while others do not.

Industrial and manufacturing plants find the ocean useful for the disposal of waste products as well. Some companies seal waste chemicals into metal containers and dump them onto the ocean floor, where they believe the containers will not disturb anything. However, the ocean floor is home to crabs, ground fish, and other marine creatures that are disrupted by these alien containers. Over time some of these containers have begun to leak, sometimes releasing radioactive waste and other chemicals harmful to the marine environment.

The majority of garbage dumped into the ocean comes from ships and boats. The number of fishing boats, naval vessels, cruise ships, cargo ships, and recreational boats taking to the sea daily is estimated to be in the millions. Many of these vessels are like floating towns. With thousands of people on board, sewage, garbage, and other waste can quickly accumulate, and space for storing it is often limited. Some ships, such as fishing and naval vessels, remain on the open ocean for weeks or months. It is therefore not surprising that many of these vessels solve their garbage problems by dumping it overboard. Though the disposal of sewage and wastewater is illegal, many seafaring vessels ignore such laws because they are difficult to enforce.

Of the types of garbage dumped, plastics are among the most dangerous to marine creatures. Plastic does not degrade. Medical equipment, such as needle syringes and wrappers, is often made of plastic and washes up on beaches months after being dumped into the deep ocean. Marine birds and mammals often mistake plastic bags and pellets for food. Though the animal feels full after eating plastic, the material offers no nutritional value, and the animal dies. Sea turtles and whales have died after eating plastic bags that became entangled in their intestines. Marine birds and seals have starved to death after plastic beverage rings got stuck around their beaks or snouts. During the late 1980's, an estimated 7.3 billion kilograms (16 billion pounds) of plastic was dumped into the ocean annually.

In December, 1988, the International Convention for the Prevention of Marine Pollution banned the dumping of plastics anywhere in the ocean. Referred to as the MARPOL treaty, it also bans dumping or littering of any kind within 5 kilometers (3 miles) of shore and provides specific regulations on the type of garbage that is legal to dump into the ocean. For example, it is illegal to dump floating plastic and packing material between 5 and 20 kilometers (3 and 13 miles) from shore. Paper, glass, metal, pottery, and food that is dumped within this zone must be ground to less than 2.5 centimeters (1 inch) in size. At distances beyond 40 kilo4meters (25 miles) from shore, only plastic is illegal to dump. The MARPOL treaty was signed by the United Kingdom, Canada, and the United States, among other countries. It calls for stiff penalties, including fines of up to US$50,000

and possible imprisonment for those in violation.

Most refuse dumped into the ocean returns to shore on the currents. Conservation groups annually clean this litter from the beaches in programs called "beach sweeps." They record the amount and types of material recovered. Plastic products continue to make up 60 percent of the debris gathered.

Lisa A. Wroble

SUGGESTED READINGS: A valuable overview of the ocean and the marine ecosystems that are disrupted by dumping is available in *Exploring Earth's Biomes: Ocean* (1996), by April Pulley Sayre. Martha Gorman's *Environmental Hazards: Marine Pollution* (1993), includes chapters on deep ocean dumping and its effect on the ocean. For thorough explanations of the many resources being destroyed by ocean dumping, see *The Wealth of Oceans* (1995), by Michael Weber and Judith A. Gradwohl, and *Endangered Animals and Habitats: The Oceans* (1998), by Lisa A. Wroble. Peter Weber's *Abandoned Seas: Reversing the Decline of the Oceans* (1993), focuses on the destruction caused by ocean dumping and how the ocean can be revitalized. The Internet also has many resources on ocean dumping. Ocean Planet (http://seawifs.gsfc.nasa.gov/ocean_planet/html/ocean_planet_overview.html) is a site based on a traveling exhibit created by the Smithsonian Institution. A section on the "Oceans in Peril" includes information on marine pollution and chemical and plastics dumping.

SEE ALSO: Dredging; Marine Mammal Protection Act; Ocean pollution; Seabed disposal.

Ocean pollution

CATEGORY: Water and water pollution

The introduction of harmful substances or energy into the ocean can have troublesome effects on people and the marine environment. Of these pollutants, 80 percent come from land-based sources, 10 percent come from dumping, and 10 percent come from maritime operations.

Five major types of pollutants can be found in the oceans. The first type, degradable waste, makes up the greatest volume of discharge and is composed of organic material that will eventually be reduced to stable inorganic compounds such as carbon dioxide, water, and ammonia through bacterial attack. Included in this category are urban wastes, agriculture wastes, food-processing wastes, brewing and distilling wastes, paper pulp mill wastes, chemical industry wastes, and oil spillage. The second type of pollutant is fertilizer, which has an effect similar to organic wastes. Nitrates and phosphorus compounds are carried into the ocean by rivers. Once in the sea, they enhance phytoplankton production, sometimes to the extent that the accumulation of dead plants on the seabed produces anoxic (lack of oxygen) conditions.

Dissipating wastes are mainly industrial wastes that lose their damaging properties after they enter the sea. Therefore, their effects are confined to the immediate area of the discharge. Some examples are heat, acid, alkalis, and cyanide. Another pollutant is particulates. These are small particles that may clog the feeding and respiratory structures of animals. They can also reduce plant photosynthesis by reducing light penetration; when settled on the ocean bottom, they may smother animals and change the nature of the seabed. Some examples are dredging spoil, powdered ash from coal-fired power stations, china clay waste, colliery waste, and clay from gravel extraction. Particulates also include larger objects, such as containers and plastic sheeting.

The last type of pollutant is conservative waste. These include categories such as heavy metals (mercury, lead, copper, and zinc), halogenated hydrocarbons, and radioactivity. These are not subject to bacterial attack and are not dissipated, but they can react with plants and animals.

POLLUTION INPUTS

Pollutants can reach the sea in many ways. Rivers that have been polluted by land runoff are the largest source of harmful substances. River water enters the sea carrying all the materials it has accumulated. Nutrients from fertilizers and livestock, as well as silt from eroding

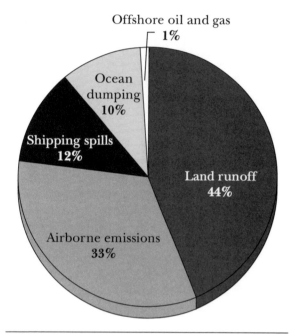

Marine Pollution Sources (percentage by weight)

Offshore oil and gas 1%
Ocean dumping 10%
Shipping spills 12%
Airborne emissions 33%
Land runoff 44%

Source: Data taken from United Nations Environment Programme, *The State of the Marine Environment.* Oxford, England: Blackwell Scientific Publications, 1991.

fields, are the leading pollutants. Herbicides and pesticides that drain from fields can fog even the clearest water. Herbicides are suspected carcinogens. Cars also contribute to runoff. They leave behind oil, grease, zinc, and copper.

Another type of input is direct outfall from pipes that empty into the sea. This pollution comes from urban and industrial wastes that are deposited directly into estuaries. During the nineteenth century rivers such as the Thames, Tees, Clyde, and Hudson became incredibly polluted by this type of input. Now, with increasing pressure to preserve inland sources of drinking water, new industries that need large quantities of water are located on the coast, increasing the level of ocean pollution. Urban areas on the coast also present a problem with untreated municipal wastes and sewage. Even though most wastes are treated, 60 percent of the world's population lives on the coast, often in fertile and productive estuarine and delta areas. This makes

The 1989 Exxon Valdez *oil spill resulted in widespread ecological damage. Although they gain media attention, such spills are a relatively minor source of ocean pollution.* (AP/Wide World Photos)

Another input comes from offshore activities such as dumping dredging spoil and sewage sludge. Dredging spoil comes from shipping channels and entrances to ports that require dredging to keep them open. Dredged material is usually barged out to sea to be dumped. Dredging spoil from harbors contains large quantities of heavy metals and other contaminants that are transferred to dumping grounds. Sewage sludge from the treatment of land-based sewage has been dumped at sea in large quantities. This sludge is contaminated with heavy metals, oils and greases, and other substances. Offshore industrial activities that can pollute the ocean include oil exploration and extraction, gravel extraction, and, in the future, extraction of minerals from the seabed. Other harmful wastes that are dumped at sea include ashes from power stations and colliery (from coal) waste, along with various dangerous materials such as radioactive wastes and old ammunition.

The last type of input comes from the air. While this type of input is complex and not well studied, scientists feel that it represents a large and major contribution to pollution. Atmospheric contaminants may exist as gases (mercury, selenium, and boron) or aerosols (most other metals). They are deposited by gas exchange at the sea surface, fall out as particles (dry deposition), or are scavenged from much of the air column by precipitation within clouds (wet deposition).

EFFECTS AND PREVENTION EFFORTS

Nutrients from sewage, treated or untreated, are increasing the rate of plant growth in coastal waters, a phenomenon known as cultural eutrophication. Sewage and agricultural runoff are high in nitrogen and phosphorus, which encourages ocean plants to grow. This results in unsightly algal scum on beaches. When the extra algae die and sink to the bottom, the resulting bacterial decomposition uses available dissolved oxygen, causing deoxygenation on the

it almost impossible to treat all waste before it enters the sea.

Ships can pollute the ocean in many ways. They often carry toxic substances such as oil, liquefied natural gas, pesticides, and industrial chemicals. Shipwrecks can lead to the release of these chemicals and result in serious damage. An example is the 1989 *Exxon Valdez* oil spill, which caused the deaths of many sea animals. Noxious or dangerous materials are frequently carried on deck as a safety precaution and can be lost overboard during storms. During routine operations, ships discharge oily ballast water, bilge water, and cargo tank washings (not always legally). They also discard much of their litter overboard.

bottom waters. In extreme cases, this can kill fish. Algae blooms often pick up toxic substances. The algae is then eaten by shellfish, which makes the shellfish unfit for human consumption.

Water-soluble compounds from crude oil and refined products include a variety of compounds toxic to a large number of marine plants and animals. Chemicals released during oil spills are toxic to a wide range of plankton, which is at special risk because it lives near the water's surface. Salt marshes and mangrove swamps trap oil from oil spills. When the plants are in bloom, the oil coats the blossoms, which rarely produce seeds. Anywhere from 10 to 100,000 seabirds die from contact with oil in the northeast Atlantic Ocean every year. Other harmful substances from shipping include antifouling agents, which are known to have major effects on the reproductive biology of shellfish and are banned in some countries.

Discarded plastic items can cause harm to the larger animals of the sea. They can trap mammals, diving birds, or turtles, causing them to drown. Also, a large amount of fishing gear is lost every year. The nets continue to snare fish long after they are lost.

Conservative pollutants are not affected by bacteria; when consumed by animals, they cannot be excreted and thus remain in the body. This process is called bioaccumulation. Animals that eat polluted plants and animals receive an enriched diet of those conservative materials, which they also are unable to excrete. The pollutants eventually accumulate to toxic levels in their bodies, leading to death. This process, called biomagnification, particularly affects the ocean predators.

Although marine pollution is still considered a major cause of concern, a number of things can be done to minimize the pollution and its effects. The greatest problems are found in coastal areas. Political solutions to coastal pollution are being proposed and implemented. Agenda 21, approved by governments attending the Earth Summit held in Rio de Janeiro in 1992, called for all coastal countries to develop integrated coastal zone management plans by the year 2000. The United Nations Law of the Sea Treaty addressed the issue of pollution on the open sea. Some countries have begun to control pollution in their estuaries and bays, and fish and shellfish are returning to formerly polluted areas. Maritime ship-based activities have considerably less impact on the ocean than in the past, mainly because of the introduction of international treaties limiting the discharge of wastes at sea. In addition, schools, aquariums, nonprofit groups, and governments are all working to educate the public about the need to prevent ocean pollution.

David MacInnes, Jr.

SUGGESTED READINGS: For a well-illustrated, general introduction to ocean pollution, see John Pernetta's *Atlas of the Ocean* (1994). Interesting and nontechnical accounts can be found in David Bulloch, *The Wasted Ocean* (1989), and K. A. Gourlay, *Poisoners of the Seas* (1988). Two readable accounts can be found in Michael Parfit and Jim Richardson, "Troubled Waters Run Deep" *National Geographic* 184 (November 1, 1993), and John G. Mitchell, "Widespread as Rain and Deadly as Poison: Our Polluted Runoff," *National Geographic* 189 (February, 1996). A technical but readable source is R. B. Clark, *Marine Pollution* (1989). A more balanced and thoughtful work is Peter Weber's *Abandoned Seas: Reversing the Decline of the Oceans* (1993).

SEE ALSO: Dredging; Minimata Bay mercury poisoning; Ocean dumping; Oil spills; Seabed disposal.

Odor control

CATEGORY: Pollutants and toxins

Odor complaints are one of the top citizen pollution concerns in many cities in Europe and the United States; about 10 percent of Americans complain about odor pollution problems. Although a few odors signal the presence of hazardous air pollutants, complaints often arise because of obnoxious nontoxic stenches. However, malodorous substances and their sources are often difficult to identify and ameliorate.

Odor pollution results from a combination of cultural expectations and the ability of people to perceive various odors. Cultural expectations play a strong role in odor pollution problems. Distinctive odors may awaken memories. For example, people who visit foreign countries may leave with unforgettable olfactory experiences, while local residents remain oblivious to what visitors deem malodorous situations. Vietnam veterans returning to visit Southeast Asia after the war often recount experiences tied to a particular odor that they have not smelled for decades.

The human nose can detect tiny concentrations of certain chemicals, rendering qualitative and quantitative analyses to identify the malodorous substances inconclusive. Mercaptans rank high on the list of odor pollution complaints, which is not surprising given the widespread industrial applications of these chemicals and the ability to smell minute concentrations. Ethyl mercaptan is detectable by smell at concentrations of fewer than 10^{-6} milligrams per liter of air; much higher concentrations are needed for detection by standard chemical tests.

A variety of chemicals emanating from many sources often contribute to an odor problem. Establishing a single cause for an odor problem is difficult because people who are able to identify a particular odor when it is found alone are often unable to identify it as a component in a complex mixture. Another factor is the concentration of a substance, which may influence the acceptability of the odor. This is particularly true of indole, which is classed as pleasant-smelling in minute concentrations but is overwhelmingly unpleasant in moderate and high concentrations. Because indole is a decomposition product of tryptophan, which is used as a chemical reagent and in the manufacture of perfumes and pharmaceuticals, it is a potential odor pollutant in the vicinity of those industrial settings.

Throughout the United States, new housing subdivisions are encroaching into areas traditionally reserved for agriculture. Recently transplanted urban dwellers may suddenly complain about a range of odors resulting from farming activities, such as mowing hay and manuring fields. The solution to this type of odor problem may be to alter the expectations of new residents or advise them to use air conditioners rather than to open their windows in an attempt to breathe "fresh air."

Within cities, odor pollution often arises because of poor methods of disposal of garbage and food residues. Large urban areas often develop task forces that combine the resources of the pollution control office, the city sewer department, and refuse collectors to ameliorate the source of stenches. Investigations of vile odors in San Francisco, California, have sometimes identified the main culprit to be aging butter discharged into the sewer system by restaurants. Odors emanating from San Francisco sewer systems are eliminated by using mobile Vactor units (similar to those used to clean up toxic spills) to remove the source of the stench and by spraying the area with disinfectant.

Many different industries have been identified as contributing to odor pollution, including food processing and paper manufacturing factories, refineries, power plants, and waste disposal operations. Pollution control officers frequently find that similar manufacturing facilities have different odor problems. Apparently, differences in effluent gases from the facilities may produce differing odor strengths, resulting in different degrees of annoyance among nearby residents.

The kraft paper industry has received considerable attention for its emissions of highly odorous and unpleasant sulfurous gases. In the manufacture of kraft paper, hydrogen sulfide (H_2S) is the major odor pollutant. However, the presence of nitric oxide (NO) enhances the unpleasant perception of H_2S, which is more readily sensed in an acid gas mixture than in an alkaline gas mixture. Carbonyl sulfide (COS) and sulfur dioxide (SO_2) also affect the perceived odor strength of the effluent gases. A kraft paper operation that controls the acidity of effluent gases may generate fewer odor pollution complaints than the industry average. Asphalt plants are another notorious odor source. When these plants are fueled by recycled oil, both the energy source and the product may contribute to odor problems. One technique sometimes employed at these facilities is the use of Ecosorb to neutralize emissions of hydrocarbons and sulfur dioxide.

Masking of an unwanted or unpleasant odor by a neutral or pleasant odor is widely practiced; individuals may apply perfumes to cloak perspiration odors, while stores and offices may use "fresh scent" dispensers in their ventilation systems. Industries have even tried releasing masking odors along with known odor pollutants, with dubious success. During atmospheric inversion conditions, all air pollutants increase in concentration, including odors. Transportation-related odors, including diesel and automobile exhaust, mingle with ozone created during photochemical smog. During these photochemical air pollution episodes, the acrid odors signal a real public health threat.

Anita Baker-Blocker

SUGGESTED READINGS: An excellent introduction to the human ability to perceive odors is given in Boyd Gibbons, "The Intimate Sense of Smell," *National Geographic* (September, 1986). *Olfaction and Taste VI* (1977), edited by J. LeMagnen and P. MacLeod, contains several relevant essays, including T. Lindvall's "Perception of Composite Odorous Air Pollutants," which provides technical information on smell. In the same volume, G. Winneke and J. Kastka, "Odor Pollution and Odor Annoyance Reactions in Industrial Areas of the Rhine-Ruhr Region," discusses industrial odor pollution problems.

SEE ALSO: Air pollution; Pulp and paper mills; Sewage treatment and disposal.

Oil crises and oil embargoes

CATEGORY: Energy

Oil crises and oil embargoes involve disruptions of oil supply patterns linking the Middle East to the industrialized West, typically during times of political tension in the Persian Gulf region. Such episodes have strengthened support in Western nations for greater domestic energy production and military policies enhancing oil security in the Persian Gulf.

Although the industrialized West depends on oil as a fuel and chemical feedstock, the majority of the world's long-term oil reserves are located in the Middle East. Therefore, the West depends on a smooth flow of oil through world markets. On at least three occasions during the last few decades of the twentieth century, political events in the Middle East led to crises in the world oil markets. These crises illustrated the West's vulnerability to supply interruptions and price increases.

WORLD OIL MARKET INTERRUPTIONS
In the early years of international oil markets, the United States was the leading producer, accounting for more than one-half of the world's oil production until the 1950's. U.S. oil production, however, peaked at 11.3 million barrels per day in 1970 and began to decline as lower-cost reserves in the Middle East were tapped. Leading oil-producing nations in the Middle East had met in 1960 to form the Organization of Petroleum Exporting Countries (OPEC), but the new organization was largely ignored by the world community. Because oil flowed without significant interruption, there was also little notice of the increasing dependence of the West on Middle Eastern oil into the early 1970's.

The security of oil supply became an immediate world issue on October 6, 1973, with an attack by Egyptian forces against Israeli positions along the Suez Canal. The United States supported Israel in the brief war that followed. In retaliation, King Faisal of Saudi Arabia ordered a 25 percent cut in Saudi oil output and a cutoff of shipments to the United States. The other Arab members of OPEC joined in the embargo and production cutbacks. World oil prices roughly quadrupled, from around three dollars per barrel to twelve dollars. In the United States, emergency conservation measures were implemented, and consumers waited in long lines for limited supplies of motor fuels. The embargo ended about six months later, on March 18, 1974.

The Western economies recovered, and oil prices were relatively stable from 1974 to 1978. Oil demand resumed its growth, and oil again flowed to the West. In 1979, however, the Iranian revolution removed much of Iran's production

Oil Crises and Oil Embargoes

YEAR	EVENT
1960	Oil-producing nations holding more than 90 percent of known oil reserves form the Organization of Petroleum Exporting Countries (OPEC) to seek higher and more stable crude oil prices.
1970	U.S. oil production peaks. After being the dominant oil producer in the first half of the twentieth century, the United States finds its production declining and begins to import an increasing fraction of its oil.
1970	Libya receives higher prices for its crude oil from Occidental Petroleum.
1973	World oil prices soar after Arab members of OPEC declare a cutback in production and an embargo on shipments bound for the United States and other allies of Israel.
1979	Oil prices double after the fall of the shah of Iran's government removes Iranian capacity from production.
1990	Oil prices briefly reach previous record highs after Iraqi forces invade Kuwait and take over Kuwaiti oil fields.
1991	Operation Desert Storm forces—led by the United States—defeat Iraqi forces, which set numerous oil well fires before retreating.

from world markets at a time when there was little excess capacity. World oil prices again rose sharply, reaching a peak of nearly thirty-five dollars per barrel by 1981. During this incident, unlike 1973-1974, there was no attempt to declare an embargo. OPEC members merely raised prices and continued to ship oil.

The high prices in the aftermath of the Iranian revolution did not persist, as a worldwide surplus of oil drove prices lower during the 1980's. A third crisis, the Persian Gulf War of 1990-1991, elevated prices so temporarily that it became known by some analysts as "an oil spike," referring to the sharp upward, then downward, movement in oil prices. The upward movement was precipitated by Iraq's invasion of Kuwait in August, 1990. World oil prices reached their highest level in more than eight years, briefly peaking near the forty-dollar-per-barrel mark. Saudi Arabia and other producers began to replace Kuwait's lost output, and prices fell, with a dramatic drop after Operation Desert Storm defeated Iraqi forces in Kuwait in January, 1991. Even the burning of Kuwaiti oil wells by Iraqi

forces was unable to keep world prices at their previous high levels.

OIL CRISIS EFFECTS

The early stages of all three crises were characterized by panic buying, with producers and consumers scrambling to assure supplies. Later research showed that the crises affected price more than physical availability. Even during the most severe of the three episodes (1973-1974), the ordered cutbacks were fewer than 10 percent of world oil supply. Further, after the embargo ended there was evidence that important quantities of supposedly embargoed oil had reached the United States and other Western nations.

During each of the three crises, Western economies entered recessions. Higher oil prices acted as a tax that transferred massive amounts of wealth from consuming nations to producing nations. Later declines in world oil prices had the opposite effects, stimulating the economies of the consuming nations.

The 1973-1974 embargo led to marked

changes in energy policy. During the embargo, U.S. president Richard M. Nixon announced Project Independence. This program was designed to achieve self-sufficiency in energy through crash programs along the lines of the Apollo moon missions and the Manhattan Project to develop the first nuclear weapons. Project Independence called for major development of nuclear and coal energy resources, but with a lower priority on conservation and environmental concerns. Similar policies continued during the administration of President Gerald R. Ford but lost urgency as the embargo's effects were overcome. The Iranian disruption of 1979-1980 occurred during the term of U.S. president Jimmy Carter, adding force to Carter's characterization of energy problems as "the moral equivalent of war." Carter called for conservation and new technology to engineer a transition as important as the transition from wood to coal during the Industrial Revolution. By the time of Operation Desert Storm in 1991, policy concern had shifted to the ability of Western nations to use military force to keep oil flowing. Iraq's initially successful invasion of Kuwait and the implied threat to Saudi Arabia quickly generated support for a military solution by a coalition of nations led by the United States.

KNOWLEDGE GAINED FROM OIL CRISES

Although the oil crises have been extensively studied, their message for the future and the policy lessons to be learned from them are uncertain. One viewpoint holds that oil crises are an early warning of resource depletion. The 1973-1974 price increases were initially seen as a signal confirming predictions that the prices of nonrenewable resources would increase as depletion approached. This "depletionist" viewpoint called for government-enforced conservation of existing resources and development of new technologies to postpone or avoid the consequences of oil depletion.

A second interpretation of the oil crises is that instead of reflecting a fundamental problem of depletion, they reflect a problem with distribution; that is, oil resources occur in patterns that are different from their patterns of use. Since large amounts of oil resources are held in the politically unstable Middle East but used in the West, cutoffs and price increases are always possible. The policy problem is viewed as a politically inspired cutoff of oil shipments. There is no threat, in this view, of a calculated economic cutoff of oil because producers would find it more profitable to sell oil at a high price than to cut it off altogether. An embargo would make sense only to demonstrate control

A 1973 oil embargo by Arab members of the Organization of Petroleum Exporting Countries (OPEC) led to fuel shortages in the United States. (Library of Congress)

of a resource and the will to sustain price increases.

A third interpretation characterizes the oil crises as a simple product of monopoly power or conspiracy. During the 1970's oil crises—with their long lines for gasoline—and the rapid 1990 price increases that followed Iraq's invasion of Kuwait, such explanations were common. Early versions of this viewpoint denied resource scarcity as a problem and called for the breakup of oil companies to end the conspiracy. Later, as Middle Eastern oil-producing countries assumed control over operations from Western oil companies, the focus shifted to policy actions to combat the OPEC monopoly.

OPEC POWER

The degree of perceived power held by the oil cartel OPEC has fluctuated considerably since its formation in 1960. OPEC nations held 90 percent of the world's then-known oil reserves. Although this guaranteed the OPEC nations a key role in the future as reserves outside OPEC were depleted, the organization was weak during its first ten years. A hint of OPEC's possible power came in 1970 when a bulldozer accident broke a pipeline carrying oil from the Persian Gulf to the Mediterranean Sea. In the aftermath of the accident, which caused tight oil supplies, Libya insisted on and received higher prices for its crude oil.

During the 1973-1974 embargo, OPEC's power was little questioned. The sharp upward movement of oil prices seemed to prove that OPEC had the ability to withhold enough oil to keep prices permanently higher. Further confirmation came with the 1979-1980 incident, when oil prices again doubled. Recessions following the oil crises made the West appear dependent on OPEC's indulgence for its future prosperity.

The collapse of oil prices after the three crises led to a different interpretation. In this view, oil prices were kept artificially low before 1970 by the rapid pumping of oil from Middle Eastern reserves by Western-based companies. The companies, in this view, were seeking profit with little regard for the long-term future of the resource, since they realized that power would soon shift away from them. After 1970, as the host nations

began to realize their power, oil prices recovered from their artificially low levels. The extraordinary oil price increases of 1973-1974, in this view, were from artificially low levels to normal levels. At the time they occurred, they had been viewed as increases from normal levels to artificially high levels. Some analysts came to view OPEC as simply announcing, rather than causing, higher oil prices.

Energy analysts have observed that oil crises and embargoes have had a common pattern. First, demand growth catches up with world capacity, leading to generally tight market conditions. Next, a precipitating event occurs in the Middle East, leading to quick price increases and disruptions of ordinary supply channels. If there is no permanent loss of supply, however, the disruption soon ends. At the time of the crisis, policy priority goes to energy security over environmental concerns. With the passage of time, however, energy security loses some of its prominence as a policy goal.

William C. Wood

SUGGESTED READINGS: A definitive history of OPEC and the oil crises of the 1970's is provided in Albert L. Danielsen's *The Evolution of OPEC* (1982). For an easy-to-read account of resource depletion in this context, see W. Jackson Davis, *The Seventh Year: Industrial Civilization in Transition* (1979). A highly influential volume on energy security is *Energy Future: Report of the Energy Project at the Harvard Business School* (1979), edited by Robert Stobaugh and Daniel Yergin. Useful perspectives and history are found in President Jimmy Carter's speech of April 18, 1977, "The Energy Problem," noted for its declaration of the energy crisis as "the moral equivalent of war." An influential opposing article is David Stockman, "The Wrong War? The Case Against a National Energy Policy," *The Public Interest* (Fall, 1978). An overview of the policy setting of the oil crises is found in James M. Griffin and Henry B. Steele's *Energy Economics and Policy* (1986). A comprehensive treatment, including material on the Persian Gulf War of 1991, is found in Roger A. Hinrichs, *Energy: Its Use and the Environment* (1996).

SEE ALSO: Energy policy; Fossil fuels; Gulf War oil burning; Oil drilling; Trans-Alaskan pipeline.

Oil drilling

Category: Energy

Drilling for oil and gas requires the use of a rotary drill rig or platform and generally involves site preparation, equipment setup, drilling, fluid circulation, and waste disposal. All of these processes have the potential to degrade the environment.

The process of drilling a hole through the rock layers that overlie a petroleum reservoir is fairly simple, at least in principle. A diamond-tipped drill bit is attached at the base of a length of vertical pipe, and the pipe is rotated. The bit grinds away the rock, producing a hole the width of the drill bit and the length of the attached pipe, or drill string. As the top of the rotating drill string approaches ground level, a new piece of pipe is added, lengthening the drill string and allowing the depth of the hole to increase. The weight of the drill string is supported by the drilling rig, which helps keep the hole straight.

Meanwhile, a fluid known as drilling mud is pumped down the hole to keep the bit cool, help keep the hole from collapsing, carry rock chips (cuttings) up the hole, and prevent overpressurized, subsurface liquid or gas from causing a blowout. Blowouts occur when subsurface zones of high pressure force the fluid in the well to flow out of the hole at a high rate, in some cases blowing the drill string completely out of the hole and destroying the drill rig. Drilling mud is pumped downhole through the interior of the drill string, out through holes in the bit at the bottom of the drill string, then back to the surface between the drill string exterior and the drill hole, a space known as the well annulus. At the surface, the cuttings are collected, and the mud is recirculated downhole. Drilling mud is carefully monitored for the correct viscosity and weight. Chemicals are added, as needed, to modify these characteristics.

At intervals during drilling, the drill string is removed, and a steel liner, or casing, is cemented into place downhole. This supports the sides of the hole and isolates the annulus from the surrounding downhole environment.

If economic amounts of oil are not found, then the well is termed a "dry hole." It must be carefully cased from top to bottom and filled with cement to effectively seal the well permanently, a procedure known as plugging and abandoning the well. If oil is found in economic amounts, it will be pumped from the well, a process known as primary production. After the oil well's output begins to significantly decline, alternative methods of production may be applied. Enhanced oil recovery requires more complicated—and expensive—methods for increasing an oil well's production. The most common method, saltwater injection, involves pumping saltwater brine, produced from the subsurface with the oil, back into adjacent wells, forcing more oil to the surface. Chemicals, such as surfactants, may also be employed to help mobilize the oil.

The environmental concerns in drilling for oil on land versus at sea are somewhat different. However, in either case, the effects of poor operations management can be devastating to the environment, particularly in ecologically sensitive areas. Site access and preparation can significantly impact the ecology of land sites, especially in forest preserves, wetlands, and tundra. Soils, as well as surface water and groundwater supplies, are at risk from oil spills and improper disposal of saltwater, drill cuttings, and drilling mud. Groundwater is also at risk from improper casing, saltwater injection, and well plugging and abandonment. Groundwater pollution may go undetected for years and affect relatively large areas of the subsurface.

At sea, much larger areas may be at risk because of down-current transport of pollutants. Environments particularly threatened by marine drilling platforms include coral reefs, oyster banks, mangrove swamps, and tidal estuaries. Disposal of drilling mud and drill cuttings threatens the benthic (sea bottom) environment surrounding the platform and the pelagic (open sea) environment down current. Improper operations can release oil and brine into the water column, and faulty casing can create oil and brine seeps on the sea floor.

In the United States, numerous federal laws regulate oil drilling, production, and transport

An offshore oil-drilling platform. (Ben Klaffke)

on land and at sea. These include the Federal Water Pollution Control Act of 1972; the Outer Continental Shelf Lands Act of 1953; the Comprehensive Environmental Response Compensation and Liability Act (CERCLA) of 1980; the Marine Protection, Research, and Santuaries Act of 1972; and the Safe Drinking Water Act of 1974. Most states have stringent permit requirements, site inspection programs, and accidental spill reporting and response programs for drilling operations. Industry compliance with these programs is generally excellent.

Oil spills from offshore platforms represent only about 1 percent of the petroleum released into the oceans each year. Natural oil seeps contribute about eight times this amount, and their contribution is steadily decreasing as onshore and offshore oil production reduces the pressure in hydrocarbon reservoirs. Marine drilling platforms, while perhaps an eyesore above water, provide habitats for marine species that otherwise would not be present—an artificial reef, of sorts. Scuba divers and fishermen alike seek out platform substructures for the lush growth of invertebrates encrusting them and the many fish that this attracts. As platforms are decommissioned, debate often ensues within the environmental community itself as to whether the structures should be completely dismantled or remain intact, allowing the marine community to benefit.

Clayton D. Harris

SUGGESTED READINGS: An excellent, general introduction to marine pollution, including petroleum, can be found in chapter 8 of *Introductory Oceanography* (1998), by Harold V. Thurman. P. K. Driessen, "Oil Rigs and Sea Life: A Shotgun Marriage that Works," *Sea Frontiers* 33 (1987), discusses evidence that, in spite of the aesthetic objections and environmental concerns, the superstructure of offshore oil rigs may enhance marine ecology in the surrounding waters. For a discussion of the history and significance of Gulf Coast oil drilling and oil spills, see the chapter "Offshore Oil: Opening Pandora's Box" in *The Gulf of Mexico* (1992), by Robert H. Gore.

SEE ALSO: Fossil fuels; Ocean pollution; Oil spills.

Oil spills

CATEGORY: Water and water pollution

Oil spills are human-made disasters that pose both short- and long-term environmental threats. Spills may result from accidental or intentional discharges of raw (unrefined) crude oil or refined petroleum products, and they may occur on land or in the ocean.

Oil spills commonly occur in both terrestrial and marine settings. Terrestrial spills affect land areas, including drainage courses and bodies of surface water impounded in lakes and ponds. Subsurface waters (groundwater) are also at risk from leakage of polluted water downward through the vadose zone (zone of aeration) to the water table (zone of saturation). Some crude oil is lost during exploratory drilling, workover operations, and tank storage. Mud pits utilized during drilling usually contain oil recovered from the well during testing or oil derived from an oil-based mud used during rotary drilling. Oil and brine released from the well are sometimes stored in unlined ponds, and these fluids can soak into the ground and kill beneficial microbes in the soil and plant life in the immediate area. Subsurface water can also be contaminated if the oil or brine migrates downward to the groundwater table. Runoff from these areas can enter streams and ponded bodies of water and kill fish and other aquatic animals as well as reduce the natural vegetation.

Terrestrial spills also occur during the loading, transportation, and offloading of petroleum from tank trucks and railroad tank cars. Petroleum pipelines, both buried and above ground, are highly vulnerable to rupture from welding defects, corrosion, earthquakes, or shifting soils. Fire is a constant danger associated with such spills. During the mid-1980's a ruptured pipeline near São Paulo, Brazil, caused the deaths of more than five hundred people and resulted in the destruction by fire of 2,500 homes in the town of Vila Socco. During the 1991 Persian Gulf War, large areas of Kuwait were devastated when Iraqi soldiers damaged oil pipelines and refineries and set fire to hundreds of oil wells. Pools of oil formed near the wells and infiltrated the porous and permeable soil and rock, thereby endangering the water supply of Kuwait City and other areas. Nearly six million barrels (2.4 billion gallons) of oil were spilled, making it the largest oil-spill disaster of all time.

The disposal of petroleum products (diesel fuel, gasoline, kerosene, jet fuel, and used motor oil) is a significant problem worldwide. Some of these products contain carcinogens such as benzene, toluene, and xylene, which require special handling. According to Jane Walker in *Oil Spills* (1993), "Each year, human beings dispose of more than 5,000 million liters [38 million barrels] of used oil on land." It has been estimated that more than 6.6 million tons, or about 45 million barrels, of petroleum and petroleum products enter the oceans per year. Of this total, 44 percent is land-derived. These inputs include coastal city contributions from refineries, wastewater outlets, and other sources (13 percent); urban runoff, including storm drains (5 percent); and river runoff (26 percent).

MARINE SPILLS

Pollution of oceans or restricted seas can occur in several ways. Oil sources include natural seeps along the ocean floor, stream runoff from the land, wastewater drainage outlets from industrial complexes, offshore drilling accidents, deliberate purging of ballast or cargo areas of ships, and accidents involving oil supertankers. The petroleum input into the marine environment from ocean-derived factors has been estimated at about 56 percent. This calculation includes transportation losses during loading and unloading (30 percent), oil spills (5 percent), offshore production losses (1 percent), atmospheric pollution (10 percent), and offshore oil seeps (10 percent).

Supertanker accidents are major events resulting from shipboard explosions, collisions with other vessels, or grounding on barriers (mostly rocks or coral reefs) because of navigation errors, mechanical failure, or inclement weather. According to Walker in *Oil Spills*, "Accidents involving giant oil tankers often produce the most dramatic oil spills, causing widespread damage to our environment." However, oil spills represent only five percent of the total input of petroleum into the oceans.

Oil slicks are especially a problem in semi-enclosed bodies of water, such as the Baltic and Mediterranean Seas. Because tides are minor in these seas, any oil spilled in the area is not easily eliminated. These places are among the world's most polluted bodies of water. When an oil spill occurs in the open ocean, the oil expelled from a vessel floats on the surface of the water because of a density contrast between the two substances. Water has a specific gravity from 1 to 1.2, while oil is lighter, with a specific gravity from 0.7 to 1.0. The lighter components of the spilled petroleum immediately begin to evaporate; also, oil-degrading bacteria in the water begin to feed on the organic deposit and multiply. Warmer water and ambient air temperatures increase the rates of bacterial growth and evaporation. The major part of petroleum dumped into the open ocean evaporates or is reduced by bacteria. With time, the oil turns into an inert, tarlike substance that becomes extremely hard; some such residues have been found with barnacles attached to them.

In areas where the spilled oil is washed ashore and subjected to intense wave action, a gooey emulsion resembling chocolate mousse sometimes forms. This foamy mass coats everything along the beach, including sand-sized particles, large boulders, aquatic animals and plants, pleasure boats, and human-made shore facilities. The polluted waters can devastate resort beaches, oyster and shrimp beds, fish hatcheries, and prime fishing grounds.

The potential damage resulting from a large oil spill in the ocean includes the loss of substantial numbers of commercial and rough fish, waterfowl, aquatic mammals, shellfish, algae, and plankton. The oil contamination reduces the amount of oxygen in the water column and results in the deaths of large numbers of fish in the polluted area. Millions of fish died as a result of the 1989 *Exxon Valdez* disaster in Alaskan waters. The fishing industry, which is extremely important to the state of Alaska, was seriously threatened by the oil spill. Although this event is generally considered a major disaster, David

Supertanker Disasters, 1967 to 1999

Year	Vessel Name	Location	Barrels of Crude Oil Lost
1967	*Torrey Canyon*	Land's End, England	100,000
1976	*Argo Merchant*	Nantucket, Massachusetts	180,952
1978	*Amoco Cadiz*	Portsall, France	1,500,000
1983	*Castillo de Bellver*	Cape Town, South Africa	2,023,810
1989	*Exxon Valdez*	Prince William Sound, Alaska	261,905
1990	*American Trader*	Southern California	6,249
1990	*Berge Broker*	North Atlantic Ocean	95,238
1990	*Mega Borg*	Gulf of Mexico	75,476
1992	*Aegean Sea*	Spain	14,821
1993	*Braer*	Shetland Islands, Scotland	619,048
1996	*Sea Empress*	Southwest Wales	380,952
1999	*New Carissa*	Oregon	9,524

Source: Data primarily adapted from Jane Walker, *Oil Spills* (1993).

McConnell reported in the *Journal of Geological Education* (1999) that the accident "ranks 53d in size in comparison to other spills worldwide."

Sea birds become vulnerable to spilled oil when their feathers are soaked with oil—they lose the ability to fly and float, as well as their natural body insulation. Many of the birds starve to death, die from exposure, or drown. More than 200,000 seabirds were killed by the *Exxon Valdez* disaster. During the Persian Gulf War, more than twenty thousand birds perished from oil-related causes. Aquatic mammals also die from the loss of their food supply, exposure to cold, or poisoning when they ingest the toxic oil. Many birds and mammals have been saved by workers at animal centers set up after large spills. The animals are washed with solvents and held until their normal body resiliency returns. Shellfish in bays and marshes can perish in oil-polluted waters from asphyxiation or toxicity. Some effects of the pollution remain for years. Shellfish in a salt marsh in Massachusetts were quarantined for more than five years after heating oil was spilled nearby in 1969, and traces of petroleum persisted for decades.

CLEAN-UP PROCEDURES

Oil spills in the open ocean have been reduced or eliminated using several techniques. Igniting the slick, if attempted shortly after the spill, can be effective; however, if the volatile components have evaporated, it is difficult to start and maintain such oil fires. In 1967, when the damaged supertanker *Torrey Canyon* began to founder and break up off the coast of England, Great Britain's air force dropped bombs in an effort to ignite the oil that remained in the vessel.

Another technique used to reduce oil slicks is the application of chemicals such as detergents in an attempt to disperse the oil droplets. This procedure is not viable if the temperature of the water is too low for the chemicals to be effective. Floating booms are sometimes used to confine the oil streamers or protect inlet areas. The floating barriers are useful if the waves are not too high. Skimmers may be used to collect floating oil during early stages of spills. Plant material (threshed hay, peat moss, or wood shavings) or pulverized rock (chalk or claystone) are sometimes spread over the slick to absorb the oil. Plant material can then be removed and burned or buried; most of the applied rock particles absorb the oil, clump together, and sink to the ocean floor. In many parts of the oceans, oil-eating microbes (mostly bacteria) utilize the petroleum as a nutrient. If the microbes are not present or exist in small numbers, engineered microbes can be introduced into the area. Under ideal temperature and ocean chemistry conditions, the bacteria will rapidly multiply and consume the oil slick. This technique is known as bioremediation.

In nearshore areas, beach sand and boulders can be steamed clean or washed with water or solvents. Pools of oil can be vacuumed into tank trucks and hauled away from the site. The recovered oil can sometimes be processed into useful products. Plant material can also be applied to crude oil along the shoreline. After the 1969 Santa Barbara oil spill, the beaches along the California coast were coated by approximately ten thousand barrels of gooey oil. Numerous volunteers helped spread fresh hay along the beachfront and later recovered the oil-soaked plant material.

Donald F. Reaser

SUGGESTED READINGS: Jane Walker, *Oil Spills* (1993), provides an excellent overview of the causes and effects of oil spills. Jon P. Davidson, Walter E. Reed, and Paul M. Davis, *Exploring Earth* (1997), describes the adverse effects of oil spills in the open ocean and along coastal areas. Carla W. Montgomery, *Environmental Geology* (1989), also describes the effects of spilled oil in the open ocean, especially the hazardous effects on waterfowl. Joel S. Watkins, Michael L. Bottino, and Marie Morisawa, *Our Geological Environment* (1975), focuses on the Santa Barbara drilling accident and related oil slick, which adversely affected part of the Southern California coast. Amos Turk, Jonathan Turk, Janet T. Wittes, and Robert Wittes, *Environmental Science* (1974), provides details on the degradation of spilled crude oil from a volatile liquid to an asphaltic residue.

SEE ALSO: *Amoco Cadiz* oil spill; *Argo Merchant* oil spill; *Braer* oil spill; *Exxon Valdez* oil spill; Mexico oil well leak; Ocean pollution; Santa Barbara oil spill; *Sea Empress* oil spill; Tobago oil spill; *Torrey Canyon* oil spill.

Old-growth forests

CATEGORY: Forests and plants

Old-growth forests consist of trees that have never been harvested by loggers. The timber industry views the large, old trees as a renewable source of fine lumber, but environmentalists see them as part of an ancient and unique ecosystem that can never be replaced.

In the 1970's scientists began studying the uncut forests of the Pacific Northwest and the plants and animals that inhabited them. One of the first results of this was the U.S. Forest Service publication *Ecological Characteristics of Old-Growth Douglas-Fir Forests* (1981). In this report, Forest Service biologist Jerry Franklin and his colleagues revealed that these forests were not just tangles of dead and dying trees, but rather a unique, thriving ecosystem made up of living and dead trees, mammals, insects, and even fungi.

The forest usually referred to as old growth occurs primarily on the western slope of the Cascade Mountains in southeast Alaska, southern British Columbia in Canada, Washington, Oregon, and Northern California. The weather there is wet and mild, ideal for the growth of trees such as Douglas fir, cedar, spruce, and hemlock. Some studies have shown that there is more biomass, including living matter and dead trees, per acre in these forests than anywhere on earth. Trees can be as tall as 90 meters (300 feet) with diameters of 3 meters (10 feet) or more and live as long as one thousand years. The forest community itself grows and changes over time, not reaching biological climax until the forest primarily consists of hemlock trees, which are able to sprout in the shade of the sun-loving Douglas fir.

One of the most important components of the old-growth forest is the large number of standing dead trees, or snags, and fallen trees, or

logs, on the forest floor and in the streams. The fallen trees rot very slowly, often taking more than two hundred years to completely disappear. During this time they are important for water storage, as wildlife habitat, and as "nurse logs" where new growth can begin. In fact, seedlings of some trees, such as western hemlock and Sitka spruce, have difficulty competing with the mosses on the forest floor and need to sprout on the fallen logs.

Another strand in the complex web of the forest consists of micorrhizal fungi, which attach themselves to the roots of the trees and enhance their uptake of water and nutrients. The fruiting bodies of these fungi are eaten by small mammals such as voles, mice, and chipmunks, which then spread the spores of the fungi in their droppings. There are numerous species of plants and wildlife that appear to be dependent on this ecosystem to survive. The most famous example is the northern spotted owl, whose endangered species status and dependence on old growth has caused great political and economic turmoil in the Pacific Northwest.

By the 1970's most of the trees on the timber industry's private lands had been cut. Their replanted forests, known as second growth, would not be ready for harvest for several decades, so the industry became increasingly dependent on public lands for their raw materials. Logging old growth in the national forests of western Oregon and Washington increased from 900 million board feet in 1946 to more than 5 billion board feet in 1986.

Environmentalists claimed that only 10 percent of the region's original forest remained, and they were determined to save what was left. The first step in their campaign was to encourage the use of the evocative term "ancient forest" to counteract the somewhat negative connotations of "old growth." Then they were given an effective tool in the northern spotted owl. This small bird was found to be dependent on old growth, and its listing under the federal Endangered Species Act in 1990 was a bombshell that caused a decade of scientific, political, and legal conflict.

Under the law, the Forest Service was required to protect enough of the owl's habitat to

Old-growth forests in Washington State became the center of controversy between loggers and environmentalists during the 1980's. (Jim West)

insure its survival. An early government report identified 7.7 million acres of forest to be protected for the bird. Later, the U.S. Fish and Wildlife Service recommended 11 million acres. In 1991 U.S. District Court judge William Dwyer placed an injunction on all logging in spotted owl habitat until a comprehensive plan could be put in place. The timber industry responded with a prediction of tens of thousands of lost jobs and regional economic disaster. In 1993 President Bill Clinton convened the Forest Summit conference in Portland, Oregon, to work out a solution. The Clinton administration's plan, though approved by Judge Dwyer, satisfied neither the industry nor the environmentalists, and protests, lawsuits, and legislative battles continued.

As the twentieth century came to an end, timber harvest levels had been significantly reduced, the Northwest's economy had survived, and additional values for old-growth forests were found: as habitat for endangered salmon and other fish, as a source for medicinal plants, and simply as a repository for benefits yet to be discovered. The decades-long controversy over the forests of the Northwest had a deep impact on environmental science as well as natural resource policy and encouraged new interest in other native forests around the world, from Brazil to Malaysia to Russia.

Joseph W. Hinton

SUGGESTED READINGS: *Secrets of the Old Growth Forest* (1988), by David Kelly and Gary Braasch, is a coffee-table book with color photographs and extensive text that provides an excellent introduction to the topic. A detailed examination of the politics of the old-growth controversy, with special emphasis on events from 1987-1996, can be found in Kathie Durbin's *Tree Huggers: Victory, Defeat, and Renewal in the Northwest Ancient Forest Campaign* (1996). *The Final Forest: The Battle for the Last Great Trees of the Pacific Northwest* (1992), by William Dietrich, studies the old-growth controversy with a balanced look at individuals affected by it, conservationists and loggers alike. *Forest Primeval: The Natural History of an Ancient Forest* (1989), by Chris Maser, is a fascinating biological biography of a hypothetical thousand-year-old forest.

SEE ALSO: Ecosystems; Endangered Species Act; Endangered species and animal protection policy; Forest Summit; Logging and clear-cutting; Northern spotted owl.

Olmsted, Frederick Law, Sr.

BORN: April 26, 1822; Hartford, Connecticut
DIED: August 28, 1903; Waverly, Massachusetts
CATEGORY: The urban environment

Frederick Law Olmsted, Sr., pioneered the nineteenth century urban park movement in the United States. (Library of Congress)

Pioneering U.S. landscape architect and urban planner Frederick Law Olmsted, Sr., left a distinct mark on the American environment from New York City to the wilds of California, from Washington, D.C., to Biltmore, one of the most elaborate family estates in the nation. He synthesized a variety of experiences in his youth and young adulthood to become one of the greatest landscape designers in the history of the United States.

As a child Frederick Law Olmsted traveled widely with his family, becoming familiar with the natural landscapes of various regions. As a young adult he operated an experimental farm and learned how the environment interacts with human endeavors to tame it. He later traveled throughout the antebellum American South as a correspondent for *The New York Times*. His observations were compiled and published as *Journeys and Explorations in the Cotton Kingdom* (1861), a discerning analysis of social conditions in the southern slave society.

In 1857 Olmsted and a friend, Calvert Vaux, entered a contest to plan a park on Manhattan Island. Olmsted and Vaux envisioned a park for the people—one that departed from the grand plans, controlled landscapes, and aristocratic enclaves that characterized Europeans parks. In 1858 Olmsted was appointed chief architect of Central Park and devoted his energies to developing a natural environment that would provide a refuge for all the city's residents, rich and poor alike.

When the Civil War began, Olmsted set aside his landscaping to organize the U.S. Sanitary Commission, where he helped reform the Union Army's medical corps. From there, in 1863, he moved to California to manage the Fremont Mariposa mining enterprise. During his time in the West, he helped preserve the Yosemite and Mariposa wilderness areas and worked on a draft of what would become the charter for the national parks system. By 1875 he was back in the East directing the work of converting Boston's Back Bay from marshes into parkland. In this and other projects—Prospect Park in Brooklyn, Riverside and Morningside Parks in Manhattan, Jackson Park in Chicago, the grounds of the Capitol Building in Washington, D.C., and the grounds of the World's Columbian Exposition of 1893 in Chicago—Olmsted clung to his desire to use and preserve natural environments for all people. At the Biltmore estate, his largest private commission, he designed a managed forest that preserved the natural North Carolina habitat instead of the artifice of a traditional aristocratic estate.

Olmsted also worked as an early city planner and advocate of suburban development. He envisioned suburbs as self-contained communities linked to the city by good roads. He used his ideas in planning a series of parks along parkways in both Boston and Buffalo. He also planned every detail of the suburb of Riverside, Illinois, one of the country's first planned communities.

Olmsted's legacy is huge. He was the predominant force behind the preservation of Niagara Falls and designed nearly eighty parks and thirteen college campuses (including Stanford University in California). He fathered the American urban park movement and pioneered many of the tenets of American regional planning. In each of his projects, his aim was to move from the merely scenic to a balance between preservation and use of the environment by all people.

Jane Marie Smith

SEE ALSO: Planned communities; Urban parks; Urban planning; Urbanization and urban sprawl; Yosemite.

Open space

CATEGORY: Land and land use

The inclusion of open space is often mandated by municipalities on newly developed tracts of land in order to encourage a more efficient use of land and more flexible development practices that respect and conserve natural resources.

Often more than 10 percent of a tract area is required to be dedicated as open space. When large tracts of land are developed for cluster residential housing, a municipality may require that 40 to 60 percent of the area be made available as open space. Land designated as open space can take the form of lawns, natural areas, recreational areas, cropland or pastures, or stormwater management areas.

A lawn is a grassed area with or without trees that may be used by residents for a variety of purposes. The lawn is regularly mowed to ensure a neat and tidy appearance. A natural area consists of native or natural vegetation undisturbed during construction and land development. Occasionally, a natural area can be created by grading and replanting acreage disturbed during development. Natural areas, however, are often required to be maintained by a designated authority. The maintenance involves preventing the proliferation of undesirable plants; clearing litter, dead trees, and brush; and keeping streams freely flowing. Pathways or walking trails may be constructed within the natural area to provide places for local residents to enjoy mild exercise or relax in a less structured setting.

Open space may also take the form of an area for recreation, such as tennis courts, swimming pools, ballfields, and other types of play areas. Safety concerns make this kind of open space expensive to construct and maintain. Most often these recreational facilities are part of a public park. Open space in the form of cropland, pasture, or acreage planted with nursery stock or orchard trees provides a visually valuable pastoral setting. Buildings and structures are usually not permitted in open spaces, but in the case of agricultural areas, flexible development procedures often permit any existing dwellings or farm-related buildings to remain.

Finally, stormwater management features are often incorporated into the open space of developed land. Detention or retention basins are integral to the control of stormwater when a tract of land has buildings and paved areas constructed upon it. Impervious materials prevent rainwater from percolating into the ground. Without taking these measures on developed land, heavy erosion is likely, and neighboring lands may be subjected to flooding.

Anthony J. Nicastro

SEE ALSO: Environmental impact statements and assessments; Land-use policy; Planned communities; Urban parks; Urban planning; Wetlands.

Organic gardening and farming

CATEGORY: Agriculture and food

Organic agriculture strives to produce healthy soils and crops through practices that replenish and maintain soil fertility. Organic farmers avoid the use of synthetic and often toxic fertilizers and pesticides.

At the beginning of agricultural history, farmers believed that plants "ate the soil" in order to grow. During the nineteenth century, German chemist Justus von Liebig discovered that plants extracted nitrogen, phosphorous, and potash from the soil. His findings dramitically changed agriculture as farmers found they could grow crops in any type of soil, even sand and water solutions, if the right chemicals were added.

Diversified family farms eventually gave way to huge specialized operations; by the end of the twentieth century, less than 2 percent of the U.S. population was directly involved in crop production. Crop yields were raised with the use of chemicals, but farms and their soil and water were not being cared for as the unique ecosystem that they are. The quality of the soil was ignored as chemicals were used to produce high crop yields to feed the 98 percent of the population not involved in farming. Over time, the organic quality of the soil was lost even though the chemicals remained in the soil.

Chemicals were also found to leach into the water supply. In 1988 the U.S. Environmental Protection Agency (EPA) found that groundwater in thirty-two states was contaminated with seventy-four different agricultural chemicals. Besides the health consequences to humans, leaching causes once-friable and fertile soils to turn hard and become nonproductive. Further, the use of chemical insecticides was proving to have toxic effects on both the foods grown and the farmworkers encountering them. About forty-five thousand accidental pesticide poisonings occur in the United States each year; in 1987 the EPA ranked pesticides as the third leading environmental cancer risk.

Organic farming developed as an attempt to return to diversified farming practices that emphasize working with nature to create a renewing, ecologically sound, and sustainable system of agriculture. Organic farmers are bound by practices that are free of synthetic chemicals and genetic engineering. Lands must be chemical free for at least three years before their products can be certified as organic. The organic certification further requires that both plants and livestock be raised organically without the use of chemicals, antibiotics, hormones, or synthetic feed additives. Organic farmers must comply with both organic regulations in their state and the 1990 Organic Foods Production Act (OFPA). In the late 1990's the U.S. Department of Agriculture (USDA), through the National Organic Standards Board (NOSB), began working to develop standards and regulations that would ensure consistent national standards for organic products. There are six areas that the NOSB is examining: crops standards; livestock and livestock products standards; processing, packaging, and labeling standards; accreditation; international issues; and materials. There is also interest in developing global standards for the importation and exportation of organic products.

SOIL FERTILITY

In order for organic farmers to produce chemical-free crops, they must rely on organic practices to ensure soil fertility and control unwanted plants and insects. Organic farmers build organic materials in the soil by the addition of green manure, compost, or animal manure.

Green manure is the term used for crops that are grown to introduce organic matter and nu-

The use of organic fertilizers helps maintain the health of soils without adding toxic substances to the environment. (Yasmine Cordoba)

trients into the soil. They are a crop raised expressly for the purpose of being plowed under rather than sold to the consumer. Green manure crops protect soil against erosion, cycle nutrients from lower levels of the soil into the upper layers, suppress weeds, and keep much-needed nutrients in the soil rather than allowing them to leach out. Legumes are an excellent green manure crop because they are able to extract nitrogen from the air and transfer it into the soil, leaving a supply of nitrogen for the next crop. The legumes have nitrogen nodules on their roots; when the legumes are tilled under and decompose, they add more nitrogen to the soil. As plants decay, they also make insoluble plant nutrients—such as carbon dioxide, and acetic, butyric, and lactic and other organic acids—available in the soil.

Much of the organic material derived from green manure comes in the form of decaying roots. Alfalfa, one type of legume, sends its roots several feet down into the soil. When the alfalfa plants are turned, the entire root system decomposes into organic material. Thus they help improve water retention and soil quality at the same time. Some examples of legumes used for green manure are sweet clover, ladino clover, alfalfa, and trefoil. Nonlegumes or grasses used for green manure include rye, redtop, and timothy grass.

Further fertility can be achieved if green manure is grazed by animals and animal manure is deposited on the soil. Left on their own in a large field, cows, sheep, and horses will choose only their favorite bits of grasses to graze and will graze the field unevenly. Organic farmers use portable fencing for strip rotational grazing, in which animals graze a strip of the field all the way down and then are moved to another strip of the field while the first pasture recovers. This process is repeated until the entire field has been grazed and fertilized. Also, when the grass is grazed, a proportion of the roots die and rot to form a good amount of humus. This humus is the stable organic material that acts as a catalyst for allowing plants to find nutrients.

Compost can work for large or small farmers but is particularly useful to the small farmer or gardener who does not have the acreage to grow a green manure crop to plow under. Composting is a natural soil-building process that began with the first plants that existed on earth and continues as a natural process today. As falling leaves and dead plants, animals, and insects decompose into the soil, they form a rich, organic layer. Compost can be made by farmers and gardeners by using alternating layers of carbohydrate-rich plant cuttings and leaves, animal manure, and topsoil and allowing it to decay and form a rich, organic matter to be added to the soil.

Animal manure is used as an organic fertilizer. Rich in nitrogen, the best animal manures are found in animals with the most protein in their diet. For example, beef cattle being fattened for market have a higher level of protein than dairy cattle that are producing milk for market. The application of manure to fields improves the structure of the soil, raises organic nitrogen content, and stimulates the growth of soil bacteria and fungi necessary for healthy soils.

CROP ROTATION

Planting the same crop year after year on the same piece of ground results in depleted soils. Crops such as corn, tobacco, and cotton remove nutrients, especially nitrogen, from the soil. In order to keep the soil fertile, a legume should be planted the following year in order to add nitrogen and achieve a balanced nutrient level. Planting a winter-cover crop such as rye grass will help protect the land from erosion and will, when plowed under in the spring, provide a nutrient-rich soil for the planting of a cash crop. Crop rotations also improve the physical condition of the soil because different crops vary in root depth, are cultivated differently, and respond to either deep or shallow soil preparation.

Traditional agriculture, or monoculture, puts a large source of the same crop in easy proximity of insects that are destructive to that particular crop. Insect offspring can multiply out of pro-

portion when the same crops are grown in the same field year after year. Since insects are drawn to the same home area, they will not be able to proliferate and thrive if the crop is changed the following year to a crop they do not eat. This is one of the reasons that organic farmers rely heavily on crop rotation as one aspect of insect control. Rotating crops can also help control weeds. Some crops and cultivation methods inadvertently allow certain weeds to thrive. Crop rotation can incorporate a successor crop that eradicates those weeds. Some crops, such as potatoes and winter squash, work as "cleaning crops" because of the different styles of cultivation that are used on them.

ORGANIC INSECT AND WEED CONTROL

Organic farmers and gardeners believe that plants within a balanced ecosystem are resistant to disease and insect infestation. Plants stressed by unfavorable growing situations will be much more susceptible to such problems. Therefore, the whole premise of organic farming is working with nature to help provide healthy, unstressed plants. Rather than using chemicals, which often create resistant generations of insects and thus the need for newer and stronger chemicals, organic farmers have found natural ways to diminish pest problems. Organic farmers strive to maintain and replenish soil fertility to produce healthy plants resistant to insects. They also try to select plant species that are more resistant to insects, weeds, and disease. Crop rotation, as already discussed, is one method of keeping infestations down. Rows of hedges, trees, or even plants that are not desirable to insects planted in and around the crop field can act as barriers to insect pests. They can also provide habitats for insect's natural enemies: birds, beneficial insects, and garden snakes.

Beneficial insects can be ordered through the mail to help alleviate insect pests. The ladybug and the praying mantis are two beneficial insects that will help rid the farm or garden of aphids, mites, mealy bugs, and grasshoppers. Just one ladybug can consume fifty or more aphids per day. Considering that most people order ten to twenty thousand ladybugs for their garden and that one ladybug can lay up to one thousand fertile eggs, their overall cost is much less than buying insecticides. Trichogramma, also available by mail order, is a small wasp that will destroy moth eggs, squash borers, cankerworms, cabbage loopers, and corn earworm.

Organic farmers and gardeners also rely on what is called an "insecticide crop." Garlic planted near lettuce and peas will deter aphids. Geraniums or marigolds grown close to grapes, cantaloupes, corn, and cucumbers will deter Japanese and cucumber beetles. Herbs such as rosemary, sage, and thyme planted by cabbages will deter white butterfly pests. Potatoes will repel Mexican bean beetles if planted near beans, and tomatoes planted near asparagus will ward off asparagus beetles. Natural insecticides such as red pepper juice can be used for ant control, while a combination of garlic oil and lemon may be used against fleas, mosquito larvae, houseflies, and other insects.

Organic farming relies on the physical control of weeds, especially through the use of cutting (cultivation) or smothering (mulching and hilling). Cultivation is the shallow stirring of surface soil to cut off small developing weeds and prevent more from growing. This cultivation can be done with tractors, wheel hoes, tillers, or, for small gardens, hand hoes. Mulch is a soil cover that prevents weeds from getting the sunlight they need for growth. Mulching with fully biodegradable materials can help build soil fertility while controlling weeds. Plastic mulches can be used on organic farms as long as they are removed from the field at the end of the growing or harvest season.

Organic farming offers a safe alternative to the use of synthetic chemicals in the production of food. A National Academy of Science study estimated that twenty thousand people each year get cancer because of pesticides alone. Growing public awareness of the effects of the traditional agricultural reliance on synthetic chemicals is reflected in a growing demand for organically produced foods.

Dion Stewart and Toby Stewart

SUGGESTED READINGS: Eliot Coleman's *The New Organic Grower* (1995) is an easy-to-read manual on organic concepts for home or market gardeners. *The Organic Garden Book* (1994), by Geoff Hamilton, is an excellent book for gardeners who wish to grow vegetables, flowers, and fruits organically. The premiere book on the detriments of the agricultural use of pesticides and other chemicals is Rachel Carson's *Silent Spring* (1962). *The Art of Natural Farming and Gardening* (1985), by Ralph Engelken and Rita Engelken, is an interesting story of two decades of organic farming in Iowa. The Board of Agriculture's National Research Council provides interesting information on organic farming in their book *Alternative Agriculture* (1989). The Community Alliance with Family Farmers provides information on organic farming and lists producers and products in the *National Organic Directory* (1998).

SEE ALSO: Integrated pest management; Pesticides and herbicides; Sustainable agriculture.

Osborn, Henry Fairfield

BORN: January 15, 1887; Princeton, New Jersey
DIED: September 16, 1969; New York, New York
CATEGORY: Population issues

Naturalist and conservationist Henry Fairfield Osborn worked to preserve endangered species and their habitats. He also demonstrated a concern about the effects of population growth on the earth and wrote two books on the subject.

Henry Fairfield Osborn was one of five children of Henry Fairfield Osborn and Lucretia Thacher Perry. The senior Osborn was a professor of paleontology at Princeton University at the time of Osborn's birth and eventually became the president of the American Museum of Natural History and founder of the New York Zoological Society. Osborn graduated from the Groton School in 1905 and Princeton University with a bachelor of arts in 1909. He then studied for one year at Cambridge University.

After finishing his year in Cambridge, Osborn travelled and held several jobs, including laying track for the railroad in Utah. During World War I he served as a captain in the 351st Field Artillery of the American Expeditionary Force. After returning from the war, Osborn worked in the investment banking business on Wall Street. He married Marjorie M. Lamond on September 8, 1914, and together they had three daughters. In 1922, as he continued working in investments, Osborn became a trustee of the New York Zoological Society (NYZS). The paleontology field trips that he had taken with his father from an early age had convinced him that his real vocation lay in natural science. Osborn left the banking field in 1935 to become the secretary of the NYZS. He assumed the presidency of the society in 1940 and stayed in that position until 1968, when his health failed. Under his direction the Bronx Zoological Park was greatly improved, and the Marine Aquarium at Coney Island was created. In connection with his fund-raising for these causes, he has been called "the greatest showman since Barnum."

Osborn used his position in the society to become a strong advocate of the preservation of endangered species and their habitats. He was also an early voice warning of the dangers of human population growth. These ideas were forcefully put forth in his two books, *Our Plundered Planet* (1948) and *The Limits of the Earth* (1953). He also pushed these ideas as a member of the Conservation Advisory Committee of the U.S. Department of the Interior and the Planning Committee of the Economic and Social Council of the United Nations. Osborn was often invited to speak at both public and scientific meetings and regularly attended national and international conservation conferences. The Conservation Foundation, which Osborn founded in 1947, was incorporated into the World Wildlife Fund (WWF) in 1990. In 1948 he was given the American Design Award for his campaign to stop humanity from fighting a losing battle against nature.

Osborn suffered a slight stroke at seventy-nine years of age, which left him with a minor speech impediment. He continued to lecture and work, however, until his death in 1969 at the age of eighty-two.

Kenneth H. Brown

SEE ALSO: Conservation; Endangered species; International Union for the Conservation of Nature/World Wildlife Fund; Population growth.

Ozone layer and ozone depletion

CATEGORY: Weather and climate

The ozone layer protects life on earth from exposure to dangerous levels of ultraviolet (UV) light. Because of the introduction of human-made chemicals into the atmosphere, the amount of ozone steadily declined during the second half of the twentieth century.

Ozone (O_3) is a molecule made of three atoms of oxygen. It is considered a trace gas since it accounts for only .000007 percent of the earth's atmosphere. Ozone concentrations are measured in terms of Dobson units, which represent the thickness of all the ozone in a column of the atmosphere if it were compressed—on average only 3 millimeters thick. Depending on where it is found in the atmosphere, ozone may have either a positive or negative impact on life. When ozone is near the earth's surface, it is a major air pollutant and a chief constituent of smog. Ozone that resides in the stratosphere—above 12 kilometers (7.5 miles)—makes up the "ozone shield," a layer that protects organisms from lethal intensities of solar UV radiation. Without this ozone shield, life on earth as it is now known would probably cease to exist.

Ozone concentration peaks between an altitude of 20 to 25 kilometers (12 to 16 miles). Within this layer, two sets of chemical reactions, both powered by UV radiation, continuously occur. In one reaction, ozone is produced; in the other, ozone is broken down into oxygen molecules and ions. Since slightly more ozone is produced than destroyed by these reactions, stratospheric ozone is constantly maintained by nature. Because UV radiation is a catalyst for both reactions, most of this radiation is used up and prevented from ever reaching earth's surface.

UV radiation is divided into three segments, depending on wavelength: UVA, UVB, and UVC. UVC rays readily kill living cells with which they come into contact. UVB rays, although less energetic than UVC rays, also damage cells. The ozone shield prevents all UVC rays from reaching the surface; however, some UVB radiation does pass through the ozone layer. It is contact with this radiation that causes sunburns, accelerates the natural aging of the skin, and has been shown to increase rates of skin cancer and cataracts. The American Cancer Society estimates that more than 700,000 new cases of skin cancer occur each year in the United States, mainly a result of UV radiation. Some researchers estimate that every 1 percent decline in ozone concentration causes a 2 percent increase in UV intensity at the earth's surface, resulting in a greater risk of skin cancer, cataracts, and immune deficiencies.

Increased UV radiation also harms plant and animal life. Studies suggest that yields from crops such as corn, wheat, rice, and soybeans drop by 1 percent for each 3 percent decrease in ozone concentration. Increased UV radiation in polar regions impairs and destroys phytoplankton, which makes up the base of the food chain. A decrease in phytoplankton would likely cause population reductions at all levels of the ecosystem. Fewer than 1 in every 100,000 molecules in the atmosphere is ozone, a ratio that both underscores and belies the critical role ozone plays in protecting human health and the global environment.

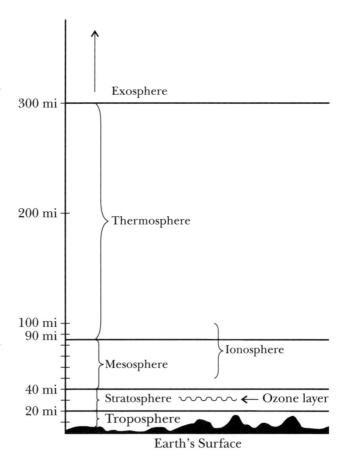

Layers of the Earth's Atmosphere

ANTARCTIC OZONE HOLE

Satellites placed in orbit in the late 1970's allowed scientists to observe concentrations of stratospheric ozone. Observations showed that during the Southern Hemisphere's spring (primarily September and October), the ozone layer above Antarctica thinned dramatically, then recovered during November. These findings were initially dismissed as being caused by instrument error; however, the measurements were confirmed by the British Antarctic Survey in 1985. During that spring, the loss of ozone exceeded 50 percent of normal concentrations. Alarmingly, continued satellite measurements indicated that each year throughout the 1980's and 1990's, the ozone concentration over Antarctica dropped to record lows, and the size of the ozone-depleted area (dubbed the "ozone hole")

grew larger. In 1986 the size of the ozone hole grew larger than the size of the Antarctic continent, and in 1993 the hole was larger than all of North America.

During an intensive Antarctic field program in 1987, extremely high levels of chlorine monoxide (CLO) were found in the stratosphere. This finding was seen by many scientists as evidence that the cause of ozone depletion was chloroflourocarbons (CFCs), anthropogenic chemicals that make excellent refrigerants, cooling fluids, and cleaning solvents. Since the 1950's millions of tons of CFCs have been produced in the United States. A huge amount of CFCs reside in the atmosphere as a result of leaking of old refrigerators and automobile air conditioners and a lack of recycling. Also, one of the

CFCs, CFC-11, was used for decades as a propellant in aerosol spray cans until it was banned in the late 1970's.

In the stratosphere, UV radiation breaks down CFCs, causing them to release chlorine (CL), a gas that readily reacts with ozone. The reaction produces oxygen (O_2) and CLO, which then combine to produce O_2 and CL. Thus, at the end of these reactions, the chlorine ion is again free to destroy another ozone molecule. It is estimated that a single chlorine ion may reside in the stratosphere for fifty years or longer and destroy hundreds of thousands of ozone molecules.

While most scientists believed the detection of stratospheric CLO was the "smoking gun" that proved that human-made CFCs were the cause of ozone depletion, many industrialists and other critics of environmental regulations asserted that the chlorine in the atmosphere was a result of natural processes such as volcanic eruptions or sea spray. This theory was finally debunked in 1995 when scientists also found hydrogen fluoride in the stratosphere. Hydrogen fluoride is not produced by any natural process; however, fluoride would be liberated from a CFC molecule when it is broken down

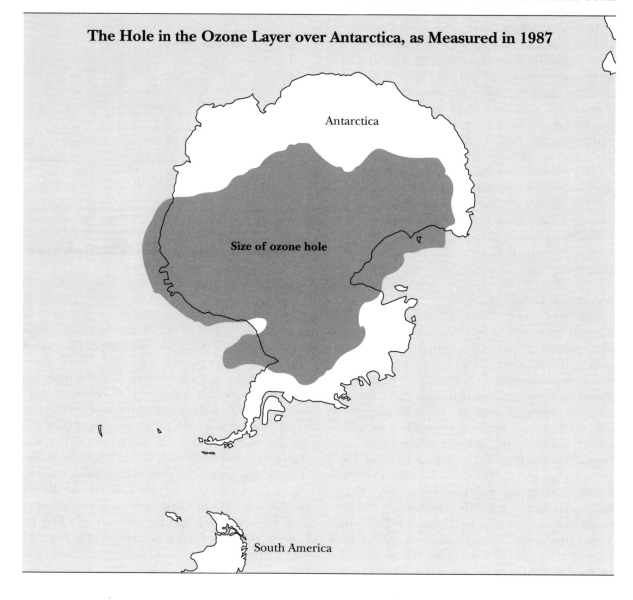

The Hole in the Ozone Layer over Antarctica, as Measured in 1987

Antarctica

Size of ozone hole

South America

by UV radiation. Scientifically, there is no longer any debate that ozone destruction is a direct result of CFCs.

Since the initial satellite observations of the 1970's and 1980's, the understanding of the complex science of the formation of the Antarctic ozone hole has greatly improved. Most CFCs are released in the Northern Hemisphere; however, they mix throughout the troposphere in about one year and then mix into the stratosphere in two to five years. A key element that enhances ozone depletion is the presence of polar stratospheric clouds. These clouds form only in the polar regions where temperatures in the stratosphere drop to below -80 degrees Celsius. The surfaces of these clouds are sites on which inactive chlorine becomes the reactive chlorine used in the ozone depletion process. The dark Antarctic stratosphere becomes so frigid during winter that these clouds are produced in abundance. As spring begins, the stratosphere is hit with solar UV radiation, accelerating the ozone depletion process. A strong vortex of winds circle the South Pole every winter, keeping the Antarctic stratosphere isolated from neighboring air masses. This allows the ozone-destructing reactions to act on a limited amount of ozone that cannot be replenished by ozone from other latitudes. The result is the appearance of the ozone hole each spring. In late spring the winds weaken, allowing ozone-rich air from other latitudes to mix into the Antarctic atmosphere.

GLOBAL OZONE DEPLETION

While the Antarctic ozone hole grabbed most of the attention, ozone concentrations declined globally during the second half of the twentieth century as a result of rising levels of atmospheric chlorine. An ozone hole over the Arctic forms in the Northern Hemisphere during the spring, although the hole is smaller in size and higher in ozone concentrations than its Antarctic counterpart. The main reason for the difference between the Arctic and Antarctic is that the winds encircling the Arctic are typically not as strong as those in the Antarctic, thus allowing ozone-rich air to constantly mix into the Arctic. Still, significant ozone depletion occurred throughout the

1990's as Arctic spring concentrations dropped about 30 percent below average levels.

Ozone levels in other parts of the world have also been in decline. At midlatitude ground stations, stratospheric ozone levels have been declining since 1970. Satellites indicate a negative trend in ozone concentrations in both hemispheres, with a net decrease of about 3 percent per decade. The depletion increases with latitude and is somewhat larger in the Southern Hemisphere. Over the United States, Europe, and Australia, 4 percent per decade is typical. A new threat to ozone followed the eruption of Mount Pinatubo in the Philippines in 1991, the century's most violent volcanic eruption. Massive amounts of sulfur aerosols were injected into the stratosphere. The aerosols function like the ice crystals in the Antarctic stratosphere by helping convert chlorine from its inactive to reactive state. During the two years following the eruption, record low levels of annual ozone concentrations were recorded: 9 percent below normal between 30 and 60 degrees north lattitude.

As expected, measurements also indicate increased UV radiation reaching the earth's surface during these same decades. During the 1990's, summer levels of UVB increased 7 percent annually over Canada, while winter levels increased 5 percent per year. At the sime time, incidents of skin cancer grew faster than any other form of cancer, with reported cases doubling between 1980 and 1998.

As evidence piled up in the 1980's that CFC production was responsible for ozone depletion, the international community decided to act. In 1987 fifty-seven nations ratified the Montreal Protocol, which put into place a framework for the phaseout of CFC production. In the following years, as ozone levels kept falling, the timetable of the phaseout was accelerated, with a total ban on CFC production to begin in January, 1996. CFCs have such a long residence time in the atmosphere, however, that even with a ban on production, the levels of CFCs in the stratosphere are not expected to return to pre-1980 levels until the middle of the twenty-first century at the earliest. Maximum ozone depletion is expected to occur between 2000 and

2020, followed by a slow recovery. UV radiation levels at the earth's surface are also expected to increase during the early part of the twenty-first century, bringing with it ensuing health and ecological effects.

The story of the ozone layer contains both positive and negative lessons of human interaction with the environment. Human production of CFCs during the latter half of the twentieth century significantly harmed the earth's protective barrier against deadly UV radiation. Human health has been affected, and unmeasurable damage has been inflicted on many ecosystems. With the earth in the balance, however, the international community decided to act, and did so forcefully. Without the Montreal Protocol, the fate of the ozone layer and life on earth would have been sealed. It is known that human activity has the power to destroy the global environment, but perhaps human activity may have the power to save it as well.

Craig S. Gilman

SUGGESTED READINGS: For a complete, nontechnical account of ozone depletion, see S. Cagin and P. Dray, *Between Earth and Sky: How CFCs Changed Our World and Endangered the Ozone Layer* (1993). For comprehensive books on several atmospheric issues written for nonscientists, see R. Somerville, *The Forgiving Air: Understanding Environmental Change* (1996), and R. Turco, *Earth Under Siege: From Air Pollution to Global Change* (1997). For in-depth analysis written for scientists, see T. J. Graedel and P. J. Crutzen, *Atmospheric Change: An Earth System Perspective* (1993).

SEE ALSO: Chlorofluorocarbons; Freon; Montreal Protocol.

P

Particulate matter

CATEGORY: Atmosphere and air pollution

Fine solid and liquid particles suspended in the atmosphere are referred to as particulates.

About 7 million tons of human-produced particulates are emitted over the United States annually; about 40 percent of this total is emitted from industrial sources, while 17 percent comes from vehicles. Oxides of nitrogen (NO_x) and oxides of sulfur (SO_x) emitted as gases from combustion are slowly converted to tiny particulates in the atmosphere. Globally, about 300 million tons of particulates are formed by conversion of industrial gases in the atmosphere; conversion of NO_x and SO_x from natural sources contributes at least this much.

Atmospheric particulates contribute to decreased atmospheric visibility. For example, during the late 1990's, the dominant haze components in the well-known smog of Los Angeles were NO_x and soot from road traffic. High concentrations of particulate matter, especially particles with diameters less than 10 micrometers, which can penetrate deep into the lung, are detrimental to health and increase cardiovascular and respiratory mortality. Acidic ultrafine particulates may cause bronchoconstriction and alveolar inflammation, which may lead to acute changes in blood coagulability. Particulates may impair lung defenses and cause physiological disturbances in gas transfer in the lung. The health threat posed by particulates with diameters less than 2.5 micrometers has led the United States to adopt a National Ambient Air Quality Standard for Fine Particulate Matter.

A study of excess mortality caused by air pollution in major European cities indicates that an increase of particulates in black smoke (largely particulates less than 1 micrometer in diameter) is associated with a 3 percent increase in daily mortality in Western Europe. An increase in particles with diameters less than 10 micrometers is associated with a 2 percent increase in daily mortality.

Particulates may be removed from the atmosphere by impaction with a solid object (a process referred to as "sedimentation"); precipitation ("washout" or "wet deposition") is a major natural mechanism for removing particulates from the atmosphere. Trees scavenge particulates, and trees can be selectively planted close to stationary sources of pollution to serve as particulate filters.

Anita Baker-Blocker

SEE ALSO: Aerosols; Air pollution; Smog.

Passenger pigeon

CATEGORY: Animals and endangered species

The passenger pigeon, at one time an abundant bird species in North America, became extinct at the beginning of the twentieth century because of human activity.

During the sixteenth century, giant flocks of passenger pigeons (*Ectopistes migratorius*) flew freely across the North American continent. This pattern changed rapidly with the increased population of the continent by Europeans. Even at the start of the nineteenth century, there were billions of passenger pigeons in North America. One hundred years later, however, there was only one left. The extinction of the passenger pigeon was caused entirely by human beings. Hunters killed millions of birds each year, but, more important, they disrupted the passenger pigeon's breeding cycle.

Passenger pigeons were most vulnerable while nesting en masse, and it was here that their fate was sealed. Flocks of pigeons nesting in trees were so dense that a hunter could kill more than one hundred birds with a single shotgun blast. In fact, passenger pigeons were so easy to shoot that they were not considered a game bird until the mid-nineteenth century. Especially high yields were achieved using nets baited with "stool pigeons," live passenger pigeons with their eyelids sewn together. The blind birds sat on stools in front of the nets in order to lure pigeons into the trap. A single net could capture more than one thousand birds.

Adult pigeons abandoned the new generation of young pigeons, or squabs, before they could fly, leaving them to wander on the ground. Much fatter than the adult birds, squabs were delicious to eat. Hunters dislodged squabs by setting fire to the bark of the trees at the base, causing them to leap from their nests to the ground. One observer recalled that the "squabs were so fat and clumsy that they would burst open on striking the ground." The senseless slaughter of squabs sealed the fate of passenger pigeons as a species.

No legislation was created to protect the passenger pigeon. In fact, the majority of laws precluded shooting in the vicinity of nesting areas so as not to interfere with the nets. It is believed that the last wild passenger pigeon was killed on March 24, 1900. One month later, while arguing for a bill to prevent interstate trade of wild birds, Iowa congressman John Fletcher Lacey noted that the wild pigeon "has entirely disappeared from the face of the earth." Passage of the Lacey Act in May, 1900, was a significant step toward protection of birds and a major precursor to the Endangered Species Act of 1970. Like the buffalo, passenger pigeons were particularly vulnerable because of their great numbers. Similarly, the expansion of railroads and influx of commercial hunters created a short-lived industry that rapidly exterminated millions of birds. Sadly, however, not enough passenger pigeons were preserved in zoos to guarantee survival of the species. Martha, the world's last passenger pigeon, died on September 1, 1914, at a zoo in Cincinnati.

Peter Neushul

SEE ALSO: Extinctions and species loss; Hunting.

People for the Ethical Treatment of Animals

DATE: established 1980
CATEGORY: Animals and endangered species

People for the Ethical Treatment of Animals (PETA) was founded in 1980 to establish and defend the rights of animals. Members of the organization protest practices that harm animals and educate the public about such abuses.

Alex Pacheco, who founded PETA with Ingrid Ward Newkirk in 1980, received initial notoriety for exposing cruelty to monkeys in a Silver Spring, Maryland, laboratory. Although PETA began its campaign with only two people during the 1980's, by the end of the 1990's it had a worldwide membership of more than 500,000 and a $10 million budget. PETA's brand of radicalism and sophisticated use of the media created a national debate. During the 1980's PETA's major concerns were vivisection, factory farming, hunting, fishing, zoos, and circuses. By the 1990's the organization's major concerns had shifted to the fur industry and the use of animals in product testing, agricultural production, and biomedical research. PETA's campaign against the fur industry convinced designers such as Giorgio Armani, Ralph Lauren, and Calvin Klein, as well as supermodels such as Cindy Crawford and Christy Turlington, not to design or model clothing made of fur.

The general principles upon which PETA was founded are simple but potent: Animals are not on earth for humans to eat or wear. Technology has enabled humans to make incredibly diverse substitutes. They further contend that animals are not on earth to use for scientific experiments. PETA advocates ending all animal abuse. Violating animal rights, they maintain, is similar to violating human rights. In fact, the former is worse because animals cannot speak for themselves.

One tactic that PETA has used in targeting animal rights violators is undercover investigations. Many of these investigations have revealed patterns of cruelty that have appalled the public. For example, PETA's Save the Whales campaign

Members of People for the Ethical Treatment of Animals gather at Procter and Gamble's headquarters in Cincinnati, Ohio, in August of 1997 to protest the company's use of animal testing. (AP/Wide World Photos)

drew widespread public support. Many people saw this as a legitimate concern and contributed generous amounts of money to the organization. One of PETA's major goals is public education through the use of graphic visual images, expert testimony, media events, seminars, workshops, and lectures. Another effective tactic is PETA's grassroots activities at colleges and universities.

PETA has been criticized for ecoterrorism and monkeywrenching, examples of which include illegal entry, vandalism, and theft of laboratory animals and equipment. Critics contend that PETA's philosophical stance is based on sentimentalism and narrow-minded dogmatism and that its practices are akin to terrorism. They charge that PETA manipulates young people's emotions through misinformation disguised as "public education." Despite active opposition from industry and research communities, however, PETA has proven itself to be one of the most effective animal rights groups in the world. It has gained mainstream support from the public and from many powerful people involved in

government, entertainment, and humanitarian organizations.

Chogollah Maroufi

SEE ALSO: Animal rights; Animal rights movement; Ecoterrorism; Hunting; Monkeywrenching; Singer, Peter; Zoos.

Pesticides and herbicides

CATEGORY: Pollutants and toxins

Pests are generally defined as unwanted organisms that interfere with human activities. Pesticides are chemicals designed to kill such undesirable organisms. The major types of pesticides in common use are insecticides (to kill insects), nematocides (to kill nematodes), fungicides (to kill fungi), herbicides (to kill weeds), and rodenticides (to kill rodents). Herbicides and insecticides make up the majority of the pesticides applied in the environment.

While the use of pesticides has mushroomed since the introduction of monocultures (the agricultural practice of growing only one crop on a large amount of acreage), the application of chemicals to control pests is by no means new. The use of sulfur as an insecticide dates back before 500 B.C.E. Salts from heavy metals such as arsenic, lead, and mercury were used as insecticides from the fifteenth century until the early part of the twentieth century, and residues of these toxic compounds are still being accumulated in plants that are grown on soil where these materials were used. In the seventeenth and eighteenth centuries, natural plant extracts such as nicotine sulfate from tobacco leaves and rotenone from tropical legumes were used as insecticides. Other natural products, such as pyrethrum from the chrysanthemum flower, garlic oil, lemon oil, and red pepper, have long been used to control insects.

In 1939 the discovery of dichloro-diphenyl-trichlorethane (DDT) as a strong insecticide opened the door for the synthesis of a wide array of synthetic organic compounds to be used as pesticides. Chlorinated hydrocarbons such as DDT were the first group of synthetic pesticides. Other commonly used chlorinated hydrocarbons include aldrin, endrin, lindane, chlordane, and mirex. Because of their low biodegradability and long persistence in the environment, the use of these compounds was banned or severely restricted in the United States after years of use. Organophosphates such as malathion, parathion, and methamidophos have replaced the chlorinated hydrocarbons. These compounds biodegrade in a fairly short time but are generally much more toxic to humans and other animals than the compounds they replaced. In addition, they are water soluble and, therefore, more likely to contaminate water supplies. Carbamates such as carbaryl, maneb, and aldicarb have also been used in place of chlorinated hydrocarbons. These compounds rapidly biodegrade and are less toxic to humans than organophosphates, but they are also less effective in killing insects.

Herbicides are classified according to their method of killing rather than their chemical composition. As their name suggests, contact herbicides such as atrazine and paraquat kill when they come in contact with the leaf surface. Contact herbicides generally disrupt the photosynthetic mechanism. Systemic herbicides such as diuron and fenuron circulate throughout the plant after being absorbed. They generally mimic the plant hormones and cause abnormal growth to the extent that the plant can no longer supply sufficient nutrients to support growth. Soil sterilants such as triflurain, diphenamid, and daiapon kill microorganisms necessary for plant growth and also act as systemic herbicides.

PESTICIDE USE

In the United States, approximately 55,000 different pesticide formulations are available, and Americans apply about 500 million kilograms (1.1 billion pounds) of pesticides each year. Fungicides account for 12 percent of all pesticides used by farmers, insecticides account for 19 percent, and herbicides account for 69 percent. These pesticides have primarily been used on four crops: soybeans, wheat, cotton, and corn. Approximately $5 billion are spent each year on pesticides in the United States, and about 20 percent of this is for nonfarm use. On a per unit of land basis, homeowners apply approximately five times as much pesticide as do farmers. On a worldwide basis, approximately 2.5 tons (2,270 kilograms) of pesticides are applied each year. Most of these chemicals are applied in developed countries, but the amount of pesticide used in underdeveloped countries is rapidly increasing. Approximately US$20 billion are spent worldwide each year, and this expenditure is expected to increase in the future, particularly in the underdeveloped countries.

The use of pesticides has had a beneficial impact on the lives of humans by increasing food production and reducing food costs. Even with pesticides, pests reduce the world's potential food supply by as much as 55 percent. Without pesticides, this loss would be much higher, resulting in increased starvation and higher food costs. Pesticides also increase the profit margin for farmers. It has been estimated that for every dollar spent on pesticides, farmers experience an increase in yield worth three to five dollars. Pesticides appear to work better and faster than alternative methods of controlling pests. These chemicals can rapidly control most pests, are cost effective, can be easily shipped and applied, and have a long shelf life as compared to alternative methods. In addition, farmers can quickly switch to another pesticide if genetic resistance to a given pesticide develops.

Perhaps the most compelling argument for the use of pesticides is the fact that pesticides have saved lives. It has been suggested that since the introduction of DDT, the use of pesticides has prevented approximately seven million premature human deaths from insect-transmitted diseases such as sleeping sickness, bubonic plague, typhus, and malaria. Perhaps even more lives have been saved from starvation because of the increased food production resulting from the use of pesticides. It has been argued that this one benefit far outweighs the potential health risks of pesticides. In addition, new pesticides are continually being developed, and safer and more effective pest control may be available in the future. In spite of all the advantages of using pesticides, their benefit must be balanced against the potential environmental damage they may cause.

ENVIRONMENTAL CONCERNS

An ideal pesticide should have the following characteristics: It should not kill any organism other than the target pest; it should in no way affect the health of nontarget organisms; it should degrade into nontoxic chemicals in a relatively short time; it should prevent the development of resistance in the organism it is designed to kill; and it should be cost effective. Since no currently available pesticides meet all of these criteria, a number of environmental problems have developed, one of which is broad-spectrum poisoning. Most, if not all, chemical pesticides are not selective. In other words, they kill a wide range of organisms rather than just the target pest. Killing beneficial insects, such as bees, lady bird beetles, and wasps, may result in a range of problems. For example, reduced pollination and explosions in the populations of unaffected insects can occur.

When DDT was first used as an insecticide, many people believed that it was the final solution for controlling many insect pests. Initially, DDT dramatically reduced the number of problem insects; within a few years, however, a

Pesticides are applied to a cotton field in Georgia. Approximately $5 billion are spent on pesticides in the United States each year. (AP/Wide World Photos)

number of species had developed genetic resistance to the chemical and could no longer be controlled with it. By the 1990's there were approximately two hundred insect species with genetic resistance to DDT. Other chemicals were designed to replace DDT, but many insects also developed resistance to these newer insecticides. As a result, although many synthetic chemicals have been introduced to the environment, the pest problem is still as great as it ever was.

Depending on the type of chemical used, pesticides remain in the environment for varying lengths of time. Chlorinated hydrocarbons, for example, can persist in the environment for up to fifteen years. From an economic standpoint, this can be beneficial because the pesticide has to be applied less frequently, but from an environmental standpoint, it can be detrimental. In addition, when many pesticides are degraded, their breakdown products, which may also persist in the environment for long periods of time, are often toxic to other organisms.

Pesticides may concentrate as they move up the food chain. All organisms are integral components of at least one food pyramid. While a given pesticide may not be toxic to species at the base, it may have detrimental effects on organisms that feed at the apex because the concentration increases at each higher level of the pyramid, a phenomenon known as biomagnification. With DDT, for example, some birds can be sprayed with the chemical without any apparent effect, but if these same birds eat fish that have eaten insects that contain DDT, they lose the ability to properly metabolize calcium. As a result, they lay soft-shelled eggs, which causes the death of most of the offspring.

Pesticides can also be hazardous to human health. Many pesticides, particularly insecticides, are toxic to humans, and thousands of people have been killed by direct exposure to high concentrations of these chemicals. Many of these deaths have been children who were accidentally exposed to toxic pesticides because of careless packaging and storage. Numerous agricultural laborers, particularly in Third World countries where there are no stringent guidelines for the handling of pesticides, have also been killed as a result of direct exposure to these chemicals. Workers in pesticide factories are also a high-risk group, and many of them have been poisoned through job-related contact with the chemicals. Pesticides have been suspected of causing long-term health problems such as cancer. Some of the pesticides have been shown to cause cancer in laboratory animals, but there is currently no direct evidence to show a cause-and-effect relationship between pesticides and cancer in humans.

D. R. Gossett

SUGGESTED READINGS: An excellent discussion of the environmental issues associated with the use of pesticides can be found in *Living in the Environment: An Introduction to Environmental Science* (1992), by G. Tyler Miller, Jr. Even though Rachel Carson's *Silent Spring* (1962) is somewhat outdated, it is an excellent dramatization of the potential dangers of pesticides and served as a major catalyst for bringing the hazards of using pesticides to the attention of the general public. Good discussions of environmental ethics, including the ethics of using agricultural chemicals, can be found in *People, Penguins, and Plastic Trees* (1995), by Christine Pierce and Donald VanDeVeer. *Environmental Issues in the 1990's* (1992), by Antionette M. Mannion and Sophia R. Bowlby, offers a good discussion of some of the environmental issues associated with the use of pesticides. *Man and Environment: A Health Perspective* (1990), by Anne Nadakavukaren, contains a good discussion on human health issues associated with the use of pesticides.

SEE ALSO: Agricultural chemicals; Agricultural revolution; Biopesticides; Dichloro-diphenyl-trichloroethane (DDT); Diquat; Green Revolution; Integrated pest management; Malathion; Organic gardening and farming.

Pets and the pet trade

CATEGORY: Animals and endangered species

The growing demand for animals, domestic and exotic, legal and illegal, has developed into a

$100 billion industry. An increasing number of dogs are being bred in puppy mills under filthy conditions to meet the growing demand for domestic pets. Meanwhile, many wealthy collectors seek rare pets as a symbol of status. However, such collecting can further threaten already endangered species. Regulations have been enacted in an attempt to stop the illegal and dangerous trafficking of animals.

Pets play an important part in the lives of humans. They provide many benefits, including companionship, love, humor, exercise, a sense of power, and outlets for displacement, projection, and nurturance. Studies have shown that talking to pets reduces stress, promotes feelings of reverie and comfort, and enhances longevity and physical health. Pets have also played therapeutic roles for the sick and elderly by helping return a sense of reality and actively involving them in caring, sacrifice, and companionship. Institutionalized elderly have shown great signs of improvement when dogs were introduced to the environment. A 1980-1981 study on people who were receiving care from the hospital-based care program in a large Veterans Administration medical center showed that pet owners fared better with health issues than patients without pets.

The increasing demand for domestic pets has led to a dramatic increase in the number of puppy mills, many of which are run by people who know little to nothing about the animals they are breeding. Such puppy mills often operate under filthy conditions characterized by piles of trash, unhealthy dogs, the use of shopping carts and crates to house animals, and the presence of rats, which often eat the same food as the dogs. Animals suffer from overcrowding, lack of veterinary care, unsanitary conditions, and incessant breeding. Many females are bred twice each year, which is usually each time they come into heat. However, this is not a healthy practice. The ignorance of the breeders often results in the delivery of malnourished and sick puppies to pet shops.

Commercial kennels are licensed and regulated by the United States Department of Agriculture, which enforces the Animal Welfare Act of 1967. However, these regulations provide only limited standards of care, and many animal rights activists have demanded strict revisions.

Illegal Trade and Regulations

The illegal trade of animals is a worldwide occurrence. The most prominent regulation of trade is the Convention on International Trade in Endangered Species (CITES), established July 1, 1975. CITES was signed by 145 countries that agreed to ban commercial international trade in endangered species and regulate and monitor trade in other threatened species. CITES categorizes animals according to three main appendices: Appendix I prevents commercial trade of live animals, their parts, or derivatives; Appendix II permits trade under certain controls; and Appendix III allows trade only with an export permit from the country listing the animal or a certificate of origin from a country that did not list it. Threatened and endangered animals fall under the three appendices based upon their probability of extinction. Species under the greatest threat of extinction are listed in Appendix I, while those least threatened are listed in Appendix III.

Thailand and Japan offer numerous examples of violation of the appendices. In one instance, Thailand falsified a quota by issuing inexact statements concerning the number of Appendix II tortoises being shipped out of the country. Investigators found that the number of tortoises in each shipment ranged from seven to seven thousand. The implementation of regulations has increased the value of endangered animals, thus encouraging smugglers to take extreme risks in order to make money. In February, 1995, a man was arrested, sentenced to two years in prison, and fined for smuggling three young gibbons—a type of ape found in southeastern Asia and the East Indies—in his luggage. Two of the gibbons died of disease, and the third was placed in a zoo. In January, 1996, a man returning to Japan from Thailand was arrested for attempting to bring in thirty-nine Appendix II tortoises. Authorities later discovered that the same man had previously made twenty-six successful smuggling trips between Thailand and Japan.

CITES does allow the transportation of animals that have been captive bred for commercial purposes, which opens a loophole through which illegal trafficking can occur. False documentation declaring that a transported animal has been captive bred has been used to get illegal animals into Japan. Another problem Japan faces is the lack of domestic reinforcement. Pet shops and other sources have continued to stock endangered species that have made their way into the country. Without enforcement of CITES, the illegal activity of buying, selling, and trading endangered animals continues to look attractive for smugglers.

UPHOLDING REGULATIONS

Two cases illustrate how the members of CITES have worked together to uphold the convention's regulations. Operation Chameleon and Operation Jungle Trade were two international projects implemented to stop trade in endangered animals; both of them were under the jurisdiction of the Fish and Wildlife Service in the United States.

In September, 1998, Operation Chameleon was instituted to stop the trade of rare reptiles throughout the world, most of which were listed in the CITES Appendix II. Among the rare reptiles confiscated were the Chinese alligator, Gray's monitor (a lizard native to the Philippines), the false gavial (a crocodile found in parts of Malaysia, Indonesia, Singapore, and southern Thailand), the spider tortoise (Madagascar), and the radiated tortoise (Madagascar). Of these, the Gray's monitor and the spider tortoise are the only reptiles that can be legally traded with a permit. The Chinese alligator can trade for $15,000, while the Gray's monitor can sell for $8,000. Adult radiated tortoises and false gavials are worth $5,000, and spider tortoises can sell for about $2,000. Investigators were successful in shutting down a major chain of the illegal reptile trade between Asia and North America. Most of the reptiles, which originated in southeast and central Asia, New Zealand, and Madagascar, and were destined for the United States.

Operation Jungle Trade sought to bring an end to the illegal trade of exotic birds across the United States-Mexico border and into several other countries, including Australia, Belize, Brazil, Costa Rica, Egypt, Ghana, Honduras, New Zealand, Panama, and South Africa. The case produced forty arrests and the seizure of more than 660 exotic birds and other animals. Most of the confiscated animals were native Mexican species listed under CITES Appendix I. During Operation Jungle Trade, law enforcement officials from Australia, Canada, New Zealand, and Panama worked together to document criminal activity and gather evidence, carrying out more than forty separate but related investigations all over the world.

U.S. REGULATIONS

The Lacey Act of 1900 and the Endangered Species Act of 1973 gave the responsibility of controlling the import and export of wildlife to the United States Department of the Interior. The United States is a member of CITES through the Endangered Species Act, in which the convention's regulations and guidelines are made law. The Department of the Interior carries out its obligations by designating specific ports of entry for wildlife, providing inspectors for each port to monitor shipments of wildlife, and issuing licenses to commercial wildlife traders. Furthermore, the Interior Department cooperates with other federal organizations and other countries to monitor and investigate possible animal trade violations.

Falling under the category of other federal agencies, the U.S. Customs Service is primarily responsible for the inspection and clearance of all goods imported to the United States. Legal support is provided by attorneys from the Wildlife and Marine Resources Section of the Environment and Natural Resources Division of the Department of Justice. The Animal and Plant Health Inspection Service of the Department of Agriculture requires all birds and specific mammals to be quarantined before entering the United States. Even though there are only ninety-two inspectors and 230 special agents, progress is being made, and the illegal trading of animals is being further regulated.

The illegal animal trade has tremendous impacts on the animals themselves. A main portion of the animal trade in Taiwan between 1985 and 1990 was in orangutans. For every one that survived transportation, three or four died. Illegal transportation is almost never accompanied by proper animal care. A similar problem arises with puppy mills. Four-week-old dogs are often shipped in crates without food or water. Many die, and if they survive they arrive at pet stores sick, diseased, and with psychological problems.

The Animal and Plant Health Inspection Service of the Department of Agriculture plays the important role of quarantining animals that come into the United States. This is a critical aspect of the importation of animals. Species from South America, for example, may carry a specific strain of a disease or virus not present in North America. Quarantines and health inspections are bypassed by illegal smugglers, raising the possibility that imported diseases could spread to other animals or humans.

Allison M. Popwell

SUGGESTED READINGS: A good look at the relationships between humans and animals can be found in Alan M. Beck and Aaron Honori Katcher, *New Perspectives on Our Lives with Companion Animals* (1983). Information on the growing pet care industry can be found in Han Greenberg's "Pet Plush Apartments," *U.S. News and World Report* (February, 1998). News releases by the United States Fish and Wildlife Service can be found on the Internet. "Federal Agents Target Illegal Bird Trade" (http://www.hbonline.com/huntingnews/fwsillbirds1.htm) contains a detailed report of Operation Jungle Trade. "Probe of International Reptile Trade Ends with Key Arrests" (http://www.fws.gov/r9extaff/pr9851.html) discusses Operation Chameleon. The American Humane Society's "Help Reform the Puppy Mills" (http://arrs.envirolink.org/projects/puppy_mill.html) provides information about puppy mills, what a puppy mill is, and how to avoid receiving a pet from a mill.

SEE ALSO: Animal rights; Captive breeding; Convention of International Trade in Endangered Species; Endangered species; Endangered Species Act; Endangered species and animal protection policy.

Pinchot, Gifford

BORN: August 11, 1865; Simbury, Connecticut
DIED: October 4, 1946; New York, New York
CATEGORY: Preservation and wilderness issues

Gifford Pinchot, the first head of the U.S. Forest Service, was also the first American forester.

Because no forestry school existed in the United States, Gifford Pinchot trained at the French Forest School in Nancy in northeastern France to become a forester. His first forestry projects were for George W. Vanderbilt at the Biltmore estate and W. Seward Webb, Vanderbilt's brother-in-law, in the Adirondack Mountains in New York.

Pinchot believed that all forest resources should be made available for wise use by humans and was critical of Yellowstone National Park and the Adirondack State Forest Preserve because they did not permit logging. In his autobi-

Gifford Pinchot, who, in 1905, became the first head of the U.S. Forest Service. (Archive Photos)

ography *Breaking New Ground* (1947), Pinchot stated that "conservation is the foresighted utilization, preservation, and/or renewal of forests, waters, lands, and minerals, for the greatest good of the greatest number for the longest time." Conservationists led by Pinchot believed in the wise use of all the nation's resources. They were not preservationists who valued nature for nature's sake. Pinchot's chief ambition was to practice forestry on government land.

In 1891 Congress passed a law allowing land to be removed from the public domain and set aside in forest reserves. At that time, the Bureau of Forestry was in the Department of Agriculture, but the forest reserves were administered through the General Land Office in the Department of the Interior. The reserves were run by political hacks holding patronage appointments. In 1897 the secretary of the interior asked Pinchot to travel west and report on the condition of the reserves. Pinchot's assignment caused the head of the Bureau of Forestry to resign, and Pinchot was appointed to replace him in 1898. To supply their son with trained foresters, Pinchot's parents endowed the Yale Forest School, which opened in 1900 as a two-year postgraduate program. Pinchot still had no authority over the forest reserves.

Pinchot had advised Theodore Roosevelt on forestry when Roosevelt was governor of New York. Roosevelt's message to Congress after President William McKinley's assassination in 1901 favored placing the forest reserves under the authority of the Department of Agriculture. The transfer occurred in 1905 when the Forest Service was organized. Pinchot boasted that the forest reserves would be run to benefit poor settlers rather than rich people, a position that ultimately led to Pinchot's downfall.

With Roosevelt's approval, Pinchot brought conservation to national attention at the National Governors' Conference in 1907, the National Conservation Commission in 1908, and the North American Conservation Conference in 1909. Pinchot's views were not supported by Roosevelt's hand-picked successor, President William Howard Taft, or by Taft's secretary of the interior, Richard Ballinger. The resulting feud caused Taft to fire Pinchot. Although Pinchot remained active in public service (he served two terms as governor of Pennsylvania), he never fully regained his position of national prominence.

Gary E. Dolph

SEE ALSO: Conservation; Conservation policy; Forest and range policy; Forest management; Forest Service, U.S.; National forests; Roosevelt, Theodore.

Planned communities

CATEGORY: The urban environment

Planned communities are distinguished from other human settlements by the fact that most design aspects are determined before the settlement is built. Among the factors considered by designers of such communities is their impact on the surrounding environment.

Ever since the development of agriculture, enclaves of people have clustered in central locations and dwelt in close proximity to one another. Such ancient cities as Babylon, Nineveh, Jerusalem, Cairo, Athens, and Rome have set the pattern for urban planners through the ages. The typical Greek or Roman city had two main thoroughfares running perpendicular to each other that intersected to form a public area at the city's center. Smaller streets running parallel to the main thoroughfares contained dwellings and shops. This configuration created the grid pattern still used by many city planners. Medieval cities were usually walled for defense and had narrow streets that curtailed the movement of invaders. These small arteries led to a central square, where inhabitants gathered to socialize and receive information. During the Renaissance, narrow passageways gave way to large boulevards that opened cities to light and space.

The first city in the United States to be built according to a plan was Washington, D.C., designated to replace New York City as the nation's capital. The city planner for Washington, Pierre Charles l'Enfant, borrowed heavily from his mentor Baron Georges-Eugène Haussmann, inventor of Paris's spokes-and-hub plan. L'Enfant superimposed diagonal thoroughfares, all of which led to the Capitol, upon a giant rectangle.

A great advance in city planning occurred in 1908 with the creation of Forest Park Gardens outside New York City. The designer of this project, Frederick Law Olmsted, Sr., had designed company towns in New England. He was committed to preserving large areas of open space in densely populated residential communities. Olmsted persuaded the commissioners of New York City to purchase large tracts of undeveloped land, which eventually became Central Park. Olmsted's influence is felt in such subsequent projects as the Sunnyside Gardens (now called Queens) in New York, and Radburn, New Jersey, a bedroom community for Manhattan created in 1929. Many later planned communities, such as Reston in Virginia, Greenbelt and Columbia in Maryland, and Westlake and Laguna Niguel in California, heeded Olmsted's call for the preservation of space for public parks, bicycle lanes, jogging paths, and other such amenities.

Contemporary city planners consistently pay careful attention to environmental concerns in their planning. Among the matters they consider are the creation of parks and open spaces, noise abatement, proximity to shopping facilities so that automotive pollution is minimized, and zoning regulations that prevent the incursion of factories, airports, and other potential polluters into residential areas.

One model community that has taken such considerations very seriously is Tapiola, 10 kilometers (6 miles) outside Finland's capital, Helsinki. Tapiola, the brainchild of Heikki von Hertzen, a banker who developed the garden-city plan, has separate roads for automobiles and bicycles, as well as an elaborate network of footpaths for pedestrians. Major roads are routed around Tapiola, minimizing vehicular traffic within the town. Tapiola's dwellings are all within 230 meters (755 feet) of a food store, so driving within the community is unnecessary. Every building within the town is heated by a central heating plant. Tapiola values people over machines and industry, although some small, nonpolluting industry is permitted within it to provide work for its inhabitants, relieving them of the need to commute.

When World War II ended in 1945, hundreds of thousands of veterans returning to the United States married and established families. They had an immediate need for affordable housing. This need was met in part by William Levitt, who planned and built communities in the Northeast. By 1950 he had turned a potato patch near New York City into Levittown, a community of 17,500 reasonably priced single-family homes that became a model of city planning for the masses. Its schools, shops, and parks were attractive, although the residences in the community were identical cookie-cutter houses. Levitt overcame this problem by using different colors of exterior paint and through landscaping, which lent individuality to the look-alike houses of the community.

With the graying of America, many new communities have been established for people over the age of fifty. Among the notable leisure communities that serve this population are Del Webb's various Sun Cities, Lake Havasu City, Leisure World, and Green Valley in Arizona. A recent offshoot of this movement has been the establishment of Manasota, a retirement community limited to gays and lesbians, outside Sarasota, Florida.

Among the most ambitious planned communities in the United States are two near Washington, D.C. Columbia, Maryland, begun in 1963 by John Wilson Rouse, consists of seven villages centered on a business area. The population of Columbia, a self-contained community, is about 110,000, many of whom are employed within the town. At about the time Rouse was establishing Columbia, Robert E. Simon, Jr., planned a community 32 kilometers (20 miles) northwest of Washington, eventually named Reston, Virginia. This community, like Columbia, consists of seven villages that accommodate 75,000 people in a broad variety of dwellings ranging from single-family homes to high-rise apartments and condominiums. Similar planned communities have been established since 1965 in other parts of the United States, including Southern California, Arizona, New York, Pennsylvania, Arkansas, Texas, Hawaii, North Carolina, and Florida.

R. Baird Shuman

SUGGESTED READINGS: E. V. Walter offers useful insights into environmental concerns in city planning in *Placeways: A Theory of the Human Environment* (1988), as do Sim Van Der Ryn and Peter Calthorpe in *Sustainable Communities: A New Design Synthesis for Cities, Suburbs, and Towns* (1986). Gurney Breckenfeld's *Columbia and the New Cities* (1971), Carlos Campbell's *New Towns: Another Way to Live* (1976), and Dennis Hardy's *From Garden Cities to New Towns* (1991) also focus on the need for environmental awareness in city planning.

SEE ALSO: Olmsted, Frederick Law, Sr.; Urban planning; Urbanization and urban sprawl.

Plastics

CATEGORY: Pollutants and toxins

Serious environmental issues are raised in both the manufacture and disposal of plastics. Most plastics are made from oil, the use of which can have a negative impact on the environment. Methods used to minimize the effect of plastic disposal include incineration, biodegradation, recycling, and source reduction.

Plastics are formally defined as materials that, at some point in their manufacture, can be molded into various shapes. Industry classifies plastics as either thermoplastic or thermosetting. Thermoplastics can be repeatedly softened by heating and molded into shapes that harden upon cooling. Thermosetting plastics are also soft when first heated and can be molded into shapes. When reheated, however, they decompose instead of softening. The major raw material for making plastics is oil. The drilling and transport of oil may cause environmental problems, among them oil spills, tanker accidents, and air pollution from oil refineries.

Many people feel the major environmental issue involving plastics is their disposal. The very properties that make plastics so useful—their stability and resistance to attack by chemicals and bacteria—make them almost indestructible. Unlike metal, wood, and paper, plastics do not corrode or decay. They have contributed unsightly litter along roadsides and can be found floating on the surfaces of the ocean everywhere. Since plastics are light, they account for much of the volume of solid waste. For example, while plastics account for only 7 percent of the total weight of solid waste generated in the United States, they make up about 20 percent of its volume.

While a number of solutions to the problem of disposing of plastics have been found and tried, each solution has led to adverse effects. Industry and government have therefore been using a combination of solutions to mitigate the effect of the disposal of plastic on the environment. One problem associated with disposal is the chemicals that must be added to some plastics during production to improve or alter their properties. These may contaminate the environment when the plastic is discarded. Plastics that are brittle and hard are often made soft and pliable by the addition of compounds called plasticizers. For example, polyvinylchloride (PVC) treated in this way can be used to make raincoats and drapes. Plasticizers, as a result of their widespread use, have been detected in virtually all soil and water ecosystems, including the open ocean.

There is no single way to solve the problem of disposal of plastics, but a number of methods have lessened their impact on the environment. The first is recycling. Although recycling does not literally dispose of plastics as do incineration or biodegradation, it does help reduce the amount of new plastic being made and the energy needed to make it. More than three-quarters of all plastics are thermoplastic and can be recycled by melting and remolding. In order for recycling to be successful, however, the plastics must first be separated according to type. Different plastics melt at different temperatures. Molding machines only work if they are fed with pure materials. In 1988 the Society of the Plastics Industry developed a uniform coding system that makes it possible to sort waste plastic. Most states require the use of this coding system. Uses are being developed for recycled plastic that take advantage of its low cost and durability. Plastic products with short service lives, such as foam, wrap, and containers, can be made into products with long lives, such as construction materials and plastic pipe. Recycled soft-drink bottles are being used to make carpeting and insulation for parkas. Disposable cups and plates are being converted to plastic "lumber."

Incineration of plastics to produce energy is another approach to disposal but has the drawback that certain plastics release toxic gases when burned. However, many people feel that if incineration is carefully monitored and controlled, it can lead to a large reduction of plastic waste, generate much-needed energy, and have little negative impact on the environment. Industry has also been seeking ways to make plastics more degradable. Certain elements are introduced into plastics during the manufacturing process to make them susceptible to bacterial attack (biodegradability) or decomposition by light. Landfills, however, are constructed to keep out light and bacteria. This prevents any material, even natural wastes, from decomposing while in a landfill. Biodegradable plastics are also not recyclable and, in fact, interfere with the recycling process.

The remaining option, source reduction, appears simple and direct. The proposal is to mitigate the problem by decreasing the quantity of plastics produced and consumed. However, even this option is far more complicated than it appears. The problem is that something else must be used to replace the plastic, and this substitution can be more expensive, dangerous to the environment, or produce more waste. Looking for ways to reduce the amount of packaging used by industry appears to be a solution with fewer problems and has been successfully used in Europe.

David MacInnes, Jr.

Workers shovel plastic bottles onto a shredder in a plastics recycling plant. Recycling is one means of reducing the impact of plastic on the environment. (Jim West)

SUGGESTED READINGS: For a well-written general introduction to plastics and environmental issues, see Carl H. Snyder's *The Extraordinary Chemistry of Ordinary Things* (1995). For an excellent introduction to the disposal of waste plastic and the controversies involved, see Nancy Wold and Ellen Feldman, *Plastics: America's Packaging Dilemma* (1990). A readable account of recycling of plastics can be found in Noel Grove, "Recycling," *National Geographic* (July, 1994).

SEE ALSO: Disposable diapers; Landfills; Ocean dumping; Recycling; Waste management.

Poaching

CATEGORY: Animals and endangered species

Poaching is the illegal or unauthorized hunting, capture, sale, or slaughter of wild animals, as well as the illegal harvest of protected plant species.

Poaching occurs for many reasons, most of which are economic. Bears and tigers are slaughtered for their meat, pelts, and internal organs, which are used in some forms of medicine in Asia; birds and monkeys are captured and sold as pets; predators are killed for obeying their natural instincts; deer and sheep are collected as trophies; and all are, at times, hunted simply for sport.

The concept of poaching dates back to the middle ages, when the lands and all they contained were controlled by the ruling class. Killing animals on these lands, even for the sole purpose of putting food on the table, was outlawed and punishable by death. Even well into the twentieth century, many poachers claimed the right to hunt out of season or over prescribed limits in order to subsist.

Nevertheless, people who poach to survive are a small minority; the rest are in it for profit. Experts liken the trade in illegal plants and animals to the drug cartels, with the profits rising dramatically each time the product changes hands. In India, a tiger carcass can bring between $100 and $300, an enormous sum in a region where the average wage is less than one dollar per day. Experienced middlemen separate the carcass into a series of valuable by-products. Almost the entire animal is used: Potions made of tiger bone are thought to promote longevity and cure rheumatism, the whiskers provide strength, and pills made from the eyes are believed to calm convulsions. An adult tiger can yield up to 23 kilograms (50 pounds) of bones, which will in turn sell for $185 per pound; its pelt is worth as much as $30,000.

The threat of extinction caused by poaching is real, and the losses are staggering. During the 1970's and 1980's, ivory poachers killed 93 percent of the elephants in Zambia's North Luangwa National Park. In 1988 poachers slaughtered 145 giant pandas for their gall bladders, eliminating one-seventh of their worldwide population in a single year. Studies have indicated that some species, such as orangutans and rhinoceroses, are being killed at a rate of two to three times the rate at which they are able to reproduce. As the number of remaining animals drops, their black market value dramatically rises. Dall sheep, a species native to the Rocky Mountains, have become so rare that hunters are willing to pay up to $100,000 for the privilege of shooting one.

As a result of poaching, species such as the Bengal tiger and the northern white rhinoceros no longer have a genetically viable population in the wild. Reduced numbers lead to inbreeding, which in turn leads to weaker, more disease-prone animals that are less likely to survive in an diminishing environment. In addition, trophy hunters typically seek the largest, healthiest animals. This not only reduces the quantity of wild animals in the gene pool but also reduces the quality of the herds. More than 38 percent of the elephants in North Luangwa National Park are genetically tuskless, compared to 2 percent in a natural elephant population, because poachers targeted the mature, tusk-bearing animals of breeding age.

The illegal slaughter also takes its toll on the animals remaining in the wild. Poachers target the largest, most mature animals, which leaves the younger, inexperienced creatures to fend for themselves before they have learned skills essential for survival. In turn, when these inexperienced animals breed, they often lack the parenting skills needed to raise their offspring to maturity.

Marine poaching carries a hidden, yet dangerous, side effect. Many types of shellfish, mollusks, and other types of marine life are declared off limits, not because of the danger of extinction, but because their environment has become contaminated by bacteria and other waste products. Shellfish, long a target of enterprising poachers who sell them to restaurants, are especially susceptible to contamination because they feed by filtering up to 38 liters (10 gallons) of water per hour.

An international ban on ivory in 1989, combined with increased law enforcement on Africa's game preserves, led to a slow but steady increase in the wild elephant population. Poaching is also being battled on the economic front, as the booming tourist trade has shown many governments and their citizens that their indigenous wildlife population may be more valuable alive. Education also plays an important role, as communities that formerly relied on poaching for survival are taught improved farming practices and methods of harvesting that do not endanger the native plants and wildlife.

A more controversial approach to conservation is "sustainable use," the limited harvest of threatened animals. The benefits are twofold. First, the local governments and communities gain economic benefit from the slaughter or harvest and are therefore less likely to engage in large-scale poaching. Second, the availability of legally obtained animals and their by-products should theoretically drive down their black market value so that poaching will no longer be worth the risk. Some countries have also begun "farming" endangered animals, not for eventual release into the wild, but for slaughter and sale.

P. S. Ramsey

SUGGESTED READINGS: For a concise, easy-to-read overview of poaching and its effects on the environment, see *Wildlife Poaching* (1994), by Laura Offenhartz Greene. A detailed account of the struggle to save the elephants in Zambia's North Luangwa National Park can be found in Mark Owen and Delia Owen, "Can Time Heal Zambia's Elephants?" *International Wildlife* 27 (May 15, 1997). To learn different perspectives on the issue of threatened plants and animals, see *Endangered Species* (1996), edited by Brenda Stalcup, part of the Opposing Viewpoints series.

Kenyan Wildlife Service employees search for poachers in Tsavo National Park, Kenya. Poaching practices threaten to drive several animal species, including elephants, into extinction. (AP/Wide World Photos)

Michael Satchell, "Wildlife's Last Chance," *U.S. News & World Report* (November 15, 1993) provides a grim overview of the "sustainable use" approach in the battle against poaching.

SEE ALSO: Endangered species; Endangered Species Act; Endangered species and animal protection policy; Extinctions and species loss; Hunting; Ivory and the ivory trade; Wildlife refuges.

Pollution permits and permit trading

CATEGORY: Pollutants and toxins

Pollution permits are an incentive-based strategy for pollution control in which companies are given the right to emit a specified amount of pollution during a specified time period. Permit holders may profitably reduce their emissions by selling unused permits to other polluters. Permit trading systems are viewed as a more efficient, less costly approach to pollution control than the "command and control" regulation traditionally used by government.

In a pollution permit trading system, a new type of property right is created: the right to produce a certain amount of pollution. A regulatory agency or special commission decides on the number of permits to be issued, based upon the toxicity, longevity, and other characteristics of the pollutant. If, as is likely, the issuing agency wants to reduce the amount of pollution, the total number of permits circulated account for less than the existing pollution. Each permit gives the holder the right to emit one unit of pollution during a specified period of time, normally one year. The unit—tons, pounds, cubic yards, or other measures—depends upon the particular pollutant. Each company would have many permits. The firm could choose to emit all the pollution covered by its permits or reduce pollution and sell its unused rights to a different company. The right to pollute would be transferrable, thus creating a market. Participation in the market would not be limited to corporations. An environmental group or governmental unit could buy and hold permits, further reducing the amount of pollution actually created.

The cost of the permits would not be set by government, but rather by the forces of supply and demand within that particular market. Permit brokers and some form of a trading exchange would develop to facilitate transactions. The demand in the market would come from new companies beginning operations, the expansion of existing companies, and those companies that face unusually steep pollution abatement costs. The supply of permits would come from firms going out of business or, more important, firms that had reduced emissions.

OPTIMAL LEVEL OF POLLUTION

At first, granting ownership to a right to pollute seems harmful to society. It seems as if society would benefit more if pollution was completely elimated rather than being traded to someone else. However, one must consider that there are costs involved in reducing pollution. From an economic perspective, zero pollution is not the optimal level of pollution. Because resources are limited, both the abatement cost—the cost of decreasing the pollution—and the monetary damage caused by the pollution must be considered. The optimal level occurs when the abatement cost and the damage cost are balanced. If the cost of reducing the pollution is greater than the damage created by the pollution, society is using too many resources to reduce that pollution. Environmentalists often object to this approach, arguing that not all environmental changes are quantifiable in dollars and that zero pollution is the optimal level of pollution.

Granting firms the ability to buy and sell pollution rights improves the overall efficiency of abatement. Each firm can decide whether it should reduce its pollution and sell more permits to others. Because firms do not use exactly the same materials and production processes, different firms have different abatement costs. Firms with abatement costs above the market price of the requisite number of permits will benefit by buying permits rather than spending their resources on pollution control. Firms with abatement costs below the market price will

Sulphur Dioxide Allowances in Circulation

SOURCE	NUMBER OF TRANSACTIONS	NUMBER OF ALLOWANCES
Initial Allocation (1995 to 2027)	8,377	283,639,201
Annual EPA Auctions (1993 to 1998)	389	1,350,000
Allowances Added to Circulation	488	10,644,708
Allowances Removed from Circulation	−963	−11,848,400
Totals	8,291	283,785,509

Source: U.S. Environmental Protection Agency Acid Rain Program, 1998.

benefit by reducing their pollution even more and selling additional permits.

A simple example may help clarify this concept. Suppose XYZ company and ABC company are similarly sized firms located near each another. Although they produce the same type of good, they use different production techniques. XYZ's production costs increase by $1,000 each time it removes 1 ton of air pollution. ABC spends $200 to remove 1 ton of pollution. Suppose the permitting agency wants to eliminate 20 tons of pollution. If each company is required to reduce pollution by 10 tons, the total cost would be $12,000 (XYZ spends $10,000, and ABC spends $2,000). Using marketable permits, however, ABC could reduce pollution by 20 tons and sell ten permits to XYZ for $500 each. XYZ spends $5,000 on the permits, which is $5,000 less than it would have spent on abatement. ABC makes money: It spent $4,000 on abatement but made $5,000 selling the permits. Society still gets the benefit of a 20-ton reduction in air pollution. Marketable permits thus allow pollution reduction to occur in the most efficient, cost-minimizing way.

A permit trading system also has lower enforcement costs than traditional regulation. Regulation requires a permanent bureaucracy to gather and analyze information, monitor activity, and enforce compliance. Given the complexity accompanying these tasks, regulation often creates litigation. Permit trading relies on the market to achieve efficiency. Corporate decision makers do not have to wait for a government agency to tell them whether they are making proper decisions and do not have to go to court if they think the agency is wrong. The market will reward them for proper decisions and punish them for poor decisions. Permit trading encourages flexibility and rewards innovation by allowing corporations to decide the appropriate technology and techniques as long as the permitted pollution level is not exceeded.

TRADING SULFUR DIOXIDE PERMITS

Responding to the criticisms aimed at regulation by economists and corporations, the United States government began considering permit trading programs in the late 1970's. The Sulfur Dioxide Allowance Program created under Title IV of the Clean Air Act amendments of 1990 is the first statutorily mandated, national market-based approach to pollution control. The 1990 Clean Air Act includes a goal of reducing sulfur dioxide by 10 million tons by the year 2010, down from the 26 million tons emitted in 1980. Sulfur dioxide is viewed as a contributing cause of the acid rain problem in Canada and the northeastern section of the United States. In the program's initial period (1995-1999), 445 out of the 2,000 electric generating plants in the United States participated. The plants involved were large coal-burning utilities in eastern and midwestern states. Each permit issued allowed the emission of 1 ton of sulfur dioxide. Emission reductions were to begin in 1995, with full compliance by 2010. The utilities were issued 8.7 million permits, based upon previous pollution of 10.9 million tons in 1980. The program had an immediate positive effect: In 1995 sulfur dioxide emissions were reduced to only 5.3 million tons.

The Environmental Protection Agency (EPA) maintains the central registry of allowances and transfers permits between accounts when a transaction is completed. As of 1999, the Allowance Program had been very successful even though the actual number of trades is much lower than forecast. Not only has the level of sulfur dioxide decreased faster than expected, but the compliance cost for industry and the enforcement cost for government is less than expected as well. It was estimated that plants needing to buy emission permits would pay between $180 and $981 per ton. Because so many plants successfully reduced emissions, the price of a permit fell to less than $100 and has generally traded around $125. To participate in the program, all firms are required to install and maintain continuous monitoring devices. Because the government does not have to perform the continuous inspections and litigation typical of traditional regulation, enforcement cost is quite modest. To ensure compliance, violators face a $2,000-per-ton fine for excess emissions, well above the $125 per ton set in the market. The fine is nondiscretionary, meaning that any violator must pay the entire fine and offset the violating emission in the following year.

CONCERNS

The success of permit trading is dependent upon several factors. There must be a limited number of allowances in circulation. Existing polluters, understanding that permits will have value in the future, want as many as possible. Thus, the first steps in the development of a trading program—deciding what formula to use to distribute permits and deciding whether to auction the permits or give them away—can stir great controversy. The eventual benefits of the tradable permit system are independent of the initial distribution. In other words, the eventual

market price will not depend upon whether the permits were initially sold or given away. The distribution process is political, not economic, in nature. This creates several problems. If permits are distributed based upon current levels of pollution, companies that have already taken steps to reduce their emissions are penalized. If permits are distributed equally, it creates different effects on very large and very small companies. Some industries may have an easier time reducing pollution than others. Some local economies are completely dependent upon only one industry, while others are more diversified. What happens if the allocation of permits harms a plant that is the major employer in an economically distressed area? These concerns create considerable political difficulty in determining how to distribute permits.

The major problem facing the development of marketable permits is that many potential pollution markets are hampered by thin markets with only a few buyers and sellers. Thin markets are troublesome because trades are so infrequent that the market price may not be immediately apparent. For example, if only ten cars were sold in Houston each year, would the eleventh buyer know the approximate cost of a car before stepping into the showroom? If a seller loses a potential buyer by setting the price too high, how long will it be before the next customer arrives? Under these circumstances, firms face considerable transaction costs as buyers and sellers independently try to determine the appropriate price. The thin market problem proved an insurmountable obstacle in several early attempts to develop permit trading programs. The Emissions Trading Program initiated by the EPA and the Fox River, Wisconsin, water pollution abatement program never developed a market. While the Sulfur Dioxide Allowance Program has fewer trades than anticipated, it has had sufficient transactions to establish a market.

Allan Jenkins

SUGGESTED READINGS: Brian McLean, director of the Acid Rain Division of the EPA, describes the early results of the Sulfur Dioxide Allowance Program in "Evolution of Marketable Permits: The U.S. Experience with Sulfur Dioxide Allowance Trading," *The International Journal of Environment and Pollution* 8 (1997). The EPA's Web site (http://www.epa.gov) has considerable information about pollution control strategies and related topics. To read about the economist's perspective on the optimal level of pollution and the benefits of a permit trading system, see Barry Field's *Environmental Economics: An Introduction* (1997). Sheldon Kamieniecki, George Gonzalez, and Robert Vos compare the incentive-based market approach to other possible approaches in *Flashpoints in Environmental Policymaking: Controversies in Achieving Sustainability* (1997). Terry Anderson and Donald Leal identify additional opportunities to bring market solutions to environmental problems in *Free Market Environmentalism* (1991).

SEE ALSO: Acid deposition and acid rain; Air pollution policy; Clean Air Act and amendments; Environmental economics.

Polychlorinated biphenyls

CATEGORY: Pollutants and toxins

Polychlorinated biphenyls (PCBs) are among the most insidious of environmental pollutants. Although their production has been banned in many countries, widespread use has led to almost universal contamination with PCBs, which are resistant to biodegradation and have a tendency to bioaccumulate.

PCBs were first synthesized in 1881 and became readily available during the 1930's. A PCB is a biphenyl molecule on which chlorine molecules substitute for two or more of the hydrogens. A related chemical, polybrominated biphenyl (PBB), contains bromine instead of chlorine. Many different kinds of PCBs form during their synthesis because each ring on the biphenyl molecule can have up to five chlorine atoms. Monsanto Company made PCBs in the United States and sold them under the trade name of Aroclor. Between 1930 and 1975, more than 570 million kilograms (1.26 billion pounds) of PCBs were made in the United States alone.

Some of the properties that made PCBs attractive chemicals were high boiling points, low water solubility, and heat resistance. These physical properties meant that PCBs were hard to burn, resisted acids and bases, and were mostly inert. Consequently, PCBs were used in many manufactured products, such as adhesives, fluorescent lights, insulation in transformers, high-pressure hydraulic fluids, plasticizers, varnishes and paints, and protective coatings for wood, metal, and concrete. Braided cotton and asbestos in electric wire insulation were also impregnated with PCBs.

The properties that made PCBs useful to industry also made them potentially long lasting and widespread pollutants. However, PCBs were not suspected of being an environmental problem because they were mostly inert, were not deliberately spread, and were not acutely toxic. One exception was an incident in Japan in 1968 when rice bran oil was accidentally contaminated with PCBs. It was difficult to show chronic PCB toxicity, although symptoms such as chloracne and low birthweight in babies born to exposed mothers were known. The U.S. Environmental Protection Agency (EPA) did not list PCBs as cancer-causing chemicals until they had long been in use.

In 1966 Sören Jensen, a Swedish chemist, reported observing traces of PCBs in animal tissue that dated as far back as the 1944. Scientists in California corroborated his results, and subsequent analyses demonstrated that PCBs were everywhere in the environment: rainwater in England, river water in Japan, seals in Scotland, eagles in Sweden, Baltic Sea cod, mussels in the Netherlands, penguin eggs in Antarctica, pelican eggs in Panama, and, most disturbingly, human hair and fat. All evidence pointed to a global spread of PCBs. The most likely route of spread was determined to be the incineration of PCB-tainted wastes. It is now virtually impossible to find samples of organic material that do not contain some trace amounts of PCBs. Even foods grown in chemical-free fields can receive trace amounts of PCBs from atmospheric deposition of incinerator exhausts.

PCBs are not very soluble in water. They are, however, extremely soluble in lipids and rapidly

A thermal treatment system is used to remove polychlorinated biphenyls from contaminated soil at the Industrial Latex Superfund site in Wallington, New Jersey, in 1999. (AP/Wide World Photos)

adsorbed through the walls of the digestive system. PCBs are almost twice as soluble in fats as dichloro-diphenyl-trichloroethane (DDT) and thousands of times more likely than other environmental pollutants to bioconcentrate in animal tissue. Jensen's results indicated that they were starting to bioaccumulate in the food chain. Scientists and environmentalists alike worried that PCBs might ultimately become a greater environmental problem than DDT because more of them had been produced, they lasted longer, and they apparently had wider distribution.

Declines in fish-eating bird populations and evidence of thin-shelled eggs began to increase after World War II. This coincided with a rapid rise in the use of chlorinated chemicals such as PCBs. PCBs cause birds to give birth to thin-shelled eggs, as does DDT, because these chemicals inhibit enzymes involved in calcium movement. The hormone estrogen controls the calcium level in breeding female birds, and PCBs stimulate enzymes that make estrogen

more soluble and readily excreted. When estrogen is low, calcium reserves are low, and little calcium is available for eggshell formation.

Several events ultimately stopped PCB and PBB production in the United States by 1975. In 1974 thousands of cattle and other farm animals in Michigan were quarantined and destroyed because they were contaminated with PBBs. The contamination occurred when a chemical company accidentally mixed bags of a fire retardant containing PBBs with bags of a magnesium oxide mix used in cattle feed. The chemicals were mixed with feed that was distributed around the state. Farmers first began noticing toxicity in their farm animals in late 1973, and contaminated milk and eggs exposed several thousand farm families to PBBs.

It has proved difficult to remove PCBs once they get into the environment. Highly chlorinated compounds, such as PCBs, are extremely resistant to biodegradation. In the late 1980's, scientists at the General Electric Research and Development facility in Schenectady, New York, began looking at sediment from sites in the upper Hudson River that were contaminated with up to 268,000 kilograms (591,000 pounds) of PCBs. The PCBs came from a capacitor manufacturing operation at Hudson Falls and Fort Edward, New York, that operated between 1952 and 1971. The scientists reported that highly chlorinated PCBs changed to lower chlorinated PCBs in anaerobic sediments. They suggested that PCBs in the environment biodegraded in a two-step process. Chlorine removal in anaerobic aquatic sediments was followed by further decomposition in aerobic environments that eventually caused total PCB destruction. More important, chlorine removal detoxified carcinogenic PCB congeners known to persist in humans.

Mark Coyne

SUGGESTED READINGS: The first widespread report of PCBs in the environment appeared in *New Scientist* 32 (December). James Lovelock, "The Electron Capture Detector and Green Politics," *LCGC: The Magazine of Separation Science* 8 (November, 1990), explains how the technology to detect PCBs in nature affected the environmental movement and the social and political response to other environmental pollutants such as chlorofluorocarbons (CFCs). It is an easy-to-read, balanced, and personal essay. Reprints are available from Aster Publishing Corp., Marketing Service, 859 Willamette St., P.O. Box 10460, Eugene, OR 97440. Carl Gustafson, "PCBs: Prevalent and Persistent," *Environmental Science and Technology* 4 (October 1970), provides a short review article that concisely describes what PCBs are, why they were made, and how they contribute to environmental problems.

SEE ALSO: Biomagnification; Bioremediation; Chloracne.

Population-control and one-child policies

CATEGORY: Population issues

Many population experts believe that the population explosion is the direct cause of such environmental problems as pollution, ozone and resource depletion, forest destruction, extinction of plant and animal species, increased violence, epidemics, and starvation. Such experts maintain that for humanity to have a positive future, effective and appropriate population-control systems, such as China's one-child program, must be implemented.

About six billion people live on earth, double the planetary population in 1960. Eighty million are added yearly. Most population experts, such as Paul Ehrlich and Anne Ehrlich, believe that such uncontrolled population growth is a major factor in most important social and environmental problems. In contrast, some researchers, such as Ben Wattenberg, a senior fellow at the conservative American Enterprise Institute, believe that the population problem has been overblown. In Europe, North America, and Japan, birthrates are below the replacement level. Even in many less-developed countries, fertility rates are plunging toward replacement levels. Loss of population endangers economic prosperity because there will be fewer producers and consumers. In contrast, those who believe that overpopulation is a real problem agree that birthrates in industrialized nations are below replacement levels but point out that birthrates in many less developed countries are far too high. Also, if the planetary carrying capacity is defined as the population load level at which all people could have their basic needs for food, clothing, housing, health care, and education satisfied, the human population load is already far beyond the earth's maximum carrying capacity.

Attention is usually focused on the billions of people and the high birthrates in the less developed countries in South America, Africa, and Asia. However, the industrialized world, about one-fifth of the world's population, consumes three-fourths of the planet's resources. In a lifetime, one U.S. citizen consumes as many resources as five hundred people in India.

Family planning programs give contraceptive information and devices to couples so they may freely choose their own family size. Other possible voluntary population-control programs include delaying marriage, sex education programs in schools, accessible abortion services, and welfare benefits favoring small families. Coercive control systems might include requiring unwed teenagers to place their children up for adoption, mandatory sterilization after two children, or compulsory abortion of pregnancies after one child. Government-supported voluntary family planning programs are popular. However, many couples choose to limit family size only after they have more than two children. Thus, family planning programs have not effectively prevented high population growth rates.

Although compulsory government control of family size is opposed in most countries, projections of increasing environmental destruction and reduced resources caused by population pressures are even causing some democratic countries to consider changing tax laws to favor single adults, childless couples, and small families. In their book *The Population Explosion*

(1990), Paul Ehrlich and Anne Ehrlich conclude that population research has found five crucial noncoercive factors that cause significantly reduced levels of pregnancy: adequate nutrition, effective sanitation, basic health care, education of women, and equal rights for women. When the status of women is no longer based on fertility, family size usually declines.

The People's Republic of China is the only nation to officially adopt a one-child policy. In the late 1970's Chinese leaders were startled to learn that China's population had surged to more than one billion people, which was more than 100 million more than previous estimates. In 1979 the government set a goal of limiting 50 percent of the nation's couples to one child and the other 50 percent to two children. Within five years, the average family size had dropped close to the replacement level of two children per family. Peer pressure was a major motivational

A Chinese child hugs her doll while her mother speaks with a family planning official. In an attempt to slow the growth of its population, China implemented a one-child policy during the 1970's only to relax it during the following decade. (AP/Wide World Photos)

force. In certain factories, notices were posted indicating which women could become pregnant. If social pressure was not successful, abortions and sterilizations were coerced. However, traditional rural families strongly resisted the one-child limit. Evidence indicating that the first child would be a girl was often followed by abortion, and female infanticide was a frequent occurrence in rural areas.

In the 1980's China's one-child policy was relaxed. The need for children to help in farming also pressured the birthrate upward again. Projections indicate that China's population will reach at least 1.5 billion before zero population growth is achieved. Even if one-child policies could be achieved in noncoercive ways, critics claim that they would result in disaster. Such a country could lose 50 percent of its population in two generations. Critics predict that economic disaster would result because there would not be enough young people in the workforce to pay the pensions of the elderly.

Lynn L. Weldon

suggested readings: A comprehensive introduction to population-control issues, as well as research and activities resources, is Paul Ehrlich and Anne Ehrlich, *The Population Explosion* (1990). Worldwatch Institute publishes six issues of *Worldwatch* magazine per year and comprehensive annual volumes of *The State of the World*. The Population Reference Bureau produces the annual *World Population Data Sheet* and materials for school programs.

see also: Ehrlich, Paul; Population-control movement; Population growth.

Population-control movement

Category: Population issues

The population-control movement is an international development strategy driven by concerns about the negative impacts of overpopulation. The movement has advocated the worldwide spread of family planning services, particularly the distribution of contraceptives.

Birth-control strategies have been used to regulate the timing and spacing of births for millennia, but the idea that humans should limit family size in order to improve quality of life originated with Thomas Malthus, an English cleric and economist who wrote *Essay on the Principle of Population* in 1798. Nineteenth century neo-Malthusians were convinced that overpopulation caused poverty and that contraceptives should be provided to the poor, a position opposed by the medical community of the United States. Physicians triumphed when the Comstock law was passed in 1873, making contraceptives and associated information illegal.

Support for birth control came from a diverse constituency, including suffragists, moral reformers, and advocates of eugenics (the science of improving human hereditary qualities). Some feminists advocated abstinence rather than "unnatural" contraceptives, but most believed that women should have the right to choose when to have a child. Margaret Sanger, a radical feminist and socialist, wrote extensively about birth control, set up clinics, and recruited physicians. Sanger also founded the American Birth Control League in 1921—later renamed the American Birth Control Federation (ABCF)—and helped organize the first World Population Conference in Geneva, Switzerland, in 1927. At the conference, many eugenicists, including Henry F. Osborn and Frederick Osborn, called for government intervention in birth control and sterilization of the "unfit."

In 1922 British sociologist and educator Sir Alexander M. Carr-Saunders wrote a book entitled *The Population Problem* that laid the basis for "transition theory," a description of how fertility and mortality rates change during modernization. The evidence from European history indicated that the pattern of high fertility and mortality that preceded nutritional and health improvements was immediately followed by lowered mortality coupled to high fertility and rapid growth. Eventually the increased costs of raising children in an industrial, market economy led to smaller families, but in the poorer regions of Africa, Asia, and South America, large families were still preferred, even after mortality declined. How to speed the process of fertility decline be-

Milestones in Population-Control Movement

YEAR	EVENT
1914	Margaret Sanger publishes *The Women Rebel*, in which she provides information about birth control; she is indicted for violating the Comstock Law.
1952	Sanger helps establish the International Planned Parenthood Federation (IPPF), which sets up birth-control programs worldwide; the Population Council is also founded.
1959	India establishes the first national family-planning program, followed by Pakistan; by 1963 policies exist in Taiwan, Korea, and Tunisia. Similar efforts probably also help fertility decrease in Singapore, Hong Kong, Puerto Rico, Jamaica, and Trinidad.
1960	Birth-control pills are available for the first time.
1966	The U.S. Congress, believing that food shortages are a result of overpopulation, uses food aid monies to finance family-planning programs in the developing world.
1969	The United Nations founds the Fund for Population Activities, signaling the beginning of its active involvement in population control.
1973	*Roe v. Wade* legalizes abortion in the United States.
1974	The World Population Conference is held in Bucharest, Romania; it is the first international population conference attended by government representatives.
1994	The World Population Conference is held in Cairo, Egypt; women's status and reproductive health are major themes in the quest for a broader, more humane approach to family planning.

came an important goal of demographers, eugenicists, and many feminists, such as Sanger and her British counterpart, Marie Stopes.

Eugenicist influence on the population control movement was evident when the ABCF targeted African American doctors for assistance in spreading the family planning message in 1939, a policy that was labeled racist by prominent African Americans. This effort followed the founding of birth control clinics in economically depressed Puerto Rico in 1935 by the government's relief agency. However, the Nazi atrocities during World War II tempered the more extreme positions of eugenicists, and the impoverishment of many educated and middle-class families during the Depression also confirmed that poverty was not necessarily a function of bad genes. Thus, birth-control information was

added as part of the services provided to the poor by the U.S. government following the New Deal, and making that information more acceptable to a wider married constituency resulted in renaming the ABCF the International Planned Parenthood Federation (IPPF) in 1942.

GENERATING SUPPORT

By the 1950's and 1960's, a number of wealthy businesspeople in the United States were waging a battle on behalf of birth control because of what they perceived as the evils of overpopulation. The major contributors were the Ford and Rockefeller Foundations, which channeled their efforts through two organizations, the IPPF and the Population Council, both established in 1952, the former by John D. Rockefeller III. Hugh Moore, founder of Dixie Cups, sent his pamphlet *The Population Bomb* to ten thousand prominent citizens to garner support during the late 1950's, a time when most U.S. families practiced family planning and new contraceptive technologies were being developed. The first birth-control pill (Enovid) was marketed in 1965.

Bringing the U.S. government into the picture was the big challenge. President Dwight Eisenhower formed the Committee to Study the U.S. Military Assistance Program in 1958 because of complaints that U.S. funds for overseas assistance put too much emphasis on military support at the expense of economic aid. Major General William H. Draper, Jr., the former director of the European Recovery Program, chaired the committee and was encouraged by Moore to study the population problem. Ansley Coale and Edgar Hoover's *Population Growth in Low Income Countries* (1958) had just been published and showed that having too many children overtaxed family resources, reducing private investment. That economic development might be impeded by rapid population growth provided Draper with the rationale for using international assistance for fertility limitation, but his report was later disavowed by President Eisenhower and the

Catholic bishops of the United States. Draper tried to convince President John Kennedy to take action, but Kennedy suggested that the private sector, particularly the Ford Foundation, fund population control.

During the 1950's, the Knowledge, Attitude, and Practice Surveys (KAP) had measured the "unmet need" for contraception and found that most women would have smaller families if they could. Dramatic publicity about overpopulation during the early 1960's alarmed U.S. voters. Thus, by 1965, during the administration of President Lyndon B. Johnson, the U.S. government began a tradition of funding population programs through the U.S. Agency for International Development (USAID) and later also through the United Nations (U.N.).

UNITED NATIONS

In 1962 the United Nations invited member states to formulate population policies and in 1966 reemphasized the connection between population and socioeconomic factors. The United Nations also reinforced individual governments' rights in determining policy and the right of families to determine number, timing, and spacing of births. However, Western and Asian countries, particularly India, Sweden, and the United States, pressured the United Nations to take on a leadership role in instituting family planning programs.

In 1967 the secretary general of the United Nations established the U.N. Trust Fund for Population Activities (UNFPA) with the goals of helping developing nations with population activities, expanding the United Nations' role in family planning, and pursuing new programs. He was assisted in his efforts by Reimert T. Ravenholt, head of USAID, who wanted contraceptives distributed worldwide in order to hasten economic development. By 1971 the UNFPA was

Women sit in the waiting room of a family planning clinic in Johannesburg, South Africa. The population-control movement has made tremendous efforts to provide family planning services to people in Africa, where population growth rates have exceeded those of Europe and North America. (AP/Wide World Photos)

a recognized hub of the U.N. system, which also included support for population activities by the International Labour Organisation, the Food and Agriculture Organization, the U.N. Educational, Scientific, and Cultural Organization (UNESCO), and the World Health Organization.

The first world conferences to study population issues were held in Rome, Italy, in 1954 and Belgrade, Yugoslavia, in 1965. However, the Third World Population Conference held in Bucharest, Romania, in 1974 was the first official government conference with an emphasis on policy rather than research. Delegates agreed that population issues needed to be addressed, but the Northern and Southern Hemispheres were strongly divided over the relative importance of population planning versus economic development. Southern countries believed that an inequitable distribution of resources and the heavy consumption patterns of wealthy northern nations contributed as much to environmental deterioration as population growth by the poor.

Ongoing Debates

The U.S. Federal Office of Population Affairs was founded in 1970 and provided federal money for family planning services (except abortions) for poor Americans. In 1972 the Commission on Population Growth and the American Future advocated population planning as important for world stability. The commission believed that efforts should include access to abortion and limits to illegal immigration. President Richard Nixon would not approve the document, and debate intensified when the Supreme Court prohibited interference with a woman's right to an abortion during the first three months of pregnancy in *Roe v. Wade* (1973). At the second Population Conference in 1984 in Mexico City, Mexico, the U.S. delegation reversed its support for family planning measures because of concerns about coercion and abortion in countries such as the People's Republic of China. However, the World Plan of Action written at Bucharest was revised by delegates representing nations in general agreement that family planning programs were useful whether economic development took place or not. By 1985 the U.S. no longer funded the IPPF or the UNFPA, a position that was not reversed until President Bill Clinton took office in 1993.

There is still no consensus on the interrelationships among population growth, environmental degradation, and economic development. Some feel population growth is a cause of poverty, while others would reverse that relationship. Women with the lowest education and wages and fewest opportunities outside the household tend to have the largest families. The perspective that gender inequality may be at the heart of the problem in the regions of the world that continue to exhibit high fertility is gaining momentum, particularly evident at the International Conference on Population and Development in Cairo, Egypt, in 1994.

Many southern nations have subsistence economies requiring labor-intensive strategies, and larger families are a rational choice where savings are difficult to put aside and children are productive assets who also provide security during old age. Children in rural India as young as six years old can look after domestic animals and younger siblings, and help with other tasks. In extended family households, the costs of raising children are also shared because access to common lands expands as the household becomes larger. However, increasing urbanization and market pressures are forcing changes in traditional ownership of land so that these shared lands are decreasing in spite of ever-larger demands that put pressure on local environments.

Few doubt the importance of family planning measures, and most southern nations support and feel responsible for providing services to individual couples. However, reformers are trying to shift attention to women's health and empowerment because the population-control approach is believed to have resulted in ethical violations and coercive abuses. Furthermore, it is not always successful where children are an important source of labor. An emphasis on women's reproductive health with more sensitivity to local context may be more effective in the long run, although some rethinking and reformulating of current policies will be required.

Joan C. Stevenson

SUGGESTED READINGS: For more on the international population-control movement, see Oscar Harkavy's *Curbing Population Growth* (1995), Kurt W. Back's *Family Planning and Population Control: The Challenges of a Successful Movement* (1989), and Betsy Hartmann's *Reproductive Rights and Wrongs: The Global Politics of Population Control* (1995). Population practices within individual countries and the activities of the United Nations are enthusiastically detailed in Stanley P. Johnson, *World Population—Turning the Tide: Three Decades of Progress* (1994). For descriptions of the U.S. origins of the movement and feminist critiques anticipating the shift in emphasis to individual women's reproductive health, see Linda Gordon, *Woman's Body, Woman's Right* (1976), and Ruth Dixon-Mueller, *Population Policy and Women's Rights: Transforming Reproductive Choice* (1993). Why women continue to have large families and the impact it has on the environment is discussed in Partha S. Dasgupta, "Population, Poverty, and the Local Environment," *Scientific American* (February, 1995). The Population Council encourages research and development on population issues and also publishes two inexpensive and accessible journals: *Studies in Family Planning* and *Population and Development Review*.

SEE ALSO: Population growth; Population-control and one-child policies; United Nations Population Conference; World Fertility Survey; Zero Population Growth.

Population growth

CATEGORY: Population issues

Population growth challenges the world's economy, social structures, and environment. The precise value of the controversial, theoretical maximum population capacity of the planet depends upon the availability of natural resources, the acceptable quality of life, the role of technology, and the values underlying the human relationship with nature.

The world population passed the six billion mark in 1999 according to the U.S. Bureau of the Census, triple the 1940 population figure. It is expected to continued to increase by about 70 to 80 million people every year for at least another generation. This population explosion is unprecedented in human history. The estimated population in 8,000 B.C.E. was about five million. It did not reach 500 million until around 1650, but then the increase accelerated. There were one billion people by the mid-nineteenth century and four billion by 1975.

Even though the fertility rate—the average number of children born per woman—declined after the late 1960's, vigorous growth will continue through the twenty-first century, and perhaps beyond. The Bureau of the Census predicts that in 2050 the population will be 9.34 billion. The United Nations projects 7.8 to 21.2 billion depending upon whether the fertility rate stabilizes or increases again. Whatever the actual total, the distribution of the population will change as people migrate to urban areas. Demographers predicted that soon after 2000, more people would live in cities than in the countryside for the first time in history. Most of the total population growth and the urban growth will occur in developing countries, particularly those in Africa.

How much increase can the environment sustain? Answers to that question are so biased by ideology, religious tenets, and interpretive methods that they vary by nearly four orders of magnitude: from 500 million to one trillion. However, most estimates fall between six and ten billion, which would mean that the carrying capacity has been exceeded or will be exceeded within one century. The degradation of the global environment—or at least the locales of the greatest population concentrations, the sea coasts—could endanger billions of people and far greater numbers of wildlife.

RESOURCES

Humans affect natural resources by displacing or consuming them. Many resources are renewable, such as forests, which can be regrown. Nevertheless, scientists fear that increased demands for such raw materials may surpass the rate of replacement. For example, increasing use of wood could cause loggers to cut down forests

World and Urban Population Growth, 1950-2020

Sources: U.S. Bureau of the Census International Data Base and John Clarke, "Population and the Environment: Complex Interrelationships," in *Population and the Environment* (Oxford, England: Oxford University Press, 1995), edited by Bryan Cartledge.

faster than new trees can grow; eventually, the amount of wood available would diminish. Moreover, spreading urban areas and agriculture could take over land that once supported forests.

Some resources are gone forever once used, leaving less or none for future consumers and necessitating replacement by other materials if the enterprises dependent upon the originals are to continue. For example, if petroleum reserves were depleted, propane or natural gas might serve as replacements for heating and vehicle fuel. Not all such nonrenewable resources are replaceable. The biodiversity of nature is the critical example, although some commentators insist that biodiversity is not a resource for exploitation but a heritage that should be safe-

guarded. In any case, proponents of biodiversity believe that the destruction of natural habitats and the attendant extinction of species obliterates much-needed new sources of food and medicines.

Species die out naturally. The species now living amount to only 2 to 4 percent of those that ever existed. Nonetheless, biologists believe that because of human expansion, pollution, and increasing energy consumption, a mass extinction is under way that is as serious as the one that killed the dinosaurs during the Jurassic period. The current extinction rate is calculated to be ten thousand times higher than the average rate before the rise of humans. Of the five to ten million species thought to exist, ten to twenty-five thousand disappear every year.

Most are unrecorded insect species, but since 1650 about 1 percent of recorded bird and mammal species have become extinct, and at least another 25 percent are threatened with extinction.

People also change the character of the land and its fresh water. Land cleared for wood or farming is subject to erosion, and soil fertility declines with heavy agricultural use, especially because of monoculture crops, overgrazing, and some irrigation methods. According to the United Nations, every year about 68 million acres of arable land turn into desert because of human enterprise; 35 percent of all cropland is threatened. Innovations in agricultural methods, known as the Green Revolution, multiplied crop yields and more than made up for the loss in arable land during the late twentieth century, but accelerating growth and desertification are expected to strain farmers' ability to feed the population. Little can be done about the diminishing supply of fresh water except conservation and antipollution measures. In 1998 500 million people lived without access to potable water; by 2025 the number is projected to rise to 2.8 billion, about one in every three people. Polluted water threatens health and agriculture, especially in developing countries where populations increase the fastest. Political scientists worry that competition among nations for declining water resources could escalate into armed conflict in Africa, Central Asia, the Middle East, and South America.

Air pollution also increases with expanding industry and transportation. Changes in atmospheric chemistry from particulates, hydrocarbons, carbon monoxide, nitrogen oxides, and sulfur oxides cause acid rain and smog in urban regions; some pollutants also trap solar energy, creating a greenhouse effect, or destroy ozone high in the atmosphere. The increase in average temperatures and ultraviolet radiation because of these effects will further threaten crops, human health, and wild flora and fauna.

POLITICS AND ECONOMICS

Proposals for dealing with population growth emphasize either restricted fertility or economic development, and sometimes both. However, nationalism and international politics complicate policies and make them difficult to implement.

Attempts to restrict fertility take three forms. A government may specify by law the number of children per couple. The Peoples Republic of China is the best-known example: Its one-child policy is intended to control the world's largest national population—1.2 billion in mid-1997. Although recognizing that China desperately needs to check growth, humanitarian groups denounce its government for violating human rights in carrying out the policy. Most observers doubt a restrictive population growth policy could be effective in a country without a dominating central government like China's.

Other nations and international organizations encourage limiting family size through two other methods. One is to teach women family planning techniques and offer them contraceptive devices or drugs. The other is to raise the educational level of women in general in order to encourage their entering the workforce. Studies in all countries show that educated, working women have fewer children on average than uneducated women who stay at home. However, great obstacles confront such efforts. Some religions prohibit or discourage contraceptives. The governments of developing nations are frequently suspicious of family planning programs financed by rich countries or the United Nations, afraid that the programs are really covert political sabotage. Most often, however, cultural traditions forbid women to compete with men intellectually or economically.

Proponents of economic development assume that resources can be further developed and more efficiently used to accommodate population growth. Criticism also plagues their efforts. Conservationists throughout the world, especially in industrialized regions such as North America and Europe, are the leading opponents, especially if the development entails exploitation of nonrenewable resources, anticipation of new technologies, or increased pollution. Third World governments join in denouncing the economies of developed countries, accusing them of overconsumption and unwillingness to control their own pollution. Wealthy nations account for only one-sixth of the world's population but control three-fourths of the world's gross product and trade. Moreover, the industrialized nations pollute far more than developing nations: The United States alone, with only 5 percent of the world's population, produces 25 percent of the carbon emissions that contribute to global warming.

Political theorists believe that the effects of population growth may be palliated with more equitable distribution of wealth and political power. They argue that less wasteful consumption will decrease pollution and husband resources, while freer trade and better distribution of land and the capital to develop it wisely will end such harmful practices as slash-and-burn agriculture. Stronger international institutions, many propose, could administer the needed economic and political changes. The twentieth century saw the globalization of markets, encouraged by such treaties as the General Agreement on Tariffs and Trade (GATT), which may help equalize use of labor and resources. However, resistance from national governments to new worldwide political institutions and economic systems is daunting.

PHILOSOPHICAL ISSUES

Deep philosophical differences also trouble the debate on population growth. The Roman Catholic Church, for instance, treats the environment and population as separate questions and teaches that contraception is immoral because it is contrary to the family's purpose to procreate. The Church vigorously objects to government interference in family life and permits birth control by rhythm method (sexual abstinence during ovulation) alone. Some radical environmentalists, by contrast, want to halt human reproduction altogether until the population returns to pre-Industrial Revolution levels or lower.

These antithetical positions reflect two fundamental questions. First, is humanity to be considered part of nature? Much Western philosophy and theology holds humanity to be superior to nature according to divine law or distinct from it because humans alone possess intelligence. On the other hand, many non-Western thinkers and environmentalists consider humanity to be an integral part of nature, or the "web of life"—species are mutually dependent upon each other, so none is superior. Second, should humans take from the environment whatever they want when they want it, or should they consume only in such a way that biodiversity is not threatened?

Even if all humankind were to agree on answers to these questions, the environment itself might settle the problems of population growth. Some scientists fear that the global environment is so intricate, complex, and fragile that if stressed too much, a "jump effect" could occur in which Earth's ability to support humans and other large organisms would decline precipitously—faster than human technology or conservation could compensate.

Roger Smith

SUGGESTED READINGS: For a full, balanced, statistic-rich analysis of population growth and sustainability, see Joel E. Cohen's *How Many People Can the Earth Support?* (1995). In *Critical Masses: The Global Population Challenge* (1994), George D. Moffett, a journalist, provides an engrossing summary of population-based problems, their effects on specific people around the world, and potential solutions. *Population and Human Survival* (1993), edited by Gary E. McCuen, contains essays on population and the environment with accompanying study questions designed to help students understand and evaluate the issues. For readers new to the topic, *Global Environmental Change: Its Nature and Impact*, by John J. Hidore, carefully defines types of change, including those caused by humanity. In *Living Within Limits: Ecology, Economics, and Population Taboos* (1993), Garrett Hardin examines the environmental, economic, and theoretical background to population pressure on nature. *Population and the Environment* (1995), edited by Bryan Cartledge, contains eight essays by scholars and theologians who consider the purpose of population expansion, ways to handle it, and its impact upon nature and human values.

SEE ALSO: *Limits to Growth, The*; Population-control movement; Population-control and one-child policies.